44 0502576 0

Nat

WITHDRAWN

D1332170

Nature's Purposes
Analyses of Function and Design in Biology

edited by Colin Allen, Marc Bekoff, and George Lauder

A Bradford Book
The MIT Press
Cambridge, Massachusetts
London, England

UNIVERSITY OF HERTFORDSHIRE
WATFORD CAMPUS LRC
WATFORD WD2 8AT 301673
BIB
ISBN 0262510979
CLASS
570 1 NAT
LOCATION
F WSL
BARCODE
4405085760

© 1998 Massachusetts Institute of Technology

All rights reserved. No part of this book may be reproduced in any form by any electronic or mechanical means (including photocopying, recording, or information storage and retrieval) without permission in writing from the publisher.

This book was set in Sabon on the Monotype "Prism Plus" PostScript Imagesetter by Asco Trade Typesetting Ltd., Hong Kong, and was printed and bound in the United States of America.

Library of Congress Cataloging-in-Publication Data

Nature's purposes: analyses of function and design in biology /
  edited by Colin Allen, Marc Bekoff, and George Lauder.
    p. cm.
  "A Bradford book."
  Includes bibliographical references (p.   ) and index.
  ISBN 0-262-01168-9 (hardcover: alk. paper). — ISBN 0-262-51097-9
  (pbk.: alk. paper)
  1. Biology—Philosophy. 2. Teleology. I. Allen, Colin.
II. Bekoff, Marc. III. Lauder, George V.
QH331.N385   1998
570'.1—dc21   97-29290
                                                          CIP

Contents

Introduction

Biology is unique among the natural sciences in licensing apparently teleological statements about design, purpose, and adaptive function. Teleological thinking originated from two views, both of which are assumed to have been discredited in physics, chemistry, and the other natural sciences. The first source of teleological ideas is the Aristotelian view (discredited by Galileo) that the motions of natural objects are explained by their possessing intrinsic purposes that are fulfilled so long as the objects are not subject to external interference. The second source, abandoned by scientists as untestable and irrelevant to scientific explanation, is the view that every phenomenon in the natural world is an instance of the direct operation of divine will. Modern biologists also cleave to the unsuitability of such Aristotelian and theistical underpinnings for teleological claims.

Nonetheless, teleology has seemed to many theorists to be indispensable to biology. To understand the complex morphological and behavioral traits of organisms it seems we must say what those traits are for, which is to give a teleological explanation of why organisms have them. Other biologists have argued that natural selection, providing, at root, statistical explanations for why organisms have the traits they do, obviates any need for teleological thinking (Dawkins 1987). If organisms possess structures that can be explained fully as having arisen by selection on randomly (in the sense of "undirected") originating variants, then what need is there for teleology, a designer, or indeed even the notion of design (Dawkins 1996)? There is an evident and uncomfortable tension within the biological sciences over the role that teleological explanations might play. Furthermore, if teleology cannot be eliminated from biology, this raises fundamental questions about the nature of biological explanation and the relationship of biology to the rest of science.

Because the account we give of the role of teleology in biology will have consequences for our conception of the nature of biology itself, the goal of accounting for "Nature's purposes" is arguably the most important foundational issue in the philosophy of biology. Given the significance of the topic, it is unsurprising that many essays by biologists and philosophers of biology have been attempts to come to terms with teleology. And although most good anthologies in the philosophy of biology include a sample of this literature (e.g., Sober 1994), no comprehensive collections have previously been available. (Rescher [1986] provides an anthology derived from conference proceedings and Buller [1998] presents an anthology that focuses on philosophical literature.) Why this should be so, when the other major topics in the philosophy of biology have been anthologized several times over, remains curious. That we have the temerity to attempt to fill this breach is perhaps partly explained by the fact that as a philosopher, an ethologist, and a comparative morphologist, each of us can feel confident enough to blame the others if our readers detect any glaring omissions. On the positive side, our different backgrounds have, we believe, allowed us to bring together different parts of the interdisciplinary literature on biological teleology. Philosophers, for instance, seem to have been largely unaware of the literature on teleology generated by workers in functional and evolutionary morphology, and we hope to have closed gaps that exist even among biologists themselves.

Our selections for this volume have been chosen to reflect attempts to understand biological teleology from within a *naturalistic* framework. Following Darwin, rather than interpreting claims about design and function in biological systems as literal or metaphorical references to the supernatural intentions and interventions of a Cosmic Designer, most modern commentators interpret teleological claims to be referring more or less circuitously to various natural processes or properties. There is disagreement at many levels over the merits of the various theories offered, but it is possible to identify a number of central themes (Allen and Bekoff 1995). At the most basic level is a dispute about the point of making teleological claims: Does attributing the function of pumping blood to a heart *explain* anything, or is it merely a shorthand description of certain attributes of the heart? Most of the selections in this anthology maintain that teleological claims in biology both describe and explain something, although opinions vary about exactly what is explained, and how. (For a defense of the view that teleological claims are merely descriptive, see Thompson 1987 and Lipton and Thompson 1988a, b.) Some authors maintain that teleological explanations are a species of causal explanation. Others argue that teleological claims apply to biological phenomena independently of any particular causal explanation, and that teleological claims are best understood as providing a framework for causal explanation.

Among those who accept some version of the claim that the point of making teleological claims is to give or to frame explanations, there is disagreement about what exactly is explained by teleological claims. Here the major division is between, on the one hand, the view that teleological claims correspond to explanations of the presence or maintenance of particular traits of organisms (e.g., the fact that hearts pump blood helps explain why organisms have hearts) and, on the other hand, the view that teleological claims explain how a particular component of a system contributes to some other capacity (e.g., the fact that hearts pump blood helps explain how organisms do a variety of things, such as getting oxygen to their tissues or locomotion). Our selections reveal considerable sympathy for both of these views, and some of our authors attempt to show how the two views may be combined in various ways.

Those who focus on explanation of the existence of traits are divided on the question of whether natural selection should feature in an account of the use of teleological language in biology. Should theorists be attempting to give a *conceptual analysis* of teleological notions that have roots prior to Darwin and outside biology, or are they free to cut ties to past uses of teleological language and to promote a new understanding of such language on grounds of its utility within Darwin's theory? A classic example of the use of teleological language in biology is William Harvey's seventeenth century claim that a function of the heart is to pump blood. If Harvey could know what hearts were for without knowing anything about Darwin's theory, then why does the proper analysis of Harvey's claim need to make any reference to the theory of evolution? And if natural selection is a key part of the understanding of teleological language in biology, should teleological explanations be understood in terms of the history of selection that has produced particular traits, or should it be understood in terms of the disposition of these traits to contribute to the future survival and reproduction of organisms subject to natural selection? The choice between these backward-looking and forward-looking alternatives is a theme of several of the papers in our selection.

We have attempted to order the articles in this anthology so that readers who are new to the topic of biological teleology are offered the best chance of being able to read and understand this literature. Because themes tend to cluster, a strictly historical ordering is not the best. But because later papers often presuppose some knowledge of the earlier literature, it is not appropriate to organize them entirely on thematic grounds. Thus we have attempted to compromise. First we present some of the early papers that address the role of past events of natural selection in grounding teleological claims. Then we present some of the early papers that present alternatives to these backward-looking views. The remainder of the selections represent various attempts to become more sophisticated about the goals of philosophical analysis, and the variety of uses of teleological language that is found in actual biological practice. We believe the anthology is suitable for an advanced undergraduate or graduate course.

Looking Backwards: Teleology as Etiology

Given the centrality of Darwin's theory of evolution in modern biology, many authors have attempted to explain the role of teleology within biology by invoking natural selection. We begin with such an account from biologist Francisco J. Ayala. Teleological explanations in biology are justified, according to Ayala, in three kinds of cases: (i) when an end-state or goal is consciously anticipated by the agent, (ii) in connection with self-regulating systems (e.g. regulation of body temperature), and (iii) in reference to structures anatomically and physiologically designed by natural selection to perform a certain function. Ultimately, Ayala believes, the first two kinds of cases are grounded in the third—design by natural selection, which is an undirected, historical process. Ayala's account of teleology is, therefore, historical (or "backward-looking"). He argues that because natural selection is a mechanistic process, the teleology found in biology is "internal"—there is no need for an external source of purpose or design in the form of a Creator—and it is devoid of any suggestion that there are preordained goals.

Among biologists, the term *teleology* is ambiguous. Some authors use it to refer only to those cases involving preordained goals (either Aristotelian internal purposes, or theistical external purposes) while others employ it generically for all uses of the notions of function, design, or adaptation. Ayala says that he would be happy to see the term restricted to cases involving preordination, and even eliminated from science altogether. But rather than haggle over terminology, it is important, he thinks, to clarify the various ways in which the term is used, and he rejects the suggestion (e.g., Mayr 1965; Pittendrigh 1958) that biologists would do better to abandon the term *teleology* and use *teleonomy* only in cases of end-directed systems with self-regulating mechanisms ("programs" in the jargon of Mayr 1988). Ayala points out that some biological phenomena are end-directed without being self-regulating.

Ayala also raises the important issue of the nature of teleological *explanation*. He explains and accepts Ernest Nagel's (1965) argument that teleological explanations may be reformulated as nonteleological, causal-mechanistic explanations. Despite this, he argues, teleological explanations are indispensable to biology because they connote that biological systems are "directively organized" whereas nonteleological explanations carry no such connotations; furthermore, teleological explanations are specifically directed at explaining the *existence* of the features of organisms.

Ayala's claim that the target of teleological explanation is the existence of biological traits is taken up in more detail by philosopher Larry Wright. Unlike Ayala, and perhaps because he comes to the topic as a philosopher and not a biologist, Wright is concerned not to prejudge any questions about the relationship between teleological language in biology and its use in other contexts. He chooses to focus on the notion of *function*, and the outcome of his investigations is, he maintains, an analysis of function that is not restricted to *biological* functions. According to his analysis (p. 71), *the function of* X *is* Z means:

a. X is there because it does Z, and
b. Z is a consequence (or result) of X's being there.

Although this schema makes no reference to natural selection, natural selection is held to satisfy it, because selected traits are present in off-spring given the consequences those traits had for the fitness of parents. Like Ayala, Wright maintains that teleological explanations are explanations of the existence of some state of affairs. Teleological explanation is a species of causal explanation, following the first clause of his analysis, but is distinguished from standard causal explanations by what he calls the "convolution" of the second clause.

Although Wright does not explicitly distinguish between trait tokens (or instances) and trait types, it is important to do so in order to make sense of his claim that his analysis fits a causal pattern. The presence of a token or instance of a trait cannot be causally explained by any direct effect of that instance, for the token's existence must predate its performance of any function. (Millikan, chapter 10, asserts that this is exactly what Wright's analysis commits him to, but Nagel, chapter 7, attributes something like the token-type distinction to Wright.) If the token-type distinction is applied, then a functional explanation can provide the "causal background" for the presence of some feature or trait through the history of traits of the same type performing the effect now designated as a function. In this way, Wright's analysis, like Ayala's, is backward-looking. Because of his use of the term *etiology*, Wright's and other backward-looking accounts of function have come to be known as *etiological* accounts within the philosophical literature.

In our third selection, philosopher Robert Brandon is critical of attempts (such as Wright's) to provide an account of teleological explanation that is not tied to any particular biological theory. Nonetheless, he accepts an account of teleological explanation in biology that, like Wright's, is backward-looking. But unlike Wright, Brandon focuses on adaptations rather than functions. Indeed, borrowing a distinction from Mayr (1961), Brandon separates functional from evolutionary questions: *functional biology* is concerned with questions about how structural elements interact, whereas *evolutionary biology* is concerned with *why* an organism has a particular trait, or *what* the trait is *for*. This use of the term *function* presents a potential for terminological confusion, as several authors use it in the context of evolutionary kinds of questions.

Brandon claims that the proper objects of what-for questions are adaptations: phenotypic traits that have evolved as a direct product of natural selection. Given that a particular trait *A* is an adaptation, the structure of a teleological explanation is that (p. 92) "trait *A*'s existence is explained in terms of effects of past instances of *A*; but not just any effects: we cite only those effects relevant to the adaptedness of possessors of *A*." The notion of adaptation is teleological, according to Brandon, precisely because it is used to answer what-for questions—a form of question that does not arise in the other natural sciences.

Many biologists are concerned that appeals to the past undermine their ability to apply teleological concepts rigorously. Especially in the case of fossils, it is extremely difficult to reconstruct the conditions in which organisms lived, and hence extremely difficult to assess claims about natural selection in the past. Brandon comments (p. 91) that the epistemological problems of assessments about past natural selection "simply show that evolutionary biology is far from being completed as a science." But many biologists would find this too cavalier. Nowhere are the problems more acute than for comparative morphologists working on fossils, and it is therefore unsurprising that they have often taken an entirely different approach to the notion of function, as several selections below will illustrate. Some philosophers, too, have rejected etiological accounts, and the papers in the next section cover some of the major alternatives.

Don't Look Back: Nonhistorical Approaches to Biological Teleology

We begin this section with morphologist Martin Rudwick's attempt to deal with the particular difficulties of attributing functions to the traits of fossilized organisms. Teleological language, he argues, is appropriate to organisms in just the same sense that it is appropriate to machines whose "*design* enables them to *function* for their intended *purpose*," which, in the case of organisms, is "the existence and survival of the individual organism" (p. 108). According to Rudwick, the inference of function from fossils is driven by our comprehension of engineering principles that constrain the possible roles that some structure might play in the life of an organism. While similarities to existing organisms may provide initial clues, Rudwick maintains that hypotheses about function must be tested by analyses of the sort provided by mechanical engineering. Rudwick maintains that this kind of functional analysis is "logically independent of the origin of the structures concerned.... thus logically unrelated to any and all evolutionary theories." Despite this, he argues, evolutionary theory, insofar as it stresses the adapted, machine-like character of organisms, encourages the attribution of function to fossilized forms. Rudwick's analysis of function is entirely ahistorical, for the properties revealed by an engineering-style analysis of a given structure are properties that it has presently and independently of the context in which the structure might have been used by an organism.

Like Rudwick, morphologists Walter Bock and Gerd von Wahlert are motivated to define *function* without reference to history. They claim that biologists have tended to use the term in two senses, both for what a structure is capable of based on its physical and chemical properties, and for its role in the life of the organism. Bock and von Wahlert suggest that function should be used only in the first sense. More precisely, they define a feature's functions as "all physical and chemical properties arising from its form ... that ... do not mention any reference to the environment of the organism" (p. 124). Function is distinguished sharply from *biological role*, "which includes all actions or uses of the ... feature by the organism in the course of its life history ... with reference to the environment of the organism" (p. 131). These definitions are intended to reform biological practice so as to eliminate the ambiguity inherent in using *function* in two senses, even though restricting function to the first sense "may seem initially strange and awkward to many biologists" (p. 126). The consequence of this reform would be, they believe, a notion of function "free ... of any form of teleology, Aristotelian or otherwise" (p. 125).

The benefit of distinguishing function from biological role lies, according to Bock and von Wahlert, in the clearer picture of biological adaptation that results. Understanding functions independently of and prior to biological role clarifies, they believe, the way in which features of organisms come to have a biological role within specific environments. Bock and von Wahlert believe that backward-looking approaches may in fact hinder understanding of the principles of evolutionary change, and they claim that (p. 162) "all evolutionary mechanisms can be studied best by working forward through time and without the aid of hindsight."

Working independently, philosopher Robert Cummins produces an account of function that shares some affinities with but also has significant differences from morphologists' attempts to come to terms with functions. Cummins' view is that the functions of a trait of an organism must be specified relative to a "capacity"—a trait's functions with respect to a capacity are those effects of the trait that help to explain how the organism has a particular capacity. For instance, functions of a pigeon's wings with respect to the pigeon's capacity for flight are to generate lift and propulsion, for this explains how the wings contribute to flight. Cummins' description of *functional analysis* can be seen as a precise statement of Rudwick's idea that function attributions are those that would be provided by an engineering analysis. Like the morphologists' accounts, Cummins' is entirely ahistorical. They also share the consequence that anything that has a function must presently be capable of carrying out that function. (This is controversial among the philosophers who defend the etiological account.) The similarities between Cummins' account and those of the previous two chapters should not be pushed too hard, however. There is no restriction on Cummins' part to properties of traits that make no reference to the environment—indeed, quite the opposite is the case, for most capacities will be specified in relation to environments.

Cummins is also concerned with broader issues in the philosophy of biology. In particular, he rejects the attempts by some philosophers (e.g., Nagel 1961 and Hempel 1965) to show that there is nothing distinctive about functional explanations in biology by assimilating such explanations to patterns of explanations found in the other natural sciences. According to Cummins, previous attempts to understand the notion of function have been misled by a pair of assumptions. One of these is the assumption that functional explanations are of the *presence* of a trait. This, Cummins thinks, misrepresents the character of evolutionary theory, according to which traits do not originate because of their functions, but rather are due to factors such as random genetic mutation. The other assumption is that the performance of a biological function by a trait increases the fitness of the organism that has that trait. Cummins suggests that a counterexample to this assumption is the fact that we would call the enablement of flight a function of pigeon's wings even if circumstances had changed so that flying no longer made a positive contribution to the survival and reproduction of pigeons. By making all functional claims explicitly relative to the analysis of a larger capacity, Cummins can explain, for example, why wings would still have the function of enabling flight under the circumstances where flight was no longer to the advantage of pigeons—because they feature in an analysis of the pigeon's capacity to fly.

The views of philosopher Ernest Nagel provide a backdrop to the arguments of several of the papers in this anthology, but we have chosen to include a 1977 paper of Nagel's for inclusion in this volume. We include this paper at this point because, as the title "Teleology Revisited" suggests, it represents a return to address various challenges presented to Nagel's earlier work. There is a wealth of material in the paper that is too great to summarize here, but one of the main features is a careful discussion of the distinction between function and goal-directedness. Nagel criticizes Mayr's conflation of being controlled by a program with being goal-directed, pointing out that being controlled by a program does not entail having a goal. (This is the flip of Ayala's point that having a goal or end does not entail having a program.)

Nagel's view is that functional statements in biology are equivalent to statements without functional terms. For example, he claims that "The function of chlorophyll in plants is to enable plants to perform photosynthesis" is equivalent to "When a plant is provided with water, carbon dioxide, and sunlight, it manufactures starch only if the plant contains chlorophyll." Nagel canvasses various alternative views about function, including those of Cummins, Wright, and Bock and von Wahlert. A major disagreement with Wright concerns Nagel's denial that functional explanations are causal explanations. Nagel contends that while *goal-ascribing* explanations are causal explanations, *functional* explanations are not, because the function of a particular trait does not account causally for the presence of that trait. Rather (p. 238), "explanations of function ascriptions make evident one role some item plays in a given system." But this is to concentrate on an effect rather than a cause—photosynthesis is an effect, not the cause, of chlorophyll.

Nagel's insistence on regarding functions as effects prefigures the forward-looking account of function provided by John Bigelow and Robert Pargetter. "In describing the function of some biological character," they begin, "we describe some presently existing item by reference to some future event or state of affairs." Thus, Bigelow and Pargetter signal explicitly their rejection of backward-looking, etiological analyses of function. After rejecting the idea that functional talk can be eliminated from biology, they strive to show how a forward-looking account of function can be explanatorily relevant yet remain faithful to the principle that there is no such thing as backward causation. They charge that etiological theories, are, in fact, explanatorily vacuous, for if functions are defined in terms of effects that cause the proliferation of traits with those functions, then it is circular to explain the presence of those traits in terms of those functions. Bigelow and Pargetter proffer an account of functions that likens them to propensities: a function of a trait is some effect of the trait that has the propensity to confer a survival advantage to the organism possessing the trait. In this way, functional explanations in biology are likened to those in other sciences that cite propensities to explain certain events (such as the propensity of sugar to dissolve in water).

An entirely different approach to function is provided by philosopher Mark Bedau, who argues that all of the standard naturalizing accounts of teleological language have a common feature: they replace the evaluative component of teleology—what things are *good* for—with a descriptive, value-neutral approach. This, Bedau thinks, is a mistake, and after criticizing the various major accounts, he distinguishes three grades of involvement for value in teleology, ranging from the mere (and possibly accidental) production of good effects (level 1), through selection on the basis of effects that are good, but not necessarily *because* they are good (level 2), to the causal role played by a "telic mechanism," such as the conscious anticipation of good effects, which selects things because they are good (level 3). Bedau argues that biological teleology does not

exemplify the highest grade of teleology: natural selection only accidentally selects the things that are good for organisms, for example, swimming is good for fish, and fins were selected for in fish because they
contribute to swimming, but fins were not selected *because* swimming is
good. (In Sober's [1984] terminology one can say that there is selection
of things that are good for organisms, but not selection *for* things that
are good.) In Bedau's terms, biological teleology is only second-grade
teleology, and it remains controversial insofar as we fail to distinguish
carefully between the second and third grades. Bedau believes that bringing the notion of value to center stage is compatible with a naturalistic
view of biological teleology, wherein "values would be real ineliminable
natural properties, subject to broadly scientific investigation" (Bedau
1991, p. 655). The naturalness of value properties remains an important and unresolved issue in ethical theory (Sayre-McCord 1998), and
although we think it unlikely that Bedau's position will hold much
immediate appeal for practicing biologists (see Allen and Bekoff 1995),
we also believe that it is important to keep all options in view.

Critical Developments

In this section we present a selection of papers that criticize, develop, and
extend the positions introduced so far. Several of these papers are also
much more explicit about the methodology of providing accounts of
teleological functions.

We begin with philosopher Ruth Millikan's defense of an account
of function that is explicitly wedded to natural selection. In her 1984
book *Language, Thought, and Other Biological Categories*, Millikan
constructed a theory of what she called "proper" functions that was
intended to support a theory of linguistic and psychological meaning in
terms of the biological functions of language and thought. Many commentators have characterized Millikan's view as a more sophisticated
version of Wright's, although in the paper included here she denies that
there is any overlap between her theory and Wright's (see her footnote
5). Whatever the merits of likening Millikan's account of function to
Wright's, her theory is backward-looking, and it is not necessarily
restricted to biological function. Her account, however, specifies much

more completely than any previous theory the structural requirements for the existence of a function. This is done in a way that is intended to correspond to the structure of Darwin's theory. These details far exceed any of those presented so far—as she say, it "consumes two chapters" of the earlier book. In the present paper she merely sketches her account before proceeding to her main objectives, which are to contrast her account with others that have been offered, and to take a more reflective look at the objectives of providing an account of function.

A central point of Millikan's paper is to dispute the idea that the philosophical project of characterizing function is one of *conceptual analysis*: the project of finding necessary and sufficient conditions for the application of terms according to (intuitions about) preestablished usage. She argues that the definition of proper function that she provides is intended instead as a *theoretical definition*, to be accepted or rejected insofar as it is part of a successful theory of biological phenomena. Seen in this way, Millikan believes, the philosopher's practice of constructing fanciful hypothetical cases to test intuitions about the applicability of definitions is of rather limited usefulness. One example that has been used frequently to challenge etiological accounts of function is that of a "structural double" (this term is from Mitchell, chapter 14; see also Neander, chapter 11)—a creature that, for instance, is indistinguishable from a human being but comes into existence by an unusual route, such as a massive coincidence of quantum effects. Many philosophers have the intuition that such a creature would have a heart whose function it is to pump blood (and Cummins' account can explain this intuition) but Millikan rejects examples of complex creatures created ex nihilo (and hence, on her account, without function) as irrelevant to the project of understanding the role that functions play in actual biological systems.

Millikan argues for the superiority of her theory over non-etiological accounts given its ability to provide the correct account of things that fail to perform their functions. She believes that alternative accounts of function, such as Cummins', yield the wrong theoretical results, for they imply (as Cummins indeed explicitly noted) that anything that has a function must be capable of carrying out that function. But as Millikan points out, it can be the function of a diseased heart to pump blood even if it is utterly incapable of doing so. The backward-looking theory of function can explain this because it is the pumping of blood by ancestral hearts that accounts, via natural selection, for the function of the present organ no matter how deformed

We have also included a paper by philosopher Karen Neander, who espouses an etiological account of function but disagrees with Millikan's claim that the philosopher's task is more like theoretical definition than conceptual analysis. On Neander's view theoretical definition and conceptual analysis are interdependent, particularly when it is scientific concepts that are being analyzed. Neander then proceeds to respond to three objections commonly raised against etiological accounts of function. The first involves Harvey's attributions of function to hearts in the absence of any knowledge of natural selection. Neander's response is that it is possible to identify things correctly, even if the basis for the identification is faulty. The second objection is that even if Darwin's theory is false, there would still be proper functions, so it is incorrect to define functions in terms of Darwinian theory. Neander's response to this objection is that if the theory of evolution by natural selection is false, this would show only that the definition is false, but it would not show that the etiological account is an incorrect analysis of biologists' current concept of function. The third objection is that of the possibility of creatures, or traits of creatures, that do not have a standard biological ancestry. Millikan rejected such examples on the grounds that they were irrelevant to theoretical definitions of function, but this option is not available to Neander because she accepts the goals of conceptual analysis. Instead, she argues against the intuition that functions could be attributed in the absence of historical knowledge. Even if the argument at the level of intuitions is not convincing, Neander believes that a direct attack on ahistorical accounts of function can be mounted. Like Millikan, she argues that none of the alternatives can account for the *normativity* of function, including the fact that biological structures are *supposed* to perform certain functions whether or not they are capable of performing that function.

Philosopher Ron Amundson and morphologist George Lauder contest what they see as the encroaching orthodoxy that etiological accounts of function are uniquely appropriate to evolutionary biology. Rather, they argue, a Cummins-style analysis should be preferred for the purposes of comparative and functional anatomy. They believe that etiological accounts place biologists in an untenable position because of the empirical difficulties involved in establishing the historical facts said to underlie function attributions. Amundson and Lauder defend Cummins' notion of function against the criticisms made by Millikan, Neander, and others. In particular, they maintain that the etiologists' criticisms of Cummins' view are based on unequal treatment of the rival accounts of function with respect to their application to ordinary, nonscientific statements about function. Amundson and Lauder detail the ways in which organismic traits are in fact identified morphologically rather than functionally, thus arguing against the claim that the identification of traits in terms of their historical function is essential to biological practice.

Philosophers Berent Enç and Fred Adams, in the next paper, compare etiological with forward-looking accounts of biological teleology and conclude that while etiological accounts have the upper hand on issues relating to the attribution of function, there is nothing to choose between the two sorts of accounts if the issue is causal explanation of particular traits of organisms. They argue for a noncausal approach to teleological explanation, according to which teleological language serves to group together phenomena involving diverse causal mechanisms. According to the Thesis of Multiple Realizability, different microlevel properties can underlie the same functional properties. This means that claims made at the functional level have a higher level of generality than claims that focus on microlevel properties. Teleology, according to Enç and Adams, is concerned with noncausal explanation at a higher level of generality than is provided by the microlevel properties.

Sandra Mitchell, a philosopher as well, also takes up the metalevel question of how accounts of function are to be understood, arguing that before assessing particular accounts of function, it is important to understand the goals of providing an account. Functions, she claims, are not concrete substances like gold, or water, and so it is misleading for Neander and Millikan to use the analysis and definition of substance terms as models for understanding questions about function. On her view, "function" is more like "cause" or "reason"—it is an abstract term that designates an explanatory structure. Instead of asking about the *meaning* of function terms, Mitchell thinks that it is more fruitful to understand the explanatory structure required to make sense of certain functional claims in both biology and cultural anthropology. This structure, Mitchell argues, requires the identification of selection and transmission mechanisms that provide the causal background to explain the presence of traits (including behaviors) in a given population.

In the context of giving this kind of explanation, Mitchell finds difficulties for Bigelow and Pargetter's forward-looking, dispositional account of function. She argues that Batesian mimicry presents a problem for their view because, for example, the coloration pattern of a poisonous butterfly (the Monarch) and its mimic (the Viceroy), have very similar dispositional properties but different biological functions: the function of the Monarch's coloration is to warn predators about poison, whereas the function of the Viceroy's is to mimic the Monarch's. Despite her objections to the view, Mitchell thinks that Bigelow and Pargetter have identified an explanatory concept that is important to biology. Specifically, their account of function corresponds to the explanatory task of saying how a trait contributes to the survival and reproduction of organisms with that trait. In contrast, the etiological account of function corresponds to the task of saying how having the trait has historically contributed to the presence of the trait. Mitchell concludes (p. 411) that "there is not a single, univocal explanatory task for which such language is employed" and that "the philosophical task is to recognize the plurality of explanatory projects, to clarify their relationships, and to explicate their structures." (See also Bekoff and Allen 1995.)

Synthesis or Pluralism?

Each of the articles in this section attempts to come to terms with the plurality of uses of the term *function* by biological scientists. As an ethologist, Robert Hinde's interest is in behavior. He begins by identifying the functions of behavior as the beneficial consequences to the organism of performing the behavior. He distinguishes between "weak" and "strong" senses of function. Hinde's notion of the weak functional meaning could be construed as an account similar to Cummins'. On Hinde's view: "A weak meaning answers the question 'what is it good for?' ... By contrast, 'function' in a strong sense attempts to answer the question 'through what consequences does natural selection act to maintain this character?'" (p. 422). Hinde's strong sense appears to be a version of the etiological account: the maintenance of the trait in the organism(s) is causally explained by reference to natural selection. But in his weak sense of function, what is explained is how a certain trait contributes to some further ability of the organism. Hinde's example is that wings contribute to the capacity for flight. This contribution can be understood independently of knowing how (or even whether) flight contributes to survival and reproductive success, and hence to the presence of the trait. In fact, however, Hinde claims that the weak and strong meanings of function represent the ends of a spectrum, rather than discrete alternatives. Because questions of function are "still too often treated as a matter for speculative asides," Hinde takes up the question of evidence for functional attributions and argues that both observational and experimental approaches are useful, although laboratory conditions tend to support claims about function only in the weak sense.

Philosopher Paul Griffiths seeks to show that the etiologically-defined proper functions of a biological trait can be identified with the functions attributed by a Cummins-style functional analysis of the fitness of the organism's ancestors. A difference for Griffiths between *proper* functions and functions more generally is that the former can be cited to explain the presence of a trait. This is the core of etiological accounts of proper function, yet Griffiths maintains that many of the available etiological theories are flawed. He criticizes Millikan's account for failing to capture the important distinction between a functional trait and a vestige. Neander's view is criticized on the ground that it misrepresents the structure of explanation by natural selection—in particular, Griffiths claims, Neander's account improperly requires an increase in the frequency of genes that code for a trait in order for that trait to have a proper function. Griffiths' own account of proper function invokes the fitness contribution made by a trait's effects that helps to provide a selective explanation for the presence of the trait in the population, and he shows how this idea must be precisely stated to preserve the biologically important distinction between traits with functions and vestigial traits. Because of the explicit role played by the theory of natural selection in his account, Griffiths denies that it is possible to give a perfectly unified account of biological function and artifact function. Nonetheless, a generalization of the notion of selection does, he believes, lead to a "selection-type" theory of artifact function. Ultimately, for Griffiths, it is some form of selection that is responsible for the presence of teleology.

A problem for etiological accounts is the difficulty of saying how far back in history one should go to determine function. That birds' wings are descended from the fins of fish does not show that wings are for swimming. Peter Godfrey-Smith, a philosopher, proposes to solve this problem by stipulating that (p. 453) "Biological functions are dispositions or effects a trait has which explain the recent maintenance of the trait under natural selection." This account is etiological and historical but attempts to draw a line limiting how far back one should go to determine function. Godfrey-Smith also views his project as (p. 454) "conceptual analysis guided more by the demands imposed by the role the concept of function plays in science, the real weight it bears, than by informal intuitions about the term's application." Many ethologists, often citing Tinbergen (1951, 1963), rely on a distinction between evolutionary explanations and functional explanations (see also Brandon, chapter 3). Godfrey-Smith regards this as a potential embarrassment for historical accounts of function. He proposes to remove the embarrassment by granting (as did Millikan 1989) that a certain amount of ahistorical usage of *function* by biologists can be captured by Cummins' account, but that the core etiological sense of function is a type of evolutionary explanation that appeals explicitly to recent propensities that, in the vast majority of cases, should be expected to have continued into the present. In this way, Godfrey-Smith believes he has steered a "principled middle course" between backward-looking and concurrent or forward-looking accounts.

Philosopher Philip Kitcher takes on the multiplicity of theories about function ascriptions, and argues that many of them—including pre-Darwinian conceptions—can be unified through the notion of design according to the schema that "the function of an entity S is *what S is designed to do*" (p. 479). According to Kitcher, either natural selection or agent intentions can underlie attributions of function, but the former is more significant for understanding function attributions in biology. He laments, however, that Tinbergen's distinction between "why" questions answered by selection history and "why" questions answered by current function is collapsed by standard etiological analyses of function that appeal to the history of selection as the source of the initial spread of a trait. Kitcher sides with Godfrey-Smith in preferring an analysis of function that looks to recent selection history. Although he regards Cummins-style functional analysis as "too liberal" because any complex system can be subjected to such an analysis, Kitcher, by way of several examples drawn from the biological literature, attempts to show that Cummins-style analysis is an important tool in the attempt to analyze the design of biological systems. Furthermore, because both natural selection and conscious intention can be sources of design, Kitcher believes that he has shown how a single account of function can cover both artifacts and biological entities.

Design

The notion of design in modern biology has a long history stretching most notably from Hume's (1779) criticisms of all versions—nonbiological as well as biological—of the "argument from design," to Paley's (1802) famous attempt to illustrate that design of any kind implies the existence of a designer, to current biological uses of the concept of natural selection to demonstrate design and function. Evident still in modern biology is the tension between two views: one of design as manifest evidence of the necessity for teleological explanations, the other of design as a result of some process (such as natural selection acting on random variation) that does not require teleology (Dawkins 1987). A pervasive component of many current explanations of biological traits, especially those deemed adaptations, is the notion of design for some function (Rose and Lauder 1996), and biologists seem to have a difficult time escaping the use of teleological language in their descriptions of design. Still, this teleological tension is evident in the papers chosen here by the contrasting views on the nature of design, its origin, and the methods advocated for the analysis of design.

We begin the section on design with a paper by George Lauder, who tackles what he calls "the problem of biological design"—the twin problems of explaining why relatively few of the theoretically possible phenotypes have actually evolved and of formulating testable explanations for why some morphological forms have not evolved. Design and (implicitly) function are distinguished from adaptation by the involvement of natural selection in defining *adaptation*: Lauder defines design (p. 508) as "the organization of biological structure in relation to function" whereas adaptation "is restricted to features that have arisen by means of natural selection." According to Lauder, the most common approach to the problem of design is to take a mechanical or biochemical approach to the relationship between form and function that sees organisms as in equilibrium with environmental forces. An alternative approach analyzes biological structures historically in an attempt to discover the intrinsic determinants of evolutionary change, that is, those features of organisms that facilitate morphological change in certain directions but inhibit changes in other directions. (For example, one might try to explain why it is that, despite independent origins, the wings of bats and the wings of birds both evolved from forelimbs rather than hindlimbs: is there something intrinsic to the vertebrate body plan that favors the evolution of forelimbs into wings and blocks the evolution of wings from hindlimbs?) Lauder claims that the problem of biological design is fundamentally historical. Although the historical approach is more difficult because it is harder to test, Lauder argues that the goals of historical morphology can be effectively carried out by selecting general (not unique) features for explanation, explicitly formulating hypotheses about the phylogenetic sequence of feature acquisition, and examining related monophyletic lineages (lineages containing all the descendants of a given ancestor).

Biologists Stephen Jay Gould and Elisabeth Vrba also have the needs of morphologists in mind as they set out to clarify the term *adaptation* and to argue that evolutionary morphologists need new labels for related concepts. They distinguish adaptation in the sense of "the historical processes of change or creation for definite functions" (p. 520) from adaptation in the sense of the utility of a trait for an organism in its current environment, for which they prefer to use the term *aptation*. Gould and Vrba describe traits of organisms which were originally adapted for some function (e.g., feathers for insulation) but then coopted for other effects (e.g., for insect catching) in the course of evolution "that sets the basis for a subsequent adaptation" (e.g., for flight). At certain intermediate stages, although the later adaptive changes have not yet occurred, the basis for these changes is present, and here Gould and Vrba suggest using term *exaptation*. Thus, feathers were initially adapted for flight and exapted for insect-catching; changes due to selection led to adaptation for insect-catching and exaptation for flight; further changes result in adaptation for flight. Exaptation may also appropriately describe nonadaptive traits (or *nonaptations*) that are coopted for a particular effect when the opportunity arises—the opportunistic nature of this process eliminates any suggestion of teleological directedness. Gould and Vrba argue that, in contrast, the commonly used *preadaptation* has awkward connotations about preordination of future adaptations, and they suggest that the term *preaptation* more accurately captures the idea that a trait may be the substrate for future adaptive development. Gould and Vrba suspect that evolutionists have not formally recognized the concept of exaptation because they have overplayed the claim that adaptation is the primary driving force in evolution, and have fallaciously drawn inferences about historical genesis from observations of current utility. In their view, exaptations that begin as nonaptations may be of far greater importance for understanding evolution.

In the next chapter, morphologist Carl Gans defends the view that adaptation is the primary basis for understanding "the match of form and function" which is advantageous to organisms, an idea that Gans traces to pre-Darwinian times. Gans nods toward Bock and von Wahlert's separation of function and biological role, but proceeds to use *function* to mean what they meant by *biological role*. He then notes the ambiguity in the term *adaptation* (also noted by Gould and Vrba, chapter 20) between, on the one hand, the present state of an organism being more or less well equipped to survive in its current environment and, on the other hand, the historical process of becoming adapted in the first sense. Gans contends that many of the worries about adaptationism—a thesis about the capacity of natural selection to account for the phenotypic properties of organisms—are derived from an inadequate appreciation of the distinction between the two senses of adaptation. Gans wisely rejects any identification of adaptation with perfection and tackles the question of how hypotheses about adaptation can be tested.

Although many authors who accept an etiological account of function treat the notions of function and design inseparably, philosopher Colin

Allen and ethologist Marc Bekoff argue that not only can these notions be separated, they should be, as their separation helps articulate important issues in the evolution of morphological and behavioral traits. The definition they offer is based on an etiological account of function, although they acknowledge that other conceptions of function have roles to play in biology. Allen and Bekoff adopt an etiological conception of function which they claim does not require any special adaptation of the trait for its function. If, for example, eagles use their wings to shade nestlings, this could be considered a function for which the wings have not been specifically designed. In contrast, when adaptive change has occurred in the trait through natural selection, making the trait more optimal or better adapted for the performance of the function, then it becomes appropriate to say that the trait is designed for a function. Thus, the glide ratio of eagle wings can be compared to the glide ratio of the wings of eagle ancestors. If an eagle's wings result in a higher glide ratio than its ancestor's wings and having a higher glide ratio provides a comparative fitness advantage, then it can be said that the eagle's wings are designed for soaring. Allen and Bekoff motivate many of their claims about design by means of direct comparison to psychological cases of design, but they insist that biological design should be analyzed separately from psychological design. They end their paper by considering the role of comparative studies in assessing claims about design, and the particular difficulties in the area of animal behavior due to the lack of fossilized information about behavior.

Acknowledgments

We are grateful to the authors and publishers for their permissions to reprint these works, and to Betty Stanton and The MIT Press for supporting this project. We would also like to thank Mike "Jimbob" Martin for his assistance in preparing the manuscript and the index.

References

Allen, C. and Bekoff, M. 1995. Function, Natural Design, and Animal Behavior: Philosophical and Ethological Considerations. *Perspectives in Ethology* 11: 1–47.

Bedau, M. 1991. Can Biological Teleology Be Naturalized? *Journal of Philosophy* 88, 647–57.

Bekoff, M. and Allen, C. 1995. Teleology, Function, Design, and the Evolution of Animal Behaviour. *Trends in Ecology and Evolution* 10(6): 253–255.

Buller, D. (ed.). 1998. *Function, Selection, and Design.* SUNY Press, Albany, New York.

Dawkins, R. 1987. *The Blind Watchmaker.* W. W. Norton, New York.

Dawkins, R. 1996. *Climbing Mount Improbable.* W. W. Norton, New York.

Hempel, C. G. 1965. *Aspects of Scientific Explanation.* Free Press, New York.

Hume, D. 1779. *Dialogues Concerning Natural Religion.* 1991 reprint edited by S. Tweyman. Routledge, London.

Lipton, P. and Thompson, N. S. 1988a. Comparative Psychology and the Recursive Structure of Filter Explanations. *International Journal of Comparative Psychology* 1(4): 215–229.

Lipton, P. and Thompson, N. S. 1988b. Response: Why Dr. Tinbergen Is more Sound than Dr. Pangloss. *International Journal of Comparative Psychology* 1(4): 238–244.

Mayr, E. 1961. Cause and Effect in Biology. *Science* 134: 1501–1506.

Mayr, E. 1965. Cause and Effect in Biology. In Lerner, D. (ed.) *Cause and Effect.* Free Press, New York.

Mayr, E. 1988. *Towards a New Philosophy of Biology.* Harvard University Press, Cambridge, MA.

Millikan, R. G. 1984. *Language, Thought, and Other Biological Categories.* MIT Press, Cambridge, MA.

Millikan, R. G. 1989. An Ambiguity in the Notion of Function. *Biology and Philosophy* 4: 172–176.

Nagel, E. 1961. *The Structure of Science: Problems in the Logic of Scientific Explanation.* Harcourt, Brace and World, New York.

Nagel, E. 1965. Types of Causal Explanation in Science. In Lerner, D. (ed.) *Cause and Effect.* Free Press, New York.

Paley, W. 1802. *Natural Theology, or Evidences of the Existence and Attributes of the Deity Collected from the Appearances of Nature.* Faulkner, London.

Pittendrigh, C. S. 1958. Adaptation, Natural Selection, and Behavior. In Roe, A. and Simpson, G. G. (eds.). *Behavior and Evolution.* Yale University Press, New Haven, CT, pp. 390–419.

Rescher, N. (ed.). 1986. *Current Issues in Teleology.* University Press of America, Lanham, MD.

Rose, M R. and Lauder, G. V. (eds.). 1996. *Adaptation.* Academic Press, San Diego.

Sayre-McCord, G. (ed.). 1988. *Essays on Moral Realism*. Cornell University Press, Ithaca, NY.

Sober, E. 1984. *The Nature of Selection*. MIT Press, Cambridge, MA.

Sober, E. (ed.). 1984. *Conceptual Issues in Evolutionary Biology*. MIT Press, Cambridge, MA.

Thompson, N. 1987. The Misappropriation of Teleonomy. *Perspectives in Ethology* 7: 259–274.

Tinbergen, N. 1951. *The Study of Instinct*. Oxford University Press, New York.

Tinbergen, N. 1963. On Aims and Methods of Ethology. *Zeitschrift für Tier-psychologie* 20: 410–429.

Sources

Ayala, F. J. (1970). Teleological Explanations in Evolutionary Biology. *Philosophy of Science* 37: 1–15. Reprinted with the permission of the author and the publisher, University of Chicago Press.

Wright, L. (1973). Functions. *Philosophical Review* 82: 139–168. Reprinted with the permission of the author and the publisher.

Brandon, R. N. (1981). Biological Teleology: Questions and Explanations. *Studies in the History and Philosophy of Science* 12: 91–105. Reprinted with the permission of the author and the publisher, Elsevier Science Publications Ltd., Oxford, England.

Rudwick, M. J. S. (1964). The Inference of Function from Structure in Fossils. *British Journal for the Philosophy of Science* 15: 27–40. Reprinted with the permission of the author and the publisher.

Bock, W., and von Wahlert, G. (1965). Adaptation the Form-Function Complex. *Evolution* 19: 269–299. Reprinted with the permission of the authors and the publisher.

Cummins, R. (1975/1984). Functional Analysis. *Journal of Philosophy* 72: 741–765. The version here is that reprinted with minor alterations in Sober (1984) *Conceptual Issues in Evolutionary Biology*, MIT Press, pp. 386–407. Reprinted with the permission of the author and the publishers.

Nagel, E. (1977). Teleology Revisited. *Journal of Philosophy* 76: 261–301. Reprinted with the permission of the publisher.

Bigelow, J. and Pargetter, R. (1987). Functions. *Journal of Philosophy* 86: 81–196. Reprinted with the permission of the authors and the publisher.

Bedau, M. (1992). Where's the Good in Teleology? *Philosophy and Phenomenological Research* 52: 781–805. Reprinted with the permission of the author and the publisher.

Millikan, R. G. (1989). In Defense of Proper Functions. *Philosophy of Science* 56: 288–302. Reprinted by permission of the author and the publisher, University of Chicago Press.

Neander, K. (1991). Functions as Selected Effects: The Conceptual Analyst's Defence. *Philosophy of Science* 58: 168–184. Reprinted by permission of the author and the publisher, University of Chicago Press.

Amundson, R. and Lauder, G. V. (1994). Function without Purpose: The Uses of Causal Role Function in Evolutionary Biology. *Biology and Philosophy* 9: 443–469. Reprinted with the permission of the authors and with kind permission from Kluwer Academic Publishers.

Enç, B. and Adams, F. (1992). Functions and Goal-Directedness. *Philosophy of Science* 59: 635–654. Reprinted with permission of the authors and the publisher, University of Chicago Press.

Mitchell, S. D. (1995). Function, Fitness and Disposition. *Biology & Philosophy* 10: 39–54. Reprinted with the permission of the author and with kind permission from Kluwer Academic Publishers.

Hinde, R. A. (1975). The Concept of Function. In Baerends, G., Beer, C. and Manning , A. eds. (1975). *Function and Evolution in Behaviour: Essays in Honor of Niko Tinbergen.* Clarendon Press, Oxford, pp. 3–15. Reprinted with the permission of the author and the publisher.

Griffiths, P. E. (1993). Functional Analysis and Proper Functions. *British Journal for the Philosophy of Science* 44: 409–422. Reprinted with the permission of the author and the publisher.

Godfrey-Smith, P. (1994). A Modern History Theory of Functions. *Noûs* 28: 344–362. Reprinted with the permission of the author and the publisher.

Kitcher, P. (1993). Function and Design. *Midwest Studies in Philosophy* 18: 379–397. Reprinted with the permission of the author and the publisher.

Lauder, G. V. (1982). Historical Biology and the Problem of Design. *Journal of Theoretical Biology* 97: 57–67. Reprinted with the permission of the author and the publisher.

Gans, C. (1988). Adaptation and the Form-Function Relation. *American Zoologist* 28: 681–697. Reprinted with the permission of the author and the publisher.

Gould, S. J. and Vrba, E. S. (1982). Exaptation—A Missing Term in the Science of Form. *Paleobiology* 8: 4–15. Reprinted with the permission of the authors and the publisher.

Allen, C. and Bekoff M. (1995). Biological Function, Adaptation, and Natural Design. *Philosophy of Science* 62: 609–622. Reprinted by permission of the authors and the publisher, University of Chicago Press.

I

Looking Backwards: Teleology as Etiology

1

Teleological Explanations in Evolutionary Biology

Francisco J. Ayala

Early in the nineteenth century, William Paley in his *Natural Theology* [6] pointed out the obvious functional design of the human eye. For Paley, it was absurd to suppose that the human eye, by mere chance, "should have consisted, first, of a series of transparent lenses (very different, by and by, even in their substance from the opaque materials of which the rest of the body is, in general at least, composed; and with which the whole of its surface, this single portion of it excepted, is covered) secondly of a black cloth or canvas (the only membrane of the body which is black) spread out behind these lenses so as to receive the image formed by pencils of light transmitted through them; and placed at the precise geometrical distance at which, and at which alone, a distinct image could be formed, namely at the concourse of the refracted rays: thirdly, of a large nerve communicating between this membrane and the brain."

The adaptive character of the structures, organs, and behavior of plants and animals is an incontrovertible fact. The vertebrate eye, with its complicated anatomy of highly specialized tissues, is obviously adapted for vision; the hand of man is made for grasping, and the bird's wing for flying. Organisms show themselves to be adapted to live where they live and the way they live. To explain the phenomenon of the adaptation of life is one of the main objectives of natural science and of natural philosophy.

Before 1859, the year Darwin published *The Origin of the Species*, the adaptation of organisms was either accepted as a fact without any explanation of its origin, or more frequently, it was attributed to the omniscient design of the Creator. God had given wings to birds so that they might fly, and had provided man with kidneys to regulate the composition of his blood. For Paley, living nature is a manifestation of the existence an wisdom of the Creator.

In *The Origin of the Species* Darwin accumulated an impressive number of observations supporting the evolutionary origin of living organisms. Moreover, and perhaps most importantly, he provided a causal explanation of evolutionary processes—the theory of natural selection. The principle of natural selection, as Darwin saw it, makes it possible to give a natural explanation of the adaptation of organisms to their environment. With *The Origin of the Species* the study of adaptation, the problem of design in nature came fully into the domain of natural science.

Darwin recognized, and accepted without reservation, that organisms are adapted to their environments, and that their parts are adapted to the functions they serve. Fish are adapted to live in water, the hand of man is made for grasping, and the eye is made to see. Darwin accepted the facts of adaptation, and then provided a natural explanation for the facts. One of his greatest accomplishments was to bring the teleological aspects of nature into the realm of science. He substituted a scientific teleology for a theological one. The teleology of nature could now be explained, at least in principle, as the result of natural laws manifested in natural processes, without recourse to an external Creator or to spiritual or nonmaterial forces. At that point biology came into maturity as a science.

1   Hereditary Variability   About the time Darwin published *The Origin of the Species*, Gregor Mendel was performing in his Augustinian monastery in Brünn (Austria) experiments with peas. The results of such experiments, published in 1866, provided the fundamental principles of heredity. The Mendelian principles remained generally unknown until 1900, when they were independently and nearly simultaneously rediscovered by three biologists. The principles of heredity were extended during the nineteen hundreds to a considerable number of species of plants and animals. A whole body of knowledge concerning heredity blossomed. The biological or synthetic theory of evolution as we know it today is a synthesis of Darwin's principle of natural selection and genetic knowledge. It is in essence a two-factor theory. Mutation is the ultimate source of hereditary variability; natural selection is the directional factor that results in organized complexes of hereditary material and in adaptation.

Heredity is the transmission, from parent to offspring, of the information that directs the development of the fertilized egg to its adult stage and controls the living activity of the organism. The hereditary information is carried in a chemical substance known as deoxyribonucleic acid (DNA). Molecules of DNA exist in discrete but complexly interacting units called genes. The genes are organized in chromosomes, which exist in sets. One or more sets of chromosomes—most frequently two in higher organisms —exist in the fertilized egg cell (zygote) from which the adult individual develops. In sexually reproducing organisms, one of the two sets of chromosomes is inherited from each parent via the sex cells.

The genes of a population are shuffled and combined in different ways every generation. In the process of genetic recombination during the formation of the sex cells (gametes), the two sets of hereditary material received by each individual from its parents are combined in different ways. The sex cells carry a single set of genes each, representing combinations in different proportions of the two sets possessed by the individual. Fertilization brings together two sex cells in the zygote from which the mature individual develops. Gametic recombination and fertilization create new combinations of genes and chromosomes every generation. These new sets of information are tested against the environment where the individual lives. Thus, genetic experimentation, so to speak, occurs in all natural populations every generation.

The sum total of genetic information in a population of sexually interbreeding individuals can be thought of as the "gene pool" of the population. The gene pool of a population is characterized by the totality of genes in the population, their combinations, and the relative frequencies of both among the individuals of the population. Evolution consists in changes in the gene pool of a population. Recombination produces new combinations of genes but by itself it does not change the gene pool. There are four known processes which can do so—mutation, random fluctuation of genetic frequencies known as "sampling errors," migration of individuals in and out of the population, and natural selection. The first three of these processes are essentially random. Although the relative importance for evolution of random genetic sampling has been questioned, it must have played a role in certain instances—in particular, when a new environment is colonized by a small number of individuals and when populations are reduced to few individuals in their usual environments by drastic environmental stresses. For our present purpose we need consider neither random sampling nor migration.

Genes are fairly stable entities but not completely so. Occasionally, mutations occur. The frequencies of mutations vary for different genes and for different organisms. It is probably fair to estimate the frequency of a majority of mutations in higher organisms between one in ten thousand, and one in a million per gene per generation. Mutations in a broad sense include not only changes in the hereditary information of single genes, but also changes in the arrangement and distribution of genes in chromosomes, and in the number of chromosomes and sets of chromosomes. Mutations have sometimes been described as "errors" in the replication of the hereditary material. Such a description may be misleading, since the alleged "errors" are the ultimate source of evolutionary change. Mutation provides the raw materials of evolution, i.e., mutation is the ultimate source of genetic variability.

Mutations are random changes of the hereditary material. They are random in the sense that they occur independently of the needs of the organism in which they happen. Most new mutations are in fact harmful to the organism. If mutation were the only factor promoting genetic change in a population, it would result in an array of freaks and finally in total disorder. The genetic information stored in the DNA of the population would ultimately disintegrate. However, there is a directive process that counteracts mutation and results in order and adaptation—natural selection. Natural selection is able to produce and to preserve the stored information transmitted by the hereditary process.

2 Natural Selection Natural selection was Darwin's major contribution to the explanation of the evolution of life. For Darwin, natural selection was primarily differential survival. The modern understanding of the principle of natural selection derives from Darwin's concept, although it is formulated in a somewhat different way. Natural selection is understood today in genetic and statistical terms as differential reproduction. Differential reproduction is a compound process, the elements of which are differential survival, differential mating success, and differential fecundity. Natural selection implies that some genes and genetic combinations are transmitted to the following generation on the average more frequently than their alternates. Such genetic units will become more common in every subsequent generation and their alternates less common. Natural selection is a statistical bias in the relative rate of reproduction of alternative genetic units.

Genes and gene combinations are the entities subject to natural selection. Genes do not exist by themselves but in organisms. Genes increase or decrease in relative frequency depending on their average effects in the organisms which carry them. The process of natural selection can be also predicated of individual organisms—and in a less precise sense, of populations of organisms as well—in the sense that some organisms leave more progeny than others. Individual organisms are not lasting, however. Genes persist in the progenies of the organisms which carry them.

Natural selection is a process determined by the environment. The selective advantage of certain genetic variants must be understood in relation to the environment where the population lives. A genetic unit which is favorably selected in one environment may be selected again in a different one. A trivial example is that wings—and therefore the genes responsible for the development of wings—may increase the reproductive success of a bird, but will probably be of no advantage, and presumably will be disadvantageous, to a deep sea fish. To speak of the environment of a population is, however, an oversimplification. The environment is highly heterogeneous both in the dimension of time and in the dimension of space. The environment of a population includes all the physical and biotic elements affecting the individuals of the population in the whole range of their geographic distribution. Small or large differences in climate, food resources, competitors, etc. exist within the spatial distribution of any population. Moreover, no environment remains constant in time. It changes from morning to night, from one season to another, from one year to the next. The reproductive fitness of a genetic variant is then the average result of the effects of that genetic unit in all the environments where the population lives. It may change from one to another generation as the biotic and physical environments of the population change. Environmental diversity and environmental change are responsible for the continuous evolution of natural populations. If life existed in only a single uniform and constant environment, evolution might conceivably have produced a genotype optimally fitted to that environment with no further change. An absolutely uniform and constant environment is an abstraction; it does not exist in nature.

Genes act in concert with other genes. The average effect of a gene in a population may vary depending on the other genes and genetic combinations existing in the population. The reproductive fitness of a genetic unit must be understood as the average effect it has on all the individuals carrying it. That average effect is likely to change as the genetic composition of the population changes from generation to generation.

The numbers of alternative genetic variants existing in a natural population is a debated question, but they vary widely for different kinds of organisms. If two variants, A_1 and A_2, of a gene exist in a population, there are in diploid organisms three possible different genotypes with respect to that gene, namely A_1A_1, A_1A_2, and A_2A_2. If the number of genes existing in two alternative forms is n, 3^n different genotypes are possible. That number becomes very large as n increases. For instance, if n equals 10, the number of possible different genotypes in nearly one hundred thousand; if n equals 20, there are more than one billion potential genotypes; and if n equals 30, there are nearly one million billion possible different genotypes. The number of possible genetic combinations in a population of diploid organisms, even in those organisms carrying relatively few alternative genetic units, is enormous. Most of them will never occur because the number of individuals in the population is much less than the number of possible different genetic combinations. Natural selection operates exclusively on the genetic combinations actually realized in the population.

If a gene or genetic combination increases on the average the reproductive success of the individuals carrying it, its frequency in the population will increase gradually. It has been shown both theoretically and experimentally that a newly arisen genetic unit will swamp the population in relatively few generations, even if the advantage over its alternative forms is moderately small.

Natural selection has been compared to a sieve which retains the rarely arising useful and lets go the more frequently arising harmful mutants. Natural selection acts in that way, but it is much more than a purely negative process, for it is able to generate novelty by increasing the probability of otherwise extremely improbable genetic combinations. Natural selection is creative in a way. It does not "create" the genetic entities upon which it operates, but it produces adaptive genetic combinations which would not have existed otherwise. The creative role of natural selection must not be understood in the sense of the "absolute" creation that traditional Christian theology predicates of the Divine act by which the universe was brought into being *ex nihilo*. Natural selection may be compared rather to a painter which creates a picture by mixing and distributing pigments in various ways over the canvas. The canvas and the pigments are not created by the artist but the painting is. It is conceivable that a random combination of the pigments might result in the orderly whole which is the final work of art. Some modern paintings look very much like a random association of materials, to be sure. But the probability of, say, Leonardo's *Mona Lisa* resulting from a random combination of pigments is nearly infinitely small. In the same way, the combination of genetic units which carries the hereditary information responsible for the formation of the vertebrate eye could have never been produced by a random process like mutation. Not even if we allow for the three billion years plus during which life has existed on earth. The complicated anatomy of the eye like the exact functioning of the kidney are the result of a nonrandom process—natural selection.

How natural selection, a purely material process, can generate novelty in the form of accumulated hereditary information may be illustrated by the following example. Some strains of the colon bacterium, *Escherichia coli*, to be able to reproduce in a culture medium, require that a certain substance, the amino acid histidine, be provided in the medium. When a few such bacteria are added to a cubic centimeter of liquid culture medium, they multiply rapidly and produce between two and three billion bacteria in a few hours. Spontaneous mutations to streptomycin resistance occur in normal, i.e., sensitive, bacteria at rates of the order of one in one hundred million (1×10^{-8}) cells. In our bacterial culture we expect between twenty and thirty bacteria to be resistant to streptomycin due to spontaneous mutation. If a proper concentration of the antibiotic is added to the culture, only the resistant cells survive. The twenty plus surviving bacteria will start reproducing, however, and allowing a few hours for the necessary number of cell divisions, several billion bacteria are produced, all resistant to streptomycin. Among cells requiring histidine as a growth factor, spontaneous mutants able to reproduce in the absence of histidine arise at rates of about four in one hundred million (4×10^{-8}) bacteria. The streptomycin resistant cells may now be transferred to an agar-medium plate with streptomycin but with no histidine. Most of them will not be able to reproduce, but about a hundred will start dividing and form colonies until the available medium is saturated. Natural selection has produced in two steps bacterial cells resistant to streptomycin and not requiring histidine for growth. The probability of the two mutational events happening in the same bacterium is of about four in ten million billion $(1 \times 10^{-8} \times 4 \times 10^{-8} = 4 \times 10^{-16})$ cells. An event of such low probability is unlikely to occur even in a large laboratory culture of bacterial cells. With natural selection cells having both properties are the common result.

Natural selection produces highly improbable combinations of genes by proceeding step-wise. The human eye did not appear suddenly in all its present perfection. It requires the appropriate integration of many genetic units, and thus it could not have resulted from a random process. Our ancestors have had for at least the last half billion years some kind of organs sensitive to light. Perception of light, and later vision, were important for their survival and reproductive success. Natural selection accordingly favored genes and gene combinations increasing the functional efficiency of the eye. Such genetic units gradually accumulated eventually leading to the highly complex and efficient human eye.

Natural selection can account for the rise and spread of genetic constitutions and therefore of types of organisms, that would never have existed under the uncontrolled action of random mutation and recombination of the hereditary materials. In this sense, although it does not create the raw materials, that is, the genes, selection is definitely creative.

3 Natural Selection and Adaptation   Evolutionary changes in the gene pool of a population frequently occur in the direction of increased adaptation. The organisms likely to leave more descendants are those whose variations are most advantageous as adaptations to the environment. Natural selection, however, occurs in reference to the environment where the population presently lives. Evolutionary adaptations are not anticipatory of the future. The environmental challenges that a population may meet in the future cannot affect in any way the reproductive fitness of the organisms in the present environment. If the population is unable to react adaptively to a new environmental challenge, the result may be extinction. The fossil record bears witness that a majority of the species living in the past became eventually extinct without issue.

The evolutionary course of a population is conditioned by the past history of the population. The genetic configuration of a population is determined by the environments where the population has lived in the past. Those genes and genetic combinations were favorably selected which increased the reproductive fitness of their carriers in the environments where the population lived. The present configuration of its gene pool sets limits to the evolutionary potentialities of a population. The only genes that may be favored by natural selection are those actually present in the population. An obvious example is the colonization of the land by organisms. The colonization of the land by plants occurred during the Silurian geological period, and by animals during the Devonian period. New and diversified environments were open to the evolution of life. New forms of plants evolved, but the basic adaptations to plant life remained in all of them. These adaptations had occurred in the past and set limits to the evolutionary potentialities of their descendants. The considerable diversification of anatomic and physiological characteristics that occurred in animals were not open to plants and vice versa.

Natural selection is thoroughly opportunistic. A new environmental challenge is responded to by appropriate adaptations in the population or results in its extinction. Adaptation to the same environment may occur in a variety of different ways. An example may be taken from the adaptations of plant life to desert climate (Dobzhansky [1]). The fundamental adaptation is to the condition of dryness which carries the danger of desiccation. During a major part of the year, sometimes for several years in succession, there is no rain. Plants have accomplished the urgent necessity of saving water in different ways. Cacti have transformed their leaves into spines, having made their stems into barrels containing a reserve of water. Photosynthesis is performed in the surface of the stem instead of in the leaves. Other plants have no leaves during the dry season, but after it rains they burst into leaves and flowers and produce seeds. A third type of adaptation exists. Ephemeral plants germinate from seeds, grow, flower, and produce seeds—all within the space of the few weeks while water is available. The rest of the year the seeds lie quiescent in the soil.

Natural selection can explain the facts of the adaptation of living organisms to their environments and to their ways of life. The account of natural selection given here is also consistent with the history of life as obtained from the fossil record and with the diversity of plants and animals existing today (Simpson [10]). The fossil record shows that the evolution of life occurred in a haphazard fashion. The phenomena of radiation, expansion, relaying of one form by another, diversification, occasional trends and extinction shown by the fossil record, are best explained by the synthetic theory of evolution. They are not compatible with a preordained plan whether imprinted from without by an omniscient Creator, or the result of the orthogenetic activity of any immanent nonmaterial force, be it called "élan vital," "radial energy" or "vital force."

4   Teleological Explanations in Biology   Nagel ([5], p. 24) has written that "the notion of teleology is neither hopelessly archaic nor necessarily a mark of superstition." The concept of teleology is in general disrepute in modern science. The main reason for this discredit is that the notion of teleology is equated with the belief that future events are active agents in their own realization. Such belief, however, is not necessarily implied in the concept of teleology. Teleological explanations are appropriate in certain areas of natural science. In particular, I shall attempt to show that teleological explanations are appropriate and indispensable in biology, and that they are fully compatible with causal accounts, although they cannot be reduced to nonteleological explanations without loss of explanatory content.

The notion of teleology arose most probably as a result of man's reflection on the circumstances connected with his own voluntary actions.

The anticipated outcome of his actions can be envisaged by man as the goal or purpose towards which he directs his activity. Human actions can be said to be purposeful when they are intentionally directed towards the obtention of a goal.

The plan or purpose of the human agent may frequently be inferred from the actions he performs. That is, his actions can be seen to be purposefully or teleologically ordained towards the obtention of the goal. In this sense the concept of teleology can be extended, and has been extended, to describe actions, objects or processes which exhibit an orientation towards a certain goal or end-state. No requirement is necessarily implied that the objects or processes tend consciously towards their specific goals, nor that there is any external agent directing the process or the object towards its end-state or goal. In this generic sense, teleological explanations are those explanations where the presence of an object or a process in a system is explained by exhibiting its connection with a specific state or property of the system to whose existence or maintenance the object or process contributes. Teleological explanations require that the object or process contribute to the existence of a certain state or property of the system. Moreover, they imply that such contribution is the explanatory reason for the presence of the process or object in the system. It is appropriate to give a teleological explanation of the operation of the kidney in regulating the concentration of salt in the blood, or of the structure of the hand obviously adapted for grasping. But it makes no sense to explain teleologically the falling of a stone, or a chemical reaction.

There are at least three categories of biological phenomena where teleological explanations are appropriate, although the distinction between the categories need not always be clearly defined. These three classes of teleological phenomena are established according to the mode of relationship between the object or process and the end-state or property that accounts for its presence. Other classifications of teleological phenomena are possible according to other principles of distinction. A second classification will be suggested below.

1. When the end-state or goal is consciously anticipated by the agent. This is purposeful activity and it occurs in man and probably in other animals. I am acting teleologically when I pick up a pencil and paper in

order to express in writing my ideas about teleology. A deer running away from a mountain lion, or a bird building its nest, has at least the appearance of purposeful behavior.

2. In connection with self-regulating or teleonomic systems, when there exists a mechanism that enables the system to reach or to maintain a specific property in spite of environmental fluctuations. The regulation of body temperature in mammals is of this kind. In general the homeostatic reactions of organisms belong to this category of teleological phenomena. Two types of homeostasis are usually distinguished by biologists—physiological and developmental homeostasis, although intermediate situations may exist. Physiological homeostatic reactions enable the organism to maintain certain physiological steady states in spite of environmental shocks. The regulation of the composition of the blood by the kidneys, or the hypertrophy of a structure like muscle due to strenuous use, are examples of this type of homeostasis. Developmental homeostasis refers to the regulation of the different paths that an organism may follow in its progression from zygote to adult.

Self-regulating systems or servo-mechanisms built by man are teleological in this second sense. The simplest example of such servo-mechanisms is a thermostat unit that maintains a specified room temperature by turning on and off the source of heat. Self-regulating mechanisms of this kind, living or man-made, are controlled by a feedback system of information.

3. In reference to structures anatomically and physiologically designed to perform a certain function. The hand of man is made for grasping, and his eye for vision. Tools and certain types of machines made by man are teleological in this sense. A watch for instance, is made to tell time, and a faucet to draw water. The distinction between this and the previous category of teleological systems in sometimes blurred. Thus the human eye is able to regulate itself within a certain range to the conditions of brightness and distance so as to perform its function more effectively.

Teleological mechanisms in living organisms are biological adaptations. They have arisen as a result of the process of natural selection. The adaptations of organisms—whether organs, homeostatic mechanisms, or patterns of behavior—are explained teleologically in that their existence is accounted for in terms of their contribution to the reproductive fitness of the population. As explained above, a feature of an organism that increases its reproductive fitness will be selectively favored. Given enough time it will extend to all the members of the population.

Patterns of behavior, such as the nesting habits of birds or the web-spinning of spiders, have developed because they favored the reproductive success of their possessors in the environments where the population lived. Similarly, natural selection can account for the presence of homeostatic mechanisms. Some processes can be operative only within a certain range of conditions. If the conditions are affected by the environment, natural selection will favor self-regulating mechanisms that maintain the system within the function range. In man death results if the body temperature is allowed to rise or fall by more than a few degrees above or below normal. Body temperature is regulated by dissipating heat in warm environments through perspiration and dilatation of the blood vessels in the skin. In cool weather the loss of heat is minimized and additional heat is produced by increased activity and shivering. Finally, the adaptation of an organ or structure to its function is explained teleologically in that its presence is accounted for in terms of the contribution it makes to reproductive success in the population. The vertebrate eye arose because genetic mutations responsible for its development arose and increased the reproductive fitness of their possessors.

There are two levels of teleology in organisms. There usually exists a specific and proximate end for every feature of an animal or plant. The existence of the feature is explained in terms of the function or end-state it serves. But there is also an ultimate goal to which all features contribute or have contributed in the past—reproductive success. The ultimate end to which all other functions and ends contribute is increased reproductive efficiency. In this sense the ultimate source of explanation in biology is the principle of natural selection.

Natural selection can be said to be a teleological process in two ways. Firstly, natural selection is a mechanistic end-directed process which results in increased reproductive efficiency. Reproductive fitness can, then, be said to be the end result or goal of natural selection. Secondly, natural selection is teleological in the sense that it produces and maintains end-directed organs and processes, when the function or end-state served by the organ or process contributes to the reproductive fitness or the organisms.

However, the process of natural selection is not at all teleological in a different sense. Natural selection does not tend in any way towards the production of specific kinds of organisms or towards organisms having certain specific properties. The over-all process of evolution cannot be said to be teleological in the sense of proceeding towards certain specified goals, preconceived or not. The only nonrandom process in evolution is natural selection understood as differential reproduction. Natural selection is a purely mechanistic process and it is opportunistic in the sense discussed above. The final result of natural selection for any species may be extinction, as shown by the fossil record, if the species fails to cope with environmental change.

The presence of organs, processes and patterns of behavior can be explained teleologically by exhibiting their contribution to the reproductive fitness of the organisms in which they occur. This need not imply that reproductive fitness is a consciously intended goal. Such intent must in fact be denied, except in the case of the voluntary behavior of man and perhaps of some animals. In teleological explanation the end-state is not to be understood as the efficient cause of the object or process that it explains. The end-state is causally—and in general temporally also—posterior.

Mayr (cf. [3], p. 42) has pointed out that the term "teleology" has been applied to two different sets of phenomena. "On one hand is the production and perfection throughout the history of the animal and plant kingdoms of ever new and ever improved DNA programs of information. On the other hand is the testing of these programs and their decoding throughout the lifetime of each individual." The behavioral activities or developmental processes of an individual are controlled by the program of information encoded in the DNA inherited by the organism from its parents. The decoding of the DNA programs of information can properly be said to be a teleological—or as Mayr prefers to call it, teleonomic—process. Teleology has also been applied to the evolution of organisms, that is, to the production and perfection of DNA codes of information. The overall process of evolution cannot be said to be teleological in the sense of directed towards the production of specified DNA codes of information, i.e., organisms. But it is my contention that it can be said to be teleological in the sense of being directed towards the production of DNA codes of information which improve the reproductive fitness of a population in the environments where it lives. The process of evolution can also be said to be teleological in that it has the potentiality of producing end-directed DNA codes of information, and has in fact resulted in teleologically oriented structures, patterns of behavior, and self-regulating mechanisms.

Three categories of teleological systems have been distinguished above, according to the nature of the relationship existing between the object or process and its end-state or goal. Another classification of teleology may be suggested in reference to the agency giving origin to the teleological mechanism. The end-directedness of living organisms and their features may be said to be "internal" teleology, while that of man-made tools and servo-mechanisms may be called "external" teleology. It might also be appropriate to refer to these two kinds of teleology as "natural" and "artificial," but the other two terms, "internal" and "external," have already been used (cf. [2], p. 193). Internal teleological systems are accounted for by natural selection which is a strictly mechanistic process. External teleological mechanism are products of the human mind, or more generally, are the result of purposeful activity consciously intending specified ends.

Living organisms, then, exhibit internal teleology, but do not in general possess external teleology. The overall process of evolution is not teleological in the external sense. Evolution can be explained without recourse to a Creator or planning agent external to the organisms themselves. There is no evidence either of any vital force or immanent energy directing the process towards production of specified kinds of organisms. The evidence of the fossil record is against any necessitating force, external or immanent, leading the process towards specified goals.

5   Teleology and Causality   Nagel ([5], p. 24, 25) has convincingly argued that "teleological explanations are fully compatible with causal accounts.... Indeed, a teleological explanation can always be transformed into a causal one." Teleological explanations can be reformulated, without loss of explicit content, to take the form of nonteleological ones. A typical teleological statement in biology is the following, "The function of gills in fishes is respiration, that is the exchange of oxygen and carbon dioxide between the blood and the external water." Statements of this kind account for the presence of a certain feature A (gills) in every member of a class of systems S (fish) which possess a certain organization C (the characteristic anatomy and physiology of fishes). It does so by declaring that when S is placed in a certain environment E (water with dissolved oxygen) it will perform a function F (respiration) only if S (fish) has A (gills). The teleological statement, says Nagel, is a telescoped argument the content of which can be unravelled approximately as follows: When supplied with water containing dissolved oxygen, fish respire; if fish have no gills, they do not respire even if supplied with water containing dissolved oxygen; therefore fish have gills. More generally, a statement of the form "The function of A in a system S with organization C is to enable S in environment E to engage in process F" can be formulated more explicitly; "Every system S with organization C and in environment E engages in function F; if S with organization C and in environment E does not have A, then S cannot engage in F; hence, S must have A." The difference between a teleological explanation and a nonteleological one is, then, one of emphasis rather than of asserted content. A teleological explanation directs our attention to "the *consequences* for a given system of a constituent part or process." The equivalent nonteleological formulation focuses attention on "some of the *conditions* ... under which the system persists in its characteristic organization and activities" ([4], p. 405).

Although a teleological explanation can be reformulated in a nonteleological one, the teleological explanation connotes something more than the equivalent nonteleological one. A teleological explanation implies that the system under consideration is directively organized. For that reason, teleological explanations are appropriate in biology and in the domain of cybernetics but make no sense when used in the physical sciences to describe phenomena like the fall of a stone. Moreover, and most importantly, teleological explanations imply that the end result is the explanatory reason for the *existence* of the object or process which serves or leads to it. A teleological account of the gills of fish implies that gills came to existence precisely because they serve for respiration.

If the above reasoning is correct, the use of teleological explanations in biology is not only acceptable but indeed indispensable. Biological organisms are systems directively organized towards reproductive fitness. Parts of organisms are directively organized towards specific ends that, generally, contribute to the ultimate goal of reproductive survival. One question biologists ask about organic structures and activities is "What for?" That is, "What is the function or role of such structure or such process?" The answer to this question must be formulated in teleological language. Only teleological explanations connote the important fact that plants and animals are directively organized systems. That such connotation—or, in Nagel's expression, "surplus meaning"—can always be expressed in nonteleological language is beside the point. As Nagel ([4], p. 423) has written questions about the value of an explanation "can be answered only by examining the effective role an explanation plays in inquiry and in the communication of ideas."

It has been noted by some authors that the distinction between systems that are goal-directed and those which are not is highly vague. The classification of certain systems as end-directed is allegedly rather arbitrary. A chemical buffer, an elastic solid or a pendulum at rest are examples of physical-systems that appear to be goal-directed. I suggest the use of the criterion of utility to determine whether an entity is teleological or not. The criterion of utility can be applied to both internal and external teleological systems. A feature of a system will be teleological in the sense of internal teleology if the feature has utility for the system in which it exists and if such utility explains the presence of the feature in the systems. Utility in living organisms is defined in reference to survival or reproduction. A structure or process of an organism is teleological if it contributes to the reproductive efficiency of the organism itself, and if such contribution accounts for the existence of the structure or process. Man-made tools or mechanisms are teleological with external teleology if they have utility, i.e., if they have been designed to serve a specified purpose, which therefore explains their existence and properties. If the criterion of utility cannot be applied, a system is not teleological. Chemical buffers, elastic solids and a pendulum at rest are not teleological systems.

The utility of features of organisms is with respect to the individual or the species in which they exist at any given time. It does not include usefulness to any other organisms. The elaborate plumage and display of the peacock serves the peacock in its attempt to find a mate. The beautiful display is not teleologically directed towards pleasing man's aesthetic sense. That it pleases the human eye is accidental, because it does not contribute to the reproductive fitness of the peacock (except, of course, in the case of artificial selection by man). The criterion of utility introduces needed objectivity in the determination of what biological mechanisms are end-directed. Provincial human interests should be avoided when using teleological explanations, as Nagel says. But he selects the wrong example when he observes that "the development of corn seeds into corn plants is sometimes said to be natural, while their transformation into the flesh of birds or men is asserted to be merely accidental" ([4], p. 424). The adaptation of corn seeds have developed to serve the function of corn reproduction, not to become a palatable food for birds or man. The role of wild corn as food is accidental, and cannot be considered a biological function of the corn seed in the teleological sense.

Some features of organisms are not adaptive nor useful by themselves. They have arisen because they are concomitant of other features that are adaptive or useful. Features of organisms may also be present because they were useful to the organisms in the past although they are no longer adaptive. Vestigial organs like the vermiform appendix of man are features of this kind. If they are neutral to reproductive fitness they may remain in the population indefinitely.

6 Teleology, Teleonomy, and Aristotle   I want to take up, very briefly, two more issues; the first is a semantic question, the second a historical one. Pittendrigh [7], Simpson [10], Mayr [3], Williams [11], and others, have proposed to use the term "teleonomic" to describe end-directed processes which do not imply that future events are active agents in their own realization, nor that things or activities are conscious agents or the product of such agents. They argue that the term "teleology" has sometimes been used to explain the animal and plant kingdoms as the result of a preordained plan necessarily leading to the existing kinds of organisms. To avoid such connotation, the authors argue, the term teleonomy should be used to explain adaptation in nature as the result of natural selection.

Although the notion of teleology has been used, and it is still being used, in the alleged sense, it is also true that other authors, like Nagel [4], [5], Goudge [2], etc., employ the term "teleology" without implying a preordained relationship of means to an end. Thus, it might originate more confusion than clarity to repudiate the notion of teleology on the grounds that it connotes an intentional relationship of means to an end. The point is that what is needed is to clarify the notion of teleology by explaining the various meanings the term may have. One may then explicitly express in which sense the term is used in a particular context.

Should the term "teleology" eventually be discarded from the scientific vocabulary, or restricted in its meaning to preordained end-directed processes, I shall welcome such event. But the substitution of a term by another does not necessarily clarify the issues at stake. It would still be necessary to explicate whatever term is used instead of teleology, whether teleonomy or any other. It may further be noted that the term "teleonomic" is commonly employed in the restricted sense of self-regulating mechanisms. There are phenomena in biology that are end-directed without being self-regulating mechanisms in the usual sense. The hand of man, for example.

Pittendrigh ([7], p. 394) has written that "It seems unfortunate that the term 'teleology' should be resurrected.... The biologists' long-standing confusion would be more fully removed if all end-directed systems were described by some other term, like 'teleonomic,' in order to emphasize that the recognition and description of end-directedness does not carry a commitment to Aristotelian teleology as an efficient causal principle." The Aristotelian concept of teleology allegedly implies that future events are active agents in their own realization. According to other authors, Aristotelian teleology connotes that there exists an overall design in the world attributable to a Deity, or at least that nature exists only for and in relation to man, considered as the ultimate purpose of creation (cf. Simpson [10], Mayr [3]).

Science, for Aristotle, is a knowledge of the "whys," the "reasons for" true statements. Of a thing we can ask four different kinds of questions: "What is it?", "Out of what is it made?", "By what agent?", "What for?" The four kinds of answers that can be elicited to these questions are his four causes—formal, material, efficient, and final. Only the third type of answer is causal in the modern scientific sense. *Aition*, the Greek term that Cicero translated "cause" (*causa*, in Latin) means literally ground of explanation, i.e., what can be answered to a question. It does not necessarily mean causality in the sense of efficient agency.

According to Aristotle, to fully understand an object we need to find out, among other things, its end; what function does it serve or what results it produces. An egg can be understood fully only if we consider it as a possible chicken. The structures and organs of animals have functions, are organized towards certain ends. Living processes proceed towards certain goals. Final causes, for Aristotle, are principles of intelligibility; they are not in any sense active agents in their own realization. For Aristotle, ends "never do anything. Ends do not act or operate, they are never efficient causes" (cf. Randall, [8], p. 128).

According to Aristotle there is no intelligent maker of the world. The ends of things are not consciously intended. Nature, man excepted, has no purposes. The teleology of nature is objective, and empirically observable. It does not require the inference of unobservable causes. (cf. Ross [9], Randall [8]) There is no God designer of nature. According to Aristotle, if there is a God, He cannot have purposes (Randall [8], p. 125).

Finally, for Aristotle, the teleology of nature is wholly "immanent." The end served by any structure or process is the good or survival of that kind of thing in which they exist. Animals, plants, or their parts do not exist for the benefit of any other thing but themselves. Aristotle makes it clear that nutritious as acorns may be for a squirrel, they do not exist to serve a squirrel's meal. The natural end of an acorn is to become an oak tree. Anything else that may happen to the acorn is accidental and may not be explained teleologically.

Aristotle's main concern was the study of organisms, and their processes and structures. He observed the facts of adaptation and explained them with considerable insight considering that he did not know about biological evolution. His error was not that he used teleological explanations in biology, but that he extended the concept of teleology to the nonliving world.

Acknowledgment

I wish to express my appreciation to Professors Th. Dobzhansky, E. Mayr, and E. Nagel, who read an earlier version of this paper and made many valuable suggestions.

References

[1] Dobzhansky, Th., "Determinism and Indeterminism in Biological Evolution," in *Philosophical Problems in Biology* (ed. E. Smith), St. John's University Press, New York, 1966, pp. 55–66.

[2] Goudge, T. A., *The Ascent of Life*, University of Toronto Press, Toronto, 1961.

[3] Mayr, E., "Cause and Effect in Biology," in *Cause and Effect* (ed. D. Lerner), Free Press, New York, 1965, pp. 33–50.

[4] Nagel, E., *The Structure of Science*, Harcourt, Brace and World, New York, 1961.

[5] Nagel, E., "Types of Causal Explanation in Science," in *Cause and Effect*, (ed. D. Lerner), Free Press, New York, 1965, pp. 11–32.

[6] Paley, W., *Natural Theology*, Charles Knight, London, 1836.

[7] Pittendrigh, C. S., "Adaptation, Natural Selection and Behavior," in *Behavior and Evolution* (eds, A. Roe and G. G. Simpson), Yale University Press, New Haven, 1958, pp. 390–416.

[8] Randall, J. H., *Aristotle*, Columbia University Press, New York, 1960.

[9] Ross, D., *Aristotle*, 5th edit., Barnes and Noble, New York, 1949.

[10] Simpson, G. G., *This View of Life*, Harcourt, Brace and World, New York, 1964.

[11] Williams, G. C., *Adaptation and Natural Selection*, Princeton University Press, Princeton, N.J., 1966.

2

Functions

Larry Wright

The notion of function is not all there is to teleology, although it is sometimes treated as though it were. Function is not even the central, or paradigm, teleological concept. But it *is* interesting *and* important; and it is still not as well understood as it should be, considering the amount of serious scholarship devoted to it during the last decade or two. Let us hope this justifies my excursion into these murky waters.

Like nearly every other word in English, "function" is multilaterally ambiguous. Consider:

(1) $y = f(x)$/ The pressure of a gas is a function of its temperature.

(2) The Apollonaut's banquet was a major state function.

(3) I simply can't function when I've got a cold.

(4) The heart functions in this way ... [something about serial muscular contractions].

(5) The function of the heart is pumping blood.

(6) The function of the sweep-second hand on a watch is to make seconds easier to read.

(7) Letting in light is one function of the windows of a house.

(8) The wood box next to the fireplace currently functions as a dog's sleeping quarters.

It is interesting to notice that the word "function" has a spectrum of meanings even within the last six illustrations, which are the only ones at all relevant to a teleologically oriented study. Numbers 3, 4, and 8 are substantially different from one another, but they are each, from a teleological point of view, peripheral cases by comparison with 5, 6 and 7, which are the usual paradigms. And even these latter three are individually distinct in some respects, but much less profoundly than the others.

Quite obviously, making some systematic sense of the logical differentiation implicit in categorizing these cases as peripheral and paradigmatic is a major task of this chapter. But a clue that we are on the right track here can be found in a symptomatic grammatical distinction present in the last six illustrations: in the peripheral cases the word "function" is itself the verb, whereas in the more central cases "function" is a noun, used with the verb "to be." And since the controversy revolves around what *the function* of something *is*, the grammatical role of "function" in 5, 6, and 7 makes them heavy favorites for the logical place of honor in this discussion.

Some Rudimentary Distinctions

1   Functions versus Goals   There seems to be a strong temptation to treat functions as representative of the set of central teleological concepts which cluster around goal-directedness. However, even a cursory examination of the usual sorts of examples reveals a very important distinction. Goal-directedness is a behavioral predicate. The *direction* is the direction of behavior. When we do speak of objects (homing torpedoes) or individuals (General MacArthur) as being goal-directed, we are speaking indirectly of their behavior. We would argue against the claim that they are goal-directed by appeal to their behavior (for example, the torpedo, or the General, did not *change course* at the appropriate time, and so forth). On the other hand, many things have *functions* (for example, chairs and windpipes) which do not behave *at all*, much less goal-directedly. And behavior can have a function without being goal-directed—for example, pacing the floor or blinking your eye. But even when goal-directed behavior has a function, very often its function is quite different from the achievement of its *goal*. For example, some fresh-water plankton diurnally vary their distance below the surface. The *goal* of this behavior is to keep light intensity in their environment relatively constant. This can be determined by experimenting with artificial light sources. The *function* of this behavior, on the other hand, is keeping constant the oxygen supply, which normally varies with sunlight intensity. There are many instances to be found in the study of organisms in which the function of a certain goal-directed activity is not some further goal of that activity, as it usually is in human behavior, but rather some natural concomitant or consequence of the immediate goal. Other examples are food-gathering, nest-making, and copulation. Clearly function and goal-directedness are not congruent concepts. There is an important sense in which they are wholly distinct. In any case, the relationship between functions and goals is a complicated and tenuous one; and becoming clearer about the nature of that relationship is one aim of this essay.

2   A Function versus the Function   Recent analyses of function, including all those treated here, have tended to focus on *a* function of something, by contrast with *the* function of something. This tendency is understandable; for any analysis of this sort aims at generality, and "a function" would seem intrinsically more general than "the function" because it avoids one obvious restriction. This generality, however, is superficial: the notion of *a* function is derivable from the notion the *the* function (more than one thing meets the criteria) just as easily as the reverse (only one thing meets the criteria). Furthermore, the notion of *a* function is much more easily confused with certain peripheral, quasi-functional ascriptions which are examined below. In short, the discussion of this chapter is concerned with *a* function of X only insofar as it is the sort of thing which would be *the* function of X if X had no others. Accordingly, I take the definite-article formulation as paradigmatic and will deal primarily with it, adding comments in terms of the indefinite-article formulation parenthetically, where appropriate.

3   Function versus Accident   Very likely the central distinction of this analysis is that between the *function* of something and other things it does which are *not* its function (or one of its functions). The function of a telephone is effecting rapid, convenient communication. But there are many other things telephones do: take up space on my desk, disturb me at night, absorb and reflect light, and so forth. The function of the heart is pumping blood, not producing a thumping nose or making wiggly lines on electrocardiograms, which are also things it does. This is sometimes put as the distinction between a function, and something done merely "by accident." Explaining the propriety of this way of speaking— that is, making sense of the function/accident distinction—is another aim, perhaps the *primary* aim of the following analysis.

4   Conscious versus Natural Functions   The notion of accident will raise some interesting and important questions across another rudimentary distinction: the distinction between natural functions and consciously designed ones. Natural functions are the common organismic ones such as the function of the heart, mentioned above. Other examples are the function of the kidneys to remove metabolic wastes from the bloodstream, and the function of the lens of the human eye to focus an image on the retina. Consciously designed functions commonly (though not necessarily) involve artifacts, such as the telephone and the watch's sweep hand mentioned previously. Other examples of this type would be the function of a door knob, a headlight dimmer switch, the circumferential grooves in a pneumatic tire tread, or a police force. Richard Sorabji has argued[1] that "designed" is too strong as a description of this category, and that less elaborate conscious effort would be adequate to give something a function of this sort. I think he is right. I have used the stronger version only to overdraw the distinction hyperboically. In deference to his point I will drop the term "designed" and talk of the distinction as between natural and conscious functions.

Of the two, natural functions are philosophically the more problematic. Several schools of thought, for different reasons, want to deny that there are natural functions, as opposed to conscious ones. Or, what comes to the same thing, they want to deny that natural functions are functions in anything like the same sense that conscious functions are. Some theologians want to say that the organs of organisms get their functions through God's conscious design, and hence these things *have* functions, but not natural functions *as opposed to* conscious ones. Some scientists, like B. F. Skinner, would *deny* that organs and organismic activity have functions *because* there is no conscious effort or design involved.

Now it seems to me that the notion of an organ's having a function— both in everyday conversation and in biology—has no strong theological commitments. Specifically, it seems to me consistent, appropriate, and even common for an atheist to say that the function of the kidney is elimination of metabolic wastes. Furthermore, it seems clear that conscious and natural functions are functions in the same sense, despite their obvious differences. Functional ascriptions of either sort have a profoundly similar ring. Compare "the function of that cover is to keep the distributor dry" with "the function of the epiglottis is to keep food out of the windpipe." It is even more difficult to detect a difference in what is being requested: "What is the function of the human windpipe?" versus "What is the function of a car's exhaust pipe?" Certainly no analysis should begin by supposing that the two sorts are wildly different, or that only one is really legitimate. That is a possible *conclusion* of an analysis, not a reasonable presupposition. Accordingly, the final major aim of this analysis will be to make sense of natural functions, both as functions in the same sense as consciously contrived ones, and as functions independent of any theological presuppositions—that is, independent of conscious purpose. It follows that this analysis is committed to finding a way of stating what it is to be a function—even in the conscious cases—that does not rely on an appeal to consciousness. If no formulation of this kind can be found despite an honest search, only then should we begin to take seriously the view that we actually mean something quite different by "function" in these two contexts.

Some Analyses of Function

The analysis of function for which I wish to argue grew out of a detailed critical examination of several recent attempts in the literature to produce such an analysis, and it is best understood in that context. For this reason, and because it will help clarify the aims I have sketched above, I will begin by presenting the kernel of that critical examination.

The first analysis I want to consider is an early one by Morton Beckner.[2] Here Beckner contends that to say something s has function F' in system s' is to say, "There is a set of circumstances in which: F' occurs when s' has s, AND F' does not occur when s' does not have s" (p. 113).[3] For example, "The human heart has the function of circulating blood" means that there is a set of circumstances in which circulation occurs in humans when they have a heart, and does not when they do not. Translated into the familiar jargon, *s* has function *F'* in *s'* if and only if there is a set of circumstances containing *s* which are sufficient for the occurrence of *F'* and which also require *s* in order to be sufficient for *F'*. Now it is not clear whether the "requirement" here is necessity or merely nonredundancy. If it is necessity, then under the most natural interpretation of "circumstances" (environment), it is simply mistaken. There are *no* circumstances in which, for example, the heart is absolutely irreplaceable: we could always pump blood in some other way. On the other hand, if the requirement here is only nonredundancy, the mistake is more subtle.

In this case Beckner's formula would hold for cases in which *s* merely *does F'*, but in which *F'* is not the function of *s*. For example, the heart is a nonredundant member of a set of conditions or circumstances which are sufficient for a throbbing nose. But making a throbbing noise is not a function of the heart; it is just something it does—accidentally. In fact, there ae even dysfunctional cases which fit the formula: in some circumstances, livers are nonredundant for cirrhosis, but cirrhotic debilitation could not conceivably be the (or a) function of the liver. So this analysis fails on the functional/accidental distinction: it includes too much.

After first considering a view essentially similar to this one, John Canfield has offered a more elaborate analysis.[4] According to Canfield: "A function of I (in S) is to do C *means* I does C and that C is done is useful to S. For example, '(In vertebrates) a function of the liver is to secrete bile' means 'the liver secretes bile, and that bile is secreted in vertebrates is useful to them'" (p. 290). Canfield recognizes that natural functions are the problematic ones, but he devotes his attention solely to those cases. He treats only the organs and parts of organisms studied by biology, to the exclusion of the consciously designed functions of artifacts. As a result of this emphasis, his analysis is, without modification, almost impossible to apply to conscious functions. But even with appropriate modifications, it turns out to be inadequate to the characterization of either conscious or natural function.

In the conscious cases, there is an enormous problem in identifying the system *S*, *in* which *I* is functioning, and *to* which it must be useful. The function of the sweep-second hand of a watch is to make seconds easier to read. It would be most natural to say that the system *in which* the sweep hand is functioning—by analogy with the organismic cases—is the watch itself; but it is hard to make sense of the easier reading's being useful to the mechanism. On the other hand, the best candidate for the system *to which* the easier reading is useful is the person wearing the watch; but this does not seem to make sense as the system *in which* the sweep hand is functioning.

The crucial difficulty of Canfield's analysis begins to appear at this point: no matter what modifications we make in his formula to avoid the problem of identifying the system *S*, we must retain the requirement the *C* be useful. This is really the major contribution of his analysis, and to abandon it is to abandon the analysis. The difficulty with this is that, for example in the watch case, it is clearly not necessary that easily read seconds be useful to the watch-wearer—or anyone else—in order that making seconds easier to read be the function of the sweep hand of that wearer's watch. My watch has a sweep-second hand, and I occasionally use it to time things to the degree of accuracy it allows: it is useful to me. Now suppose I were to lose interest in reading time to that degree of accuracy. Suppose my life changed radically so that nothing I ever did could require that sort of chronological precision. Would that mean the sweep hand on my particular watch no longer has the function of making seconds easier to read? Clearly not. If someone were to ask what the sweep hand's function was ("What's it do?" "What's it there for?") I would still have to say it made seconds easier to read, although I might yawningly append an autobiographical note about my utter lack of interest in that feature. Similarly, the function of that button on my dashboard is to activate the windshield washer, even if all it does is make the mess on the windshield worse, and hence is not useful at all. That would be its *function* even if I never took my car out of the garage—or if I broke the windshield.

It is natural at this point to attempt to patch up the analysis by reducing the requirement that *C* be useful to the requirement that *C usually* be useful. But this will not do either, because it is easy to think of cases in which we would talk of something's having a function even though doing that thing was quite *generally* of no use to anybody—for example, a machine whose function was to count Pepsi Cola bottle caps at the city dump; or MIT's ultimate machine of a few years back, whose only function was to turn itself off. The source of the difficulty in all of these cases is that what the thing in question (watch, washer button, counting machine) was *designed* to do has been left out of the calculation. And, of course, in these cases, if something is designed to do X, then doing X is its function even if doing X is generally useless, silly, or even harmful. In fact, intention is so central here that it allows us to say the function of *I* is to do C, even when *I* cannot even *do* C. If the windshield washer switch comes from the factory defective, comes from the factory defective, and is never repaired, we would still say that its *function* is to activate the washer system; which is to say: that is what it was *designed* to do.

It might appear that this commits us to the view that natural and consciously contrived functions cannot possibly be the same sort of function. If conscious intent is what *determines* the function artifact has got, there is no parallel in natural functions. I take this to be mistaken, and will show why later. For now it is only important to show, from this unique vantage, the nature of the most formidable obstacle to be overcome in unifying natural and conscious functions.

The argument thus far has shown that meeting Canfield's criteria is not necessary for something to be a function. It can easily be shown that meeting them is also not sufficient. We are always hearing stories about the belt buckles of the Old West or on foreign battlefields which save their wearers' lives by deflecting bullets. From several points of view that is a very useful thing for them to do. But that does not make bullet deflection the function—or even *a* function—of belt buckles. The list of such cases is endless. Artifacts do all kinds of useful things which are not their functions. Blowouts cause you to miss flights that crash. Noisy wheel bearings cause you to have the front end checked over when you are normally too lazy. The sweep hand of a watch might brush the dust off the numbers, and so forth.

All this results from the inability of Canfield's analysis to handle what we took to be one of the fundamental distinctions of function talk: accidental versus nonaccidental. Something can do something useful purely by accident, but it cannot have, as its function, something it does only by accident. Something that *I* does by accident cannot be the function of *I*. The cases above allow us to begin to make some fairly clear sense of this notion of accident, at least for artifacts. Buckles stop bullets only by accident. Blowouts only accidentally keep us off doomed airplanes. Sweep hands only accidentally brush dust, if they do it at all. And this brings us back to the grammatical distinction I made at the outset when I divided the list of illustrations into "central" and "peripheral" ones. When something does something useful by accident rather than design, as in these examples, we signal the difference by a standard sort of "let's pretend" talk. Instead of using the verb "to be" or the verb "to have," and saying the thing in question *has* such and such a function, or saying that *is* its function, we use the expression "functioning as." We might say the belt buckle *functioned as* a bullet shield, or the blowout *functioned as* divine intervention, or the sweep hand *functions as* a dust brush. Canfield's analysis does not embrace this distinction at all.

So far I have shown only that Canfield's formula fails to handle conscious functions. This means it is incapable of showing natural functions to be functions in the same full-blooded sense as conscious ones, which is indeed serious; but that, it might be argued, really misses the point of his analysis. Canfield is not interested in conscious functions. He would be happy just to handle natural functions. For the reasons set down above, however, I am looking for an analysis which will *unify* conscious and natural functions, and it is important to see why Canfield's analysis cannot produce that unification. Furthermore, Canfield's analysis has difficulties in handling natural functions that closely parallel the difficulties it has with conscious functions; which is just what we should expect if the two are functions in the same sense.

For example, it is absurd to say with Pangloss that the function of the human nose is to support eyeglasses. It is absurd to suggest that the support of eyeglasses is even one of its functions. The function of the nose has something to do with keeping the air we breathe (and smell) warm and dry. But supporting a pince-nez, just as displaying rings and warpaint, is something the human nose does, and is useful to the system having the nose: so it fits Canfield's formula. Even the heart throb, our paradigm of nonfunction, fits the formula: the sound made by the heart is an enormously useful diagnostic aid, not only as to the condition of the heart, but also for certain respiratory and neurological conditions. More bizarre instances are conceivable. If surgeons began attaching cardiac pacemakers to the sixth rib of heart patients, or implanting microphones in the wrists of CIA agents, we could then say that these were useful things for the sixth rib and the wrist (respectively) to do. But that would not make pacemaker-hanging a function of the sixth rib, or microphone concealment a function of the human wrist.

There seems to be the same distinction here that we saw in conscious functions. It makes perfectly good sense to say the nose *functions as* eyeglass support; the heart, through its thump, *functions as* a diagnostic aid; the sixth rib *functions as* a pacemaker hook in the circumstances described above. This, it seems to me, is precisely the distinction we make when we say, for example, that the sweep-second hand *functions as* a dust brush, while denying that brushing dust is one of the sweep hand's functions. And it is here that we can make sense of the notion of accident in the case of natural functions: it is merely fortuitous that the nose supports eyeglasses; it is happy chance that the heart throb is diagnostically significant; it would be the merest serendipity if the sixth rib were to be a particularly good pacemaker hook. It is (would be) only *accidental* that (if) these things turned out to be useful in these ways.

Accordingly, we have already drawn a much stronger parallel between natural functions and conscious functions than Canfield's analysis will allow.

Thus far I have ignored Canfield's analysis of usefulness: "[In plants and animals other than man, that C is done is useful to S means] if, *ceteris paribus*, C were not done in S, then the probability of that S surviving or having descendants would be smaller than the probability of an S in which C is done surviving or having descendants" (p. 292). I have ignored it because its explicit and implicit restrictions make it even more difficult to work this analysis into the unifying one I am trying to produce. Even within its restrictions (natural functions in plants and animals other than man), however, the extended analysis fails for reasons very like the one we have already examined. Hanging a pacemaker on the sixth rib of a cardiovascularly inept lynx would be useful to that cat in precisely Canfield's sense of "useful": it would make it more likely that the cat would survive and/or have descendants. Obviously the same can be said for the diagnostic value of an animal's heart sounds. So usefulness—even in this very restricted sense—does not make the right function/accident distinction: some things do useful things which are not their functions, or even one of their functions.

The third analysis I wish to examine is a more recent one by Morton Beckner.[5] This analysis is particularly interesting for two reasons. First, Beckner is openly (p. 160) trying to accommodate both natural and conscious functions under one description. Second, he wants to avoid saying things like (to use his examples), "A function of the heart is to make heart sounds" and "A function of the Earth is to intercept passing meteorites." So his aims are very like the ones I have argued for: to produce a unifying analysis, and one which distinguishes between functions and things done by accident. And since the heart sound is useful, and intercepting metoeorites could be (perhaps already is), Beckner would probably agree in principle with the above criticism of Canfield.

Beckner's formulation is quite elaborate, so I will present it in eight distinct parts, clarify the individual parts, and then offer an illustration before going on to raise difficulties with them collectively as an analysis of the concept of function. That formulation is:

P has function *F* in *S* if and only if[6]

1. *P* is a part of *S* (in the normal sense of "part").

2. *P* contributes to *F*. (*P*'s being part of *S* makes the occurrence of *F* more likely.)

3. *F* is an activity in or of the system *S*.

4. *S* is structured in such a way that a significant number of its parts contribute to the activities of other parts, and of the system itself.

5. The parts of *S* and their mutual contributions are identified by the same conceptual scheme which is employed in the statement that *P* has function *F* in system *S*.

6. A significant number of critical parts (of *S*) and their activities definitionally contribute to one or more activities of the whole system *S*.

7. *F* is or contributes to an activity *A* of the whole system *S*.[7]

8. *A* is one of those activities of *S* to which a significant number of critical parts and their activities definitionally contribute.

Two points of clarification must be made at once. First, the notion of "the same conceptual scheme" in number 5 is obscure in some respects, and the considerable attention devoted to it by Beckner does not help very much. In general all one can say is that *P*, *F*, and the other parts

UNIVERSITY OF HERTFORDSHIRE LRC

and activities of S must be *systematically* related to one another. But in practice the point is easier to make. For example, if we wish to speak of removing metabolic wastes as the function of the human kidney, the relevant conceptual scheme contains other human organs, life, and perhaps ecology in general, but not atoms, molecular bonds, and force fields. The second point concerns the "definitional contribution" in number 6. A part (or activity) makes a definitional contribution to an activity if that contribution is part of what we mean by the word which refers to that part (or activity). For example, part of what we mean by "heart" in a biological or medical context is "something which pumps blood": we would allow considerable variation in structure or appearance and still call something a heart if it served that function. Beckner illustrates how all these steps work together, once again using the heart.

It is true that a function of the heart is to pump blood. The heart does pump blood; the body is a complex system of parts that by definition aid in certain activities of the whole body, such as locomotion, self-maintenance, copulation; the concepts "heart" and "blood' are recognizably components of the scheme we employ in describing this complex system; and blood-pumping does contribute to activities of the whole organism to which many of its organs, tissues and other parts definitionally contribute. (p. 160)

There are several difficulties with this analysis. They appear below, roughly in order of increasing severity.

First, Beckner's problems with the system S are in some ways worse than Canfield's; for Beckner explicitly wants to include artifacts, and in addition he says much more definite things about the relationship among P, F, and S. So in this case, when we say the function of a watch's sweep hand is making seconds easier to read, we must not only find a system *of* which the sweep hand is a part, and *in* or *of* which "making seconds easier to read" is an activity, but this activity must be or contribute to one to which a number of the system's critical parts definitionally contribute. In the case of natural functions of the organs and other parts of organisms, the system S is typically a natural unit, easy to subdivide from the environment: the organism itself. But for the conscious functions of artifacts, such systems, if they can be found at all, must be hacked out of the environment rather arbitrarily. With no more of a guide than Beckner has given us, there is nothing like a guarantee that we can always find such a system. Accordingly, when our minds boggle—as I take it they do in trying to conceive of "making seconds easier to read" being an activity at all, much less one meeting all of the other conditions of this analysis—we have to say that analysis is at best too obscure to be applicable to such cases, and is perhaps just mistaken.

A second difficulty stems directly from the first. It is not at all clear that functions—even natural functions—have to be activities at all, let alone activities of the sort required by Beckner. Making seconds easier to read is an example, but there are many others: preventing skids in wet weather, keeping your pants up, or propping open my office door. All of these things are legitimate functions (of tire treads, belts, and doorstops, respectively); none is an activity in any recognizable sense.

Third, we noticed in our discussion of Canfield that something could do a useful thing by accident, in the appropriate sense of "accident." Similarly, a part of a system meeting all of Beckner's criteria might easily make a contribution to an activity of that system also quite by accident. For example, an internal-combustion engine is a system satisfying Beckner's criteria for *S*. If a small nut were to work itself loose and fall under the valve-adjustment screw in such a way as to adjust properly a poorly adjusted valve, it would make an accidental contribution to the smooth running of that engine. We would never call the maintenance of proper valve adjustment the *function* of the nut. If it got the adjustment right it was just an accident. But on Beckner's formulation, we would have to call that its function. The nut does keep the valve adjusted; the engine is a complex system of parts that by definition aid in certain activities of the whole body, such as generation of torque and self-maintenance (lubrication, heat dissipation); the concepts "nut," "valve," and "valve adjustment" are components of the scheme we employ in describing this complex system; and proper valve adjustment does contribute to the smooth running of the (whole) engine, which is an activity to which many of the other parts of the engine definitionally contribute (flywheel, connecting rod, exhaust ports).

The final difficulty is also related to one we raised for Canfield's analysis. There we noticed that if an artifact was explicitly designed to do something, *that* usually *determines* its function, irrespective of how well or badly it does the thing it was supposed to do. An analogous point can be made here. If X was designed to do Y, then Y is X's function regardless of what contributions X does in fact make or fail to make. For example, the *function* of the federal automotive safety regulations is to make driving and riding in a car safer. And this is so even if they actually have just the opposite effect, through some psychodynamic or automotive quirk.

So in spite of their enormous differences, this analysis and Canfield's fail for very similar reasons: problems with the notion of system S, failure to rule out some accidental cases, and general inability to account for the obvious role of design.

There have been several other interesting attempts in the recent literature to provide an analysis of function. Most notable are those by Carl Hempel,[8] Hugh Lehman,[9] Richard Sorabji,[10] Francisco Ayala,[11] and Michael Ruse.[12] The last two of these do a somewhat better job on the function/accident distinction than the ones we have examined. But other than that, a discussion of these analyses would be largely redundant on the discussions of Beckner and Canfield. So I think we have gone far enough in clarifying the issues to begin constructing an alternative analysis.

An Alternative View

The treatments we have so far considered have overlooked, ignored, or at any rate failed to make one important observation: that functional ascriptions are—intrinsically, if you will—explanatory. Merely saying of something, X, that it has a certain function, is to offer an important kind of explanation of X. The failure to consider this, or at least take it seriously, is, I think, responsible for the systematic failure of these analyses to provide an accurate account of functions.

There are two related considerations which urge this observation upon us. First, the "in order to" in functional ascriptions is a teleological "in order to." Its role in functional ascriptions (the heart beats in order to circulate blood) is quite parallel to the role of "in order to" in goal ascriptions (the rabbit is running in order to escape from the dog). Accordingly, we should expect functional ascriptions to be explanatory in something like the same way as goal ascriptions.[13] When we say that the rabbit is running in order to escape from the dog, we are explaining *why* the rabbit is running. If we say that John got up early in order to study, we are offering an explanation of his getting up early. Similarly in the functional cases. When we say that the distributor has that cover in order to keep the rain out, we are explaining *why* the distributor has that cover. And when we say the heart beats in order to pump blood, we are ordinarily taken to be offering an explanation of why the heart beats. This last sort of case represents the most troublesome problem in the logic of function, but it must be faced squarely, and once faced, I think its solution is fairly straightforward.

The second consideration which recommends holding out for the explanatory status of functional ascriptions is the contextual equivalence of several sorts of requests. Consider:

1. What is the function of X?
2. Why do Cs have Xs?
3. Why do Xs do Y?

In the appropriate context, each of these is asking for the function of X, "What is the function of the heart?" "Why do humans have a heart?" "Why does the heart beat?" All are answered by saying, "To pump blood," in the context we are considering. Questions of the second and third sort, being "Why?" questions, are undisguised requests for explanations. So in this context functional attributions are presumed to be explanatory. And why-form function requests are by no means bizarre or esoteric ways of asking for a function. Consider:

Why do porcupines have sharp quills?
Why do (some) watches have a sweep-second hand?
Why do ducks have webbed feet?
Why do headlight bulbs have two filaments?

These are rather ordinary ways of asking for a function. And if that is so, then it is ordinarily supposed that a function explains why each of these things is the case. The function of the quills is why porcupines *have* them, and so forth.

Moreover, the kind of explanatory role suggested by both of these considerations is not the anemic, "What's it good for?" sort of thing often imputed to functional explanations. It is rather something more substantial than that. If to specify the function of quills is to explain why porcupines *have* them, then the function must be the reason they *have* them. That is, the ascription of a function must be explanatory in a rather strong sense. To choose the weaker interpretation, as Canfield does,[14] is once again to run afoul of the function-accident distinction. For, to use this example, if "Why do animals have livers?" is a request for a function, it cannot be rendered, "What is the liver good for?" Livers are good for many things which are not their functions, just like anything else. Noses are good for supporting eyeglasses, fountain pens are good for cleaning your fingernails, and livers are good for dinner with onions. No, the *function* of the liver is that *particular* thing it is good for which explains why animals have them.

Putting the matter in this way suggests that functional ascription-explanations are in some sense etiological, concern the causal background of the phenomenon under consideration. And this is indeed what I wish to argue: functional explanations, although plainly not causal in the usual, restricted sense, do concern how the thing with the function *got there*. Hence they *are* etiological, which is to say "causal" in an extended sense. But this is still a very contentious view. Functional and teleological explanations are usually *contrasted with* causal ones, and we should not abandon that contrast lightly: we should be driven to it.

What drives us to this position is the specific difficulty the best-looking alternative accounts have in making the function/accident distinction. We have seen that no matter how useful it is for X to do Z, or what contribution X's doing Z makes within a complex system,[15] these sorts of consideration are never sufficient for saying that the function of X is Z. It could still turn out that X did Z only by accident. But all of the accident counterexamples can be avoided if we include as part of the analysis something about how X came to be there (where-ever): namely, that it is there *because it does Z*—with an etiological "because." The buckle, the heart, the nose, the engine nut, and so forth were not there *because* they stop bullets, throb, support glasses, adjust the valve, and all the other things which were falsely attributed as functions, respectively. Those pseudofunctions could not be called upon to explain how those things *got* there. This seems to be what was missing in each of those cases.

In other words, saying that the function of X is Z is saying at least that

(1) X is there *because* it does Z.
    or
    Doing Z is the *reason* X is there.
    or
    That X does Z is *why* X is there.

where "because," "reason," and "why" have an etiological force. And it turns out that "X is there because it does Z,"[16] with the proper understanding of "because," "does", and "is there" provides us with not only a necessary condition for the standard cases of functions, but also the kernel of an adequate analysis. Let us look briefly at those key terms.

"Because" is, of course, to be understood in its explanatory rather than evidential sense. It is not the "because" in, "It is hot because it is red." More important, "because" is to be taken (as it ordinarily is anyway) to be indifferent to the philosophical reasons/causes distinction. The "because" in, "He did not go to class because he wanted to study" and in, "It exploded because it got too hot" are both etiological in the appropriate way.[17] And finally, it is worth pointing out here that in this sense "A because B" does not require that B be either necessary or sufficient for A. Racing cars have airfoils because they generate a downforce (negative lift) which augments traction. But their generation of negative lift is neither necessary nor sufficient for racing cars to have wings: they could be there merely for aesthetic reasons, or they could be forbidden by the rules. Nevertheless, if you want to know why they are there, it is because they produce negative lift. All of this comes to saying "because" here is to be taken in its ordinary, conversational, causal-explanatory sense.

Complications arise with respect to "does" primarily because on the above condition "Z is the function of X" is reasonably taken to entail "X does Z." Although in most cases there is no question at all about what it is for X to do Z, the matter is highly context-dependent and so perhaps I should mention an extreme case, if only as notice that we should include it. In some contexts we will allow that X does Z even though Z never occurs. For example, the button on the dashboard activates the windshield washer system (that is what it does, I can tell by the circuit diagram) even though it never has and never will. An unused organic or organismic emergency reaction might have the same status. All that seems to be required is that X be *able* to do Z under the appropriate conditions; for example, when the button is pushed or in the presence of a threat to safety.

The vagueness of "is there" is probably what Beckner and Canfield were trying to avoid by introducing the system S into their formulations. It is much more difficult, however, to avoid the difficulties with the system S than to clarify adequately this more general placemarker. "Is there" is straightforward and unproblematic in most contexts, but some illustrations of importantly different ways in which it can be rendered might be helpful. It can mean something like "is where it is," as in, "Keeping food out of the windpipe is the reason the epiglottis is where it is." It can mean "C's have them," as in "animals have hearts because they pump blood." Or it can mean merely "exists (at all)," as in, "Keeping snow from drifting across roads (and so forth) is why there are snow fences."

Now, saying that (1), understood in this way, should be construed as a necessary condition for taking Z to be the function of X, is merely to put in precise terms the moral of our examination of the function/accident distinction. We saw above that the accident counterexamples could not meet this requirement. On the other hand, this condition *is* met in all of the center-of-the-page cases. This is quite easy to show in the conscious cases. When we say the function of X is Z in these cases, we are saying that at least some effort was made to get X (sweep hand, button on dashboard) where it is precisely because it does Z (whatever). Doing Z is the reason X is there. *That* is why the effort was made. The reason the sweep-second hand is there is that it makes seconds easier to read. It is there *because* it does that. Similarly, rifles have safeties because they prevent accidental discharge.

It is only slightly less obvious how natural functions can satisfy (1): We can say that the natural function of something—say, an organ in an organism—is the reason the organ is there by invoking natural selection. If an organ has been naturally differentially selected for by virtue of something it does, we can say that the reason the organ is there is that it does that something. Hence we can say animals have kidneys *because* they eliminate metabolic wastes from the blood-stream; porcupines have quills *because* they protect them from predatory enemies; plants have chlorophyll *because* chlorophyll enables plants to accomplish photosynthesis; the heart beats *because* its beating pumps blood. And each of these can be rather mechanically put in the "reason that" form. The reason porcupines have quills is that they protect them from predatory enemies, and so forth.

It is easy to show that this formula does not represent a sufficient condition for being a function, which is to say there is something more to be said about precisely what it is to be a function. The most easily generable set of cases to be excluded is of this kind: oxygen combines readily with hemoglobin, and that is the (etiological) reason it is found in human bloodstreams. But there is something colossally fatuous in maintaining that the function of that oxygen is to combine with hemoglobin, even though it is there because it does that. The function of the oxygen in human bloodstreams is providing energy in oxidation reactions, not combining with hemoglobin. Combining with hemoglobin is only a means to that end. This is a useful example. It points to a contrast in the notion of "because" employed here which is easy to overlook and crucial to an elucidation of functions.

As I pointed out above, if producing energy is the function of the oxygen, then oxygen must be there (in the blood) because it produces energy. But the "because" in, "It is there because it produces energy," is importantly different from the "because" in, 'It is there because it combines with hemoglobin." They suggest different *sorts* of etiologies. If carbon monoxide (CO), which we know to combine readily with hemoglobin, were suddenly to become able to produce energy by appropriate (nonlethal) reactions in our cells and further, the atmosphere were suddenly to become filled with CO, we could properly say that the reason CO was in our bloodstreams was that it combines readily with hemoglobin. We could not properly say, however, that CO was there because it produces *energy*. And that is precisely what we could say about oxygen, on purely evolutionary-etiological grounds.

All of this indicates that it is the nature of the etiology itself which determines the propriety of a functional explanation; there must be specifically functional etiologies. When we say the function of X is Z (to do Z) we are saying that X is there because it does Z, but with a further qualification. We are explaining how X came to be there, but only certain kinds of explanations of how X came to be there will do. The causal/functional distinction is a distinction *among* etiologies; it is not a contrast between etiologies and something else.

This distinction can be displayed using the notion of a causal consequence.[18] When we give a functional explanation of X by appeal to Z ("X does Z"), Z is always a consequence or result of X's being there (in the sense of "is there"sketched above).[19] So when we say that Z is the function of X, we are not only saying that X is there because it does Z, we are also saying that Z is (or happens as) a result or consequence of X's being there. Not only is chlorophyll in plants *because* it allows them to perform photosynthesis, photosynthesis is a *consequence* of the chlorophyll's being there. Not only is the valve-adjusting screw there *because* it allows the clearance to be easily adjusted, the possibility of easy adjustment is a *consequence* of the screw's being there. Quite obviously, "consequence of" here does not mean "guaranteed by." "Z is a consequence of X," very much like "X does Z" earlier, must be consistent with Z's not occurring. When we say that photosynthesis is a consequence of chlorophyll, we allow that some green plants may never be exposed to light, and that all green plants may at some time or other not be exposed to light. Furthermore, this consequence relationship does not mean that whenever Z *does* occur, happen, obtain, exist, and so forth, it is as a consequence of X. There is room for a multiplicity of sufficient conditions, overdetermined or otherwise. Other things besides the adjusting screw may provide easy adjustment of the clearance. This (the inferential) aspect of consequence, as that notion is used here, can be roughly captured by saying that there are circumstances (of recognizable propriety) in which X is nonredundant for Z. The aspect of "consequence" of central importance here, however, is its asymmetry. "A is a consequence of B" is in virtually every context incompatible with "B is a consequence of A." The source of this asymmetry is difficult to specify, and I shall not try.[20] It is enough that it be clearly present in the specific cases.

Accordingly, if we understand the key terms as they have been explicated here, we can conveniently summarize this analysis as follows:

The function of X is Z means

(2) (a) X is there because it does Z,

    (b) Z is a consequence (or result) of X's being there.

The first part, (a), displays the etiological form of functional ascription-explanations, and the second part, (b), describes the convolution which distinguishes functional etiologies from the rest. It is the second part, of course, which distinguishes the combining with hemoglobin from the producing of energy in the oxygen-respiration example. Its combining with hemoglobin is emphatically not a consequence of oxygen's being in our blood; just the reverse is true. On the other hand, its producing energy *is* a result of its being there.

The very best evidence that this analysis is on the right track is that it seems to include the entire array of standard cases we have been considering, while at the same time avoiding several very persistent classes of counterexamples. In addition to this, however, there are some more general considerations which urge this position upon us.[21] First, and perhaps most impressive, this analysis shows what it is about functions that is teleological. It provides an etiological rationale for the functional "in order to," just as recent discussions have for other teleological concepts. The role of the consequences of X in its own etiology provides functional ascription-explanations with a convoluted forward orientation which precisely parallels that found by recent analyses in ascription-explanations employing the concepts goal and intention.[22] In a functional explanation, the consequences of X's being there (where it is, and so forth) must be invoked to explain why X is there (exists, and so forth). Functional characterizations, by their very nature, license these explanatory appeals. Furthermore, as I hinted earlier, (b) is often simply implicit in the "because" of (a). When this is so, the "because" is the specifically teleological one sometimes identified as peculiarly appropriate in functional contexts. The peculiarly functional "because" is the normal etiological one, except that it is limited to consequences in this way. The request for an explanation as well will very often contain this implicit restriction, hence limiting the appropriate replies to something in terms of this "because"—that is, to functional explanations. "Why is it there?" in some contexts, and "What does it do?" in most, unpack into, "What consequences does it have that account for its being there?"

The second general consideration which recommends this analysis is that it both accounts for the propriety of, and at the same time elucidates the notion of, natural selection. To make this clear, it is important first to say something about the unqualified notion of selection, from which natural selection is derived. According to the standard view, which I will accept for expository purposes, the paradigm cases of selection involve conscious choice, perhaps even deliberation. We can then understand other uses of "select" and "selection" as extensions of this use: drawing attention to specific individual *features* of the paradigm which occur in subconscious or nonconscious cases. Of course, the range of extensions arrays itself into a spectrum from more or less literal to openly metaphorical. Now, there is an important distinction within the paradigmatic, conscious cases. I can say I selected something, *X*, even though I cannot give a reason for having chosen it: I am asked to select a ball from among those on the table in front of me. I choose the blue one and am asked why I did. I may say something like, "I don't know; it just struck me, I guess." Alternately, I could without adding much give something which has the form of a reason: "Because it is blue. Yes, I'm sure it was the color." In both of these cases I want to refer to the selection as "mere discrimination," for reasons which will become apparent below. On the other hand, there are a number of contexts in which another, more elaborate reply is possible and natural. I could say something of the form, "I selected *X* because it does *Z*," where *Z* would be some possibility opened by, some advantage that would accrue from, or some other result of having (using, and so forth) *X*. "I chose American Airlines because its five-across seating allows me to stretch out." Or "They selected DuPont Nomex because of the superior protection it affords in a fire."[23] Let me refer to selection by virtue of resultant advantage of this sort as "consequence-selection." Plainly, it is this kind of selection, as oppose to mere discrimination, that lies behind conscious functions: the consequence *is* the function. Equally plainly, it is specifically this kind of selection of which *natural* selection represents an extension.

But the parallel between natural selection and conscious consequence-selection is much more striking than is sometimes thought. True, the presence or absence of volition is an important difference, at least in some contexts. We might want to say that *natural* selection is really *self-*

selection, nothing is *doing* the selecting; give the nature of X, Z, and the environment, X will *automatically* be selected. Quite so. But here the above distinction between kinds of conscious selection becomes crucial. For consequence-selection, by contrast with mere discrimination, deemphasizes volition in just such a way as to blur its distinction from natural selection on precisely this point. Given our criteria, we might well say that X *does* select itself in conscious consequence-selection. By the very nature of X, Z, and our criteria (the implementation of which may be considered the environment), X will automatically be selected.[24] The cases are very close indeed.

Let us now see how this analysis squares with the desiderata we have developed. First, it is quite clearly a unifying analysis: the formula applies to natural and conscious functions indifferently. Both natural and conscious functions are functions by virtue of their being the reason the thing with the function "is there," subject to the above restrictions. The differentiating feature is merely the *sort* of reason appropriate in either case: specifically, whether a conscious agent was involved or not. But in the functional-explanatory context which we are examining, the difference is minimal. When we explain the presence or existence of X by appeal to a consequence Z, the overriding consideration is that Z must be or create conditions conducive to the survival or maintenance of X. The exact *nature* of the conditions is inessential to the possibility of this form of explanation: it can be looked upon as a matter of mere etiological detail, nothing in the essential form of the explanation. In any given case something could conceivably get a function through either sort of consideration. Accordingly, this analysis begs no theological questions. The organs of organisms could logically possibly get their functions through God's conscious design; but we can also make perfectly good sense of their functions in the absence of divine intervention. And in either case they would be functions in precisely the same sense. This of course was accomplished only by disallowing explicit mention of intent or purpose in accounting for conscious functions. Nevertheless, the above formula can account for the very close relationship between design and function which the previous analyses could not. For, excepting bizarre circumstances, in virtually all of the usual contexts, X was designed to do Z simply entails that X is there because it results in Z.

Second, this analysis makes a clear and cogent distinction between function and accident. The things X can be said to do by accident are the things it results in which cannot explain how it came to be there. And we have seen that this circumvents the accident counterexamples brought to bear on the other analyses. It is merely accidental that the chlorophyll in plants freshens breath. But what it does for plants when the sun shines is no accident—that is why it is there. Furthermore, in this sense, "X did Z accidentally" is obviously consistent with X's doing Z having well-defined causal antecedents, just like the normal cases of other sorts of accident (automobile accidents, accidental meetings, and so forth). Given enough data it could even have been predictable that the belt buckle would deflect the bullet. But such deflection was still in the appropriate sense accidental: that is not why the buckle was there.

Furthermore, it is worth noting that something can get a function—either conscious or natural—*as the result of* an accident of this sort. Organismic mutations are paradigmatically accidental in this sense. But that only disqualifies an organ from functionhood for the first—or the first few—generations. If it survives by dint of its doing something, then that something becomes its function on this analysis. Similarly for artifacts. For example, if an earthquake shifted the rollers of a transistor production-line conveyor belt, causing the belt to ripple in just such a way that defective transistors would not pass over the ripple, while good transistors would, we cold say that the ripple was *functioning as* a quality control sorter. But it would be incorrect to say that the ripple *had* the function of quality control sorting. It does not *have* a function at all. It is there only be accident. Sorting can, however, *become* its function if its sorting ability ever becomes a reason for preserving the ripple: if, for example, the company decides against repairing the conveyor belt *for that reason*. This accords nicely with Richard Sorabji's comment that in conscious cases, saying the function of X is Z requires at least "that some efforts are or would if necessary be made" to obtain Z from X.[25]

Third, the notion of something having more than one function is derivative. It is obtained by substituting something like "partly because"[26] for "because" in the formula. Brushing dust off the numbers is one of the functions of the watch's sweep-second hand if that feature is *one* of the (restricted, etiological) reasons the sweep hand is there. Similarly in the case of natural functions. If two or three things that livers do all contribute to the survival of organisms which have livers, we must appeal to all three in an evolutionary account of why those organisms have livers. Hence the liver would have more than one function in such organisms: we would have to say that each one was *a* function of the liver.

Happily, the analysis I am here proposing also accounts for the undoubted attractiveness of the other analyses we have examined. Beckner's first analysis is virtually included in this one under the rubric, "X does Z." The rest of the formula can be thought of as a qualification to avoid some rather straightforward counter-examples which Beckner himself is concerned to circumvent in his more recent attempt. Canfield's "usefulness" is even easier to accommodate: the usefulness of something, Z, which X does is *very usually* an informative way of characterizing why X has survived in an evolutionary process, or the reason X was consciously constructed. The important point to notice is that this is only *usually* the case, not necessarily: not all useful Z's can explain survival and some things are constructed to do wholly useless things. As for Beckner's most recent analysis, the complex, mutually contributory relationship among parts central to it is precisely the sort of thing often responsible for the survival and reproduction of organisms on one hand, and for the construction of complex mechanisms on the other. Again the valuable features of that analysis are incorporated in this one.

There is still one sort of case in which we clearly want to be able to speak of a function, but which offends the letter of this analysis as it stands. In several contexts, some of which we have already examined, we want to be able to say that X has the function Z, even though X cannot be said to do Z. X is not even *able* to do Z under the requisite conditions. In the cases of this sort I have already mentioned (the defective washer switch and ineffective governmental safety regulations), it has seemed necessary to italicize (emphasize, underline) the word "function" in order to make its use plausible and appropriate. This is a logical flag: it signals that a special or peculiar contrast is being made, that the case departs from the paradigms in a systematic but intelligible way. Accordingly, an analysis has to make sense of such a case as a variant.

On the present analysis, the italic type signals the dropping of the (usually presumed) second condition. X does *not* result in Z, although, paradoxically, doing Z *is* the reason X is there. Of course, in the abstract, this sound fatuous. But we have already seen cases in which it is natural and appropriate. That *is* the reason X (switch, safety regulations) is there. And a slightly more defensive formulation of (2) will include them directly: a functional ascription-explanation accounts for X's being there by appeal to X's resulting in Z. These cases *do* appeal to X's resulting in Z to explain the occurrence of X, even though X does *not* result in Z. So the form of the explanation is functional even in these peculiar cases.

Interestingly, this account even handles the exotic fact that these italicized functions of X can cease being even italicized functions without dispensing with or directly altering X. (Something that X did not do can stop being its function!) For example, if the ineffective safety regulations were superseded by another set, and were merely left on the books through legislative sloth or expediency, we would no longer even say that had the (italicized) *function* of making driving less dangerous. But, of course, that would no longer be the reason they were there. The explanation would then have to appeal to legislative sloth or expediency. This is usually done with verb tenses: that *was* its function, but is not any longer; that was why it was there at one time, but is not why it is still there. A similar treatment can be given vestigial organs, such as the vermiform appendix in humans.

Notes

1. Richard Sorabji, "Function," *Philosophical Quarterly*, 14 (1964), 290.

2. Morton Beckner, *The Biological Way of Thought* (New York, 1959), chap. 6.

3. Beckner gives an alternative formulation in which we can speak of *activities* as having functions, instead of *things*. I have abbreviated it here for convenience and clarity. The logical points are the same.

4. John Canfield, "Teleological Explanations in Biology," *British Journal for the Philosophy of Science*, 14 (1964).

5. Morton Beckner, "Function and Teleology," *Journal of the History of Biology*, 2 (1969).

6. As before, Beckner gives an alternative formulation so that we can speak either of a thing or of an activity having a function. My treatment will be limited to things, but again the logical points are the same.

7. Beckner seems to suggest (p. 160, top) that *F* must *be* an activity of the whole system *S*, which, of course, would conflict with part of 3. But his illustration, reproduced below, suggests the phrasing I have used here.

8. Carl Hempel, "The Logic of Functional Analyses," in L. Gross, ed., *Symposium on Sociological Theory* (New York, 1959).

9. Hugh Lehman, "Functional Explanations in Biology," *Philosophy of Science*, vol. 32 (1956).

10. Sorabji, *op. cit.*

11. Francisco J. Ayala, "Teleological Explanation in Evolutionary Biology," *Philosophy of Science*, vol. 37 (1970).

12. Michael E. Ruse, "Function Statements in Biology," *Philosophy of Science*, vol. 38 (1971).

13. This is not to abandon, or even modify, the previous distinction between functions and goals: the point can be made in this form only *given* the distinction. Nevertheless, support is provided for the analysis I am presenting here by the fact that the "in order to" of goal-directedness can be afforded a parallel treatment. For that parallel treatment see my paper "Explanation and Teleology," in the June 1972 issue of *Philosophy of Science*.

14. Canfield, p. 295.

15. It is sometimes urged that this sort of thing is all a teleological explanation is asserting; this is all "why?" asks in these contexts.

16. I takes the other forms to be essentially equivalent and subject, *mutatis mutandis*, to the same explication.

17. Of course, it follows that the notion of a *reason* offered in one of the alternative formulations is the standard conversational one as well: the reason it exploded was that it got too hot.

18. The qualification "causal" here serves merely to indicate that this is not the purely inferential sense of "consequence." I am not talking about the result or consequence of an argument—e.g., necessary conditions for the truth of a set of premises. The precise construction of "consequence" appropriate here will become clear below.

19. It is worth recalling here that "is there" can only sometimes, but not usually, be rendered "exists (at all)." So, contrary to many accounts, what is being explained, and what *Z* is the result of, can very often *not* be characterized as "that *X* exists" *simpliciter*.

20. It is often claimed that the asymmetry is temporal, but there are many difficulties with this view. Douglas Gasking, in "Causation and Recipes," *Mind* (October 1955), attempts to account for it in terms of manipulability, with some success. But manipulability is even less generally applicable than time order, so, as far as I know, the problem remains.

21. The following considerations are intended primarily as support for the entire analysis considered as a whole. Since (*a*) has already been examined extensively, however, I have biased the argument slightly to emphasize (*b*).

22. The primary discussions of this sort I have in mind are those in Charles Taylor's *Explanation of Behavior* (London: Routledge and Kegan Paul, 1964) and the literature to which it has given rise.

23. Of course the advantage is not always stated explicitly: "I chose American because of its five-across seating." But for it to be selection of the sort described here, as opposed to mere discrimination, something like an advantage must be at least implicit.

24. This is a version of the old problem about the tension between rationality and freedom.

25. Sorabji, *op. cit.*, p. 290.

26. Again, it is worth pointing out that "partly" here does not indicate that "because," when *not* so qualified, represents a sufficient condition relationship. It merely serves to indicate that more than one thing plays an explanatorily relevant role in this particular case. More than one thing must be mentioned to answer adequately the functional "why?" question in this context. But that answer, as usual, need not provide a sufficient condition for the occurrence of *X*.

3

Biological Teleology: Questions and Explanations

Robert N. Brandon

Why seek a teleological explanation? In adult *Homo sapiens* there are marked morphological differences between the sexes. Why is this? Answer: Different sex specific hormones work during ontogenetic development to produce these differences. Is this answer satisfying? That depends on the question one's really asking. One might be asking what's behind these hormonal differences, what's it all for. Whether or not this question is interesting or answerable, it is not answered by the above bit on hormones. One might want more.

This paper is about the more one might want, teleological explanations. In it I give an account of such explanations and the theories that provide them by placing these theories in the general category of evolutionary theories and then by distinguishing teleological theories from nonteleological evolutionary theories. Although the resultant account shows similarities to some existing accounts, I hope that my approach sheds new light on this subject.[1]

I should also remark on the "theory-ladenness" of my account. It is deeply committed to a certain view of evolutionary theory and the evolutionary process. If this view turns out to be mistaken then teleological explanations as I characterize them would have to be rejected. This is in contrast to what seems to be the current trend in the philosophical literature. This trend is to give accounts of *biological* explanations not tied to any specific biological theory.[2] I can't see the interest in an account indifferent between divine creation and Darwinian evolution. My account does not aim at that level of abstraction.

1   Background

In the past philosophers of science have not been overconcerned with taxonomic approaches to scientific theories. Some taxonomic distinctions have been made by philosophers in the positivist tradition,[3] but not the distinction which forms the focus of this paper. The positivists characteristically took a synchronic view of scientific theories since they were explicitly concerned only with confirmational matters. More recently philosophers have been much concerned with the diachronic development of scientific theories.[4] The approach of this paper is diachronic to the extent that it relates theories to the prior questions they address.

What is a scientific theory? That's a hard question to answer although it is easy enough to give examples of theories. Here I will not attempt a full and complete answer to that question, rather I want to describe and develop what I take to be a fruitful way to think about scientific theories.

Not all scientific questions call for explanations as answers. Some do, some don't. For example, "What is the mass of the Sun?" and "Here and now to what extent is intelligence inherited genetically in humans?" don't. But "How does a warbler navigate when migrating?" and "Why is the sun spheroid?" do. ("How" and "why" are in English fairly reliable linguistic indicators of questions calling for explanations and "what" is a decent indicator of questions not calling for explanation. But the distinction is not a syntactic one. More on this later.) Theories can be thought of as sets of statements structured to provide answers for scientific questions calling for explanations. This of course doesn't say all one might want to know about theories but it does suggest looking at theories from the point of view of the questions they answer.

The suggestion is this: As the explanatory structures of science, theories should be classified according to the kinds of explanations they give. (Note that this suggestion is not at all plausible for larger units of science, such as fields and disciplines.) But how are the relevant distinctions among explanations to be made? Recall that I aim to relate theories to the questions they serve to answer; so I suggest this: Important differences in kinds of explanations reflect and are reflected in differences in the kinds of questions to which these explanations are answers.

This will be the guiding principle for what is said in the remainder of this paper on classifying scientific theories.[5]

The main distinction to be drawn in this paper is one between two types of evolutionary theories. Before doing this some characterization of the general class of evolutionary theories is in order. Evolutionary theories are those that give evolutionary explanations to evolutionary questions or problems. Clearly evolutionary problems call for diachronic explanations. We can say that "an evolutionary problem ... is one which calls for an answer in terms of the time development of items" in the domain in question (Shapere, 1974, p. 190). Thus based on this characterization we can make an exhaustive taxonomic division between evolutionary and non-evolutionary theories.[6] This division is only of minor interest since the latter category is presumably quite heterogeneous.

But within biology this distinction, first drawn by Ernst Mayr in 1961, is of considerable interest. Mayr distinguishes *functional biology* and *evolutionary biology*.[7]

The functional biologist is vitally concerned with the operation and interaction of structural elements, from molecules up to organs and whole individuals. His ever-repeated question is "How?" How does something operate, how does it function? ... The evolutionary biologist differs in his method and in the problems in which he is interested. His basic question is "Why."[8]

We see that Mayr draws this distinction as we would have him do so, on the basis of differences in questions asked. He points out that the same phenomenon, e.g. a warbler starting its migration from New Hampshire on August 25, can generate quite different question. One set of questions seeks what Mayr calls the *proximate* causes of the start of migration ("What is the mechanism and how is it triggered?"). The other seeks the evolutionary, or *ultimate*, causes ("How and why has this mechanism evolved?"). (Mayr also points out that the same linguistic expression can and often is used to ask these different kinds of questions. In this case, "Why did the warbler start its southward migration on August 25?", can be used to ask both the functional and evolutionary questions.) Mayr, quite correctly, points out that this division holds for almost all biological phenomena. That is, we can always ask both functional and evolutionary questions of a given biological phenomenon. Furthermore answers to both are necessary for a complete understanding of the phenomenon (Mayr, 1961, p. 1502).

(The functional—evolutionary division in biology is historically, as well as philosophically, interesting. A detailed history of the relation between the two has yet to be written. However the complementary relation between the two has not always been recognized. This plus the fact that the same expression can be used to ask both functional and evolutionary questions has led to great confusion. For example, in 1932 the geneticist T. H. Morgan was willing to answer quite dogmatically, "Hormones!" to those who sought an evolutionary answer to "Why are there sex differences?" [Morgan, 1932, pp. 152–169].)

Mayr's distinction marks the fundamental division of biology. It is of much greater interest than the general distinction between evolutionary and non-evolutionary theories since Mayr's class of functional biological theories is fairly homogeneous. But the focus of this paper is a division within the region of evolutionary theories (in general, not just in biology), a division between teleological and nonteleological evolutionary theories. I've discussed Mayr's distinction for two reasons: (1) It helps to locate and put into perspective the general class of evolutionary theories; and (2) It illustrates the approach to be used. (Some would term this an *erotetic* approach, one based on the logic of the questions to be answered.)

2   Evolutionary Questions

We've all seen either in monster movies or museums the large bony plates along the back of the dinosaur *Stegosaurus*. The question, "What was that line of bony plates for?" suggests itself. Similarly we might wonder about the song of humpback whales, the color patterns of penguins, the lack of leaves in cacti, or the nastiness of rats in crowded conditions. We might verbalize our questions using "Why," "How come" or "What for." I'll refer to such questions as *what-for*-questions. (Again "what for" is just a linguistic indicator of a kind of question, it is not definitive of the kind. A syntactic characterization of *what-for*-questions is not to be expected.) *What-for*-questions are, apparently, teleological. Intuitively they are legitimately asked of any organic feature. Whether or not this is the case remains to be seen.

In biology conceptions of teleology have had a checkered history of respectability, yet they haven't died because almost all naturalists have felt that some sort of answers to teleological questions are necessary for a full understanding of biological phenomena. Today biologists, and most philosophers of biology, agree that Darwinian answers to *what-for-*questions are legitimate.[9] Despite this agreement, no one has adequately clarified the conceptual foundations of this question answering. I will try to do so shortly by delimiting the things of which these questions can be asked and by explicating the theoretic means available for answering them. But first I want to contrast these biological questions with other evolutionary questions.

Evolutionary theories answer questions about items in their domain in terms of the time development of those items. The following are some nonbiological questions which call for evolutionary answers: Why do the orbits of the planets in our solar system lie more or less on the same plane? Why are stars spherical? Why is hydrogen more abundant in the Universe than uranium? Like the biological questions we will soon examine, these questions cannot be satisfactorily answered without appeal to the evolutionary processes which produced the items in question. Notice, however, that we are not tempted to formulate these questions using "what for." We don't ask "What is the spherical shape of stars for?" Why this difference? It won't do to say that biologists, unlike cosmologists, are being careless with language. Teleological talk has received careful, albeit inconclusive, scrutiny from biologists. What's careless, I claim, is failing to mark a distinction between the products of physical and biological evolution (a failure characteristic of ancient conceptualizations of the world).

We've noticed a difference in the questions biological and cosmological or physical evolutionary theories answer. Can this apparent distinction withstand critical examination? The full answer to this will not be forthcoming until I give my account of biological evolutionary explanations and show just how they answer *what-for-*questions; but if this distinction is sound there must be some differences in the relevant evolutionary processes. So let's turn out attention to those processes.

I'll use the term "physical evolution" as a catch-all to include among other things chemical evolution, geologic evolution, stellar evolution and small and large scale cosmic evolution. Physical evolution can be neatly characterized by the phrase "survival of the stable."[10] This type of evolution requires only that there be variation among the items in the evolving milieu and that there be differential survival of the items. (This characterization does not rule out catastrophic events, which, of course, do occur.) The items that tend to stick around the longest are, of course, the most stable. In a population of genetically evolving organisms there is variation and differential survival. It follows that biological or genetic evolution is a special case of physical evolution.[11] The specific features of biological evolution are of interest to us.

Consider a painting made up of splotches of different types of paint. These different paints fade and eventually disappear at different rates. The painting may be said to evolve. This is a fairly typical case of physical evolution. (What makes this case less complicated than others is that the elements, *i.e.* the splotches, neither break up nor combine to form new elements.) In biological evolution differential survival is a bit more complicated. In our painting case the elements which vary have different survival rates. In biological evolution the relevant elements of variation, whether whole organisms or genes,[12] are not particularly long-lived and their actual life span is usually not very important. To speak at the genetic level, a gene does not increase in frequency in a population by outliving its rivals. A particular gene, that is a particular DNA molecule, "lives" at most as long as the organism in which it is located. A gene increases in frequency by replicating itself faster and better (*i.e.* more accurately) than its rivals. Again at the organismic level, longevity is only one factor in determining the evolutionary success of an organism. One must live long enough to reproduce; but longevity is a factor only insofar as it contributes to reproductive success. Thus in biological evolution differential survival has been replaced by differential reproduction. (Crystal formation seems to occupy an intermediate ground. If physical evolution has "evolved into" biological evolution, such intermediate processes are expected.)

The second feature I want to point out has to do with the causes of differential reproduction. There are two possible causes: chance and natural selection. Put another way, when an organism a has more offspring than organism b in a particular environment E this is either due to a being luckier than b (*i.e.* chance) or due to a being better adapted than b to E.[13] The latter cause is thought to predominate and is the one relevant to the present discussion. Typically, though not always, natural selection takes on the character it has because of limitations on environmental resources. Darwin termed the effect of such limitations the "struggle for existence." His reasoning was as follows: Natural rates of fecundity are such that all populations of organisms potentially increase in number geometrically; all environmental resources are finite; therefore all populations eventually reach a point where resource demand exceeds supplies; when this happens there is a struggle for existence.

The major constraint on any evolutionary process is time. Both of the features of biological evolution I have just described speed up the evolutionary process. In our painting case a small difference in the physical stability of two splotches makes for only a small difference in their length of survival. But in biological evolution a small difference in the traits of organisms can have a significant effect. This amplification of effect takes place in two steps: First, when there is a severe limitation on environmental resources a small biological difference can be the difference between reproductive success and failure. Second, even when heritable differences in adaptedness (*i.e.* expected value of reproductive success) are small the effect over a number of generations is an acceleration of the rate of evolution. This is due to the multiplicative nature of reproduction.

Let me review. In the first section of this paper I suggested a certain approach to classifying scientific theories. I suggest that theories should be classed according to the kind of questions they answer. In this section I have presented a *prima facie* case for saying that the neo-Darwinian theory of evolution answers a kind of question, *what-for*-questions, which some other evolutionary theories, *e.g.* cosmological theories, do not answer. That *what-for*-questions constitute a significant type of question, *vis-à-vis* a taxonomy of scientific theories, is not implausible. More support for this will come shortly. I have pointed out that there is a substantial difference between biological evolution and other simpler types of physical evolution. But this is not at all surprising. I covered this in some detail for two reasons: First, the difference in processes seems to be a precondition of my drawing the proposed distinction between evolutionary theories; second, some attention to the relevant processes is required if one is to appreciate the difference between types of evolutionary theories. But the unsurprising fact that there are differences in evolutionary processes hardly substantiates the claim that there is an important taxonomic division among evolutionary theories. I will substantiate this claim by showing just how the neo-Darwinian theory of evolution answers *what-for*-questions and by distinguishing these questions from other questions calling for evolutionary answers.

3   A Coherent Conception of Adaptation

First we need to delimit the proper objects of *what-for*-questions in biology. They are *adaptations*. Here "adaptation" is a technical term as yet undefined. Defining it will require a fairly lengthy digression.

In an earlier paper (Brandon, 1978b) I pointed out that relative adaptedness, the relation of one organism being better adapted than another (of the same interbreeding population) to a particular environment, is the fundamental concept of neo-Darwinian theory. It was shown that this relation requires a peculiar type of definition, what I called a schematic definition. Informally the schematic definition states that *a* is better adapted than *b* in *E* if and only if *a* is better able than *b* to survive and reproduce in *E*. (This notion of ability is cashed out in terms of expected values of offspring.) Particular instances of the schema concern particular populations in particular environments. Such instances might identify adaptedness with height, coloration, metabolic rates, or more likely some complex of such properties. One should note that adaptedness is never identified with actual reproductive rates, though a high correlation between the two is the norm.

Biologists have almost always made the mistake of defining natural selection as differential reproduction. Any sensible biologist could be convinced that this is wrong since on that definition one cannot distinguish evolution by natural selection from evolution by chance (or random-drift as it is often called). Evolutionists know that natural selection is a nonrandom process; however, differential reproduction need not be. Natural selection should be defined as differential reproduction which is due to the adaptive superiority of those organisms leaving more offspring.

The word "adaptation" can be used to refer to a process and often is so used in biological contexts. The process of adaptation can sensibly be identified with evolution by natural selection. It is by this process that populations change in response to environmental changes and it is this process which produces what Darwin called "organs of extreme perfection and complication."[14]

Loosely speaking an adaptation is anything which is the product of the process of adaptation. We need to state this much more precisely. What sorts of things can be adaptations? We could talk about gene complexes being adaptations. For instance we could say that the gene complex in Viceroy butterflies which codes for their wing coloring pattern is an adaptation. This pattern closely mimics the pattern of Monarch butterflies, a species quite distasteful to butterfly predators. Clearly this gene complex is the result of evolution by natural selection, *i.e.* the process of adaptation. So it is reasonable to talk of that gene complexes as an adaptation, but it is only because we can talk of the effect or product of that gene complex as an adaptation. Remember that natural selection acts directly on phenotypes not on genes or genotypes; so adaptedness is an organismic property not a genetic property. The core notion of adaptation is phenotypic.

It is not useful to talk of whole phenotypes as adaptations. If we are right in thinking that natural selection has been the major directive force of evolution then all phenotypes are to a large extent adaptations (*i.e.* products of the process of adaptation).

Talk of whole phenotypes as adaptations is uninformative because they all are, more or less, products of the process of adaptation. Is the same true of parts of phenotypes? That is, are all phenotypic traits adaptations?[15] I have not just asked whether all phenotypic traits are useful or beneficial to their possessors. Some writers have held the view that a trait is an adaptation if it is beneficial to its possessor.[16] This view divorces adaptation from the evolutionary process and so is without interest. (Although the condition that a trait be beneficial to its possessor is generally—but not always—necessary for it *to evolve* by natural selection, it is neither a necessary nor sufficient condition for a trait *to be* an adaptation.) I mention it only to discard it. The question before us is whether all traits have the same type of causal history. The question is a complicated one. My discussion of it will not deal with all these complications completely, but will suffice for present purposes.

It should be obvious that there are many examples of traits which aren't gene-linked at all and so not adaptation. These need not concern us. There are, however, two interesting categories of traits which aren't adaptations. They are: epiphenomenal traits, and traits due to chance. Let's consider the latter first. A trait may appear full-blown with one mutation. Such a trait is clearly not an adaptation regardless of its effect on the adaptedness of its possessor. (Suppose this trait does increase the adaptedness of its possessors and thereby increases in frequency in the population. It will then become an adaptation.) Its appearance is ultimately explained by the chance mutation. More interestingly, traits may actually evolve, *i.e.* be the product of an evolutionary process, and still not be adaptations. The process is evolution by random drift.

Random drift is just the random fluctuation of alternative genes. When can such fluctuations have significant evolutionary effects? There are two cases. The first is when the alternatives are selectively neutral. The theory of such evolution, often called non-Darwinian evolution, is well developed (see King and Jukes, 1969), although there is considerable controversy over how prevalent such evolution is in nature.[17] The other case is when sampling accidents have a significant effect due to small population size. In this case the alternative genes need not be selectively neutral; an alternative which renders its possessors less adapted than organisms with its rivals may in fact reach fixation by this process.[18]

Traits due to chance are not just logical possibilities. In particular, if the generally accepted theory of speciation is correct then random drift in small populations has had tremendous evolutionary effect.[19] In fact major differences among species are largely due to chance, natural selection just fine-tunes these given differences to fit local environments.

The second category of traits that aren't adaptations I label epiphenomenal traits. This is a rather heterogeneous category. It includes traits which have evolved, not on their own merit, as it were, but due to their connection to other evolving traits. The two simplest types of connections are gene linkage and pleiotropic connections. When a not particularly deleterious gene is closely linked on a chromosome to a gene being strongly selected for, it can "hitchhike" its way to prominence. Pleiotropic connections are similar but more permanent. Pleiotropic genes are genes which affect more than one part of the phenotype (or more than one trait). For example,[20] a gene may code for an enzyme which helps detoxify a poisonous substance common in its environment. It does so by converting the poison to an insoluble pigment. Two new traits will become common in the population. One, the ability to handle the toxic substance, is an adaptation. The other, the resultant color of the organisms having the first property, is not an adaptation. Its presence is not explained by what it does but rather by its pleiotropic connection to an adaptation.

What is the basis of this distinction? In the given environment there is a causal relationship between the ability to handle the toxic substance and adaptedness; there is no such causal relationship between color and adaptedness in *this* environment. (Of course in another environment color might have a close causal connection with adaptedness.) To assert that something is an adaptation is to make a cause—historical statement.

A more nebulous type of epiphenomenal connection is illustrated by the following examples. A beating heart makes sounds. Although nowadays heart sounds are a diagnostic aid and so beneficial to the people producing them, it is highly doubtful that heart sounds played any role in the evolution of the human heart. Thus we don't want to call the making of heart sounds an adaptation. Humans and other organisms would have been just as well off, it seems, with an organ that circulated blood silently. Such an organ is conceivable, but probably not an open possibility for any of the organisms in question. But if that's so then we can hardly say that an organism would be just as well off without heart sounds: he wouldn't be as well off, he would be dead. Consider a second example. Humans, some of them at least, have the ability to do meta-logic. Critics of evolutionary approaches to epistemology might claim that this ability has had no possible evolutionary significance and so no evolutionary explanation of its presence can be given. But it may be that the ability to do meta-logic is an epiphenomenal by-product of a generalized intellectual apparatus which is a human adaptation. (I think that the epiphenomenal connections illustrated in the two preceding examples are less objective and more a product of how we decide to describe phenotypic features than gene-linked or pleiotropic connections. But this is irrelevant to present concerns.)

We have seen that not all traits are adaptations, that is, not all traits have the same type of causal history. But this notion of adaptation is useless unless we can differentiate adaptations from other traits. How can we do this? Recall our example of pleiotropy where a gene specifies an enzyme which detoxifies a poison by converting it to a pigment, thus darkening the color of the organisms. This gene becomes much more frequent in the population and so both the mechanism to handle this poison and darker color become more frequent. How can we tell which if either of these phenotypic features are adaptations? First suppose it is clear from the population genetic date that the gene is increasing in frequency due to natural selection. (By "population genetic data" I mean the data concerning gene frequencies and actualized genotypes—*i.e.* in what configurations the genes actually appear. At least in some cases the population genetic data differentiate natural selection from random drift.) This alone does not differentiate coloration from detoxification. From the population genetic data we know that, on average, individuals containing the gene for this enzyme are better adapted to their environment than individuals lacking the gene. But that does not tell us *why* they are better adapted. We know why one's better adapted than another when we understand the causal—ecological relationships relevant to differential reproduction. In our hypothetical case the only relevant variations is in ability to handle the poison. So one organism a is better adapted than b in the environment E if and only if a can detoxify more of the poison than b in E. (This if-and-only-if statement must be understood as a lawlike statement supporting counterfactuals rather than as a purely truth-functional statement.)[21] Why is an organism with this gene better adapted than an organism lacking the gene to their environment? Not because he's darker colored (the relevant causal relationships have nothing to do with coloration), but rather because he can detoxify a poison common in their environment. Thus the ability to detoxify the poison is an adaptation, the dark color is not.

Some biologists might complain that my example is too simple, and that in real cases we can't even analyze present ecological situations with the detail and accuracy required to make such a distinction, much less ecological situations of the distant past. The point would be well taken. But as I've pointed out, insofar as the notion of relative adaptedness can be made operational we can distinguish adaptations from other types of traits. And as I've pointed out elsewhere (Brandon, 1978b), relative adaptedness is the fundamental concept of evolutionary theory. Thus our notion of an adaptation being a product of evolution by natural selection is no worse off than evolutionary theory itself. The practical epistemological problems that do accompany the concept of adaptation simply show that evolutionary biology is far from being completed as a science.

4 Teleological Explanations

We have delimited the proper objects of *what-for*-questions in biology: they are adaptations. We've also seen just what adaptations are: they are phenotypic traits which have evolved as a direct product of natural selection. With this information it is now easy to show how *what-for*-question are answered.

When a biologist asks a *what-for*-question of some organic feature he postulates that it is an adaptation. (In so doing he may, of course, be wrong.) He then tries to discover the trait's "adaptive significance,"[22] that is, why it was selected for. When dealing with contemporary organisms he usually proceeds by observing just how the trait interacts with the rest of the phenotype and elements of the environment, with a view to uncovering how it contributes to the adaptedness of its possessor. If he succeeds in uncovering this he then extrapolates into the past, and presents a plausible scenario of how the trait evolved due to its effect on the adaptedness of its possessors. If he is dealing with extinct organisms his procedure is not much different, he just has more reconstruction to do. Ideally the biologist should have complete knowledge of the evolutionary dynamic[23] of the population for the relevant time period and a complete knowledge of the changing ecological situation of the population for that period. Of course, this ideal state is never attained, but we do count the less than certain plausible scenarios just characterized as evolutionary explanations.

Put abstractly, a *what-for*-question asked of adaptation *A* is answered by citing the effects of past instances of *A* (or precursors of *A*) and showing how these effects increased the adaptedness of *A*'s possessors (or the possessors of *A*'s precursors) and so led to the evolution of *A*.

The sense in which *what-for*-questions and their answers are teleological can now be clarified. Put cryptically, trait *A*'s existence is explained in terms of what *A* does. More fully, *A*'s existence is explained in terms of effects of past instances of *A*; but not just any effects: we cite only those effects relevant to the adaptedness of possessors of *A*. Note how this differs from other evolutionary explanations. For instance, one can give an evolutionary explanation of the darker coloration of the members of the population in our pleiotropy example. But one can't do this by citing the effects of the darker color. By hypothesis the darker color had no effects of evolutionary significance.

5 Conclusion

I have distinguished *what-for*-questions from other questions demanding evolutionary answers. *What-for*-questions demand teleological explanations as answers, teleological in the sense just specified. For our purposes the neo-Darwinian theory of evolution is best thought of as a supertheory or collection of theories. There is nothing teleological about the theory of evolution by random drift or theories of speciation. But what might be called the neo-Darwinian theory of adaptation or the theory of evolution by natural selection is teleological; it answers a different kind of question and so is to be distinguished from other types of evolutionary theories.

Have I created a taxonomic category for one theory? The theory of adaptation is the best developed, and perhaps the only developed teleological scientific theory; but nothing I've said precludes the possibility of others.[24]

One point of drawing the distinction between teleological and non-teleological evolutionary explanations was to shed new light on the former by contrasting them with the latter. If I've been successful in this then the taxonomic division is of philosophic interest. The reason I chose to use the word "teleological" (rather than some such word as "teleonomic") to describe adaptation-explanations is that I think they do satisfactorily answer teleological questions.[25] Put another way, I think that the unique sort of understanding yielded by adaptation-explanations is not to be confused with the understanding yielded by other types of explanations. All this is best seen and appreciated by considering real examples. I'll mention just one.

It is well known that in some species birds regulate their clutch size (*i.e.* the number of eggs laid). This is a mysterious fact; it becomes more mysterious once one begins to understand evolution. After all, shouldn't these birds be trying to produce as many eggs as possible? One explanation, popular until recently, is that birds regulate their clutch size for the good of the species. Although rarely stated explicitly the argument is that overpopulation is a bad thing for these species and so species or group selection has produced species that do regulate their population size. This regulation is an adaptation for the species. The clutch size regulation by the individual birds is just the mechanism by which the species regulates its size.[26]

An alternative explanation is also an adaptation-explanation. It states that there is an optimal clutch size for birds in a given population. Laying either fewer or more than the optimal number of eggs usually results in leaving fewer viable offspring in the next generation. This has been experimentally confirmed (see Lack, 1954). Thus birds regulate their clutch size for "their own good" or rather because doing so increases their adaptedness.

Given the relative ineffectiveness of group selection and the sufficiency of the individual level adaptation-explanation, biologists almost universally reject the first explanation and accept the latter.[27] Consider how this changes our understanding of the phenomenon. Consider how incomplete our understanding of this would be without either adaptation-explanation. That is to say, phenomena such as this cry out for teleological explanation.

Notes

1. See Wimsatt (1972), Wright (1976) or Williams (1977) for accounts similar in some respects.

2. For two extreme examples of this trend see Wright (1976, especially p. 97) and Purton (1979, especially p. 11). A nice palliative to this trend is Williams (1977).

3. For examples see Hempel (1965).

4. Notable examples being Lakatos (1970), Shapere (1974) and (1977), and Laudan (1977).

5. To my knowledge no one previously has explicitly formulated and used this principle in work on explanation. However Bromberger's work, (1963) and (1966), and Shapere's work, (1974) and (1977), as well as Suppe's discussion of their work (1977) are suggestive. One should note how radically different this approach is from the Hempelian approach; an approach which for the most part ignores and obscures the relation between questions and explanations.

6. This is similar to but not the same as Shapere's compositional-evolutionary distinction [see Shapere (1974) and (1977)]. Shapere's distinction is not exhaustive and is of greater interest since his compositional category is not characterized by exclusion.

7. Mayr (1961). To my knowledge no philosopher has given Mayr's distinction its due attention.

8. I should immediately draw attention to one potential source of confusion. The quoted material indicates why Mayr chose to use the word "functional." However there also is a considerable literature on functional ascriptions and/or functional explanations in biology. On most accounts (all plausible accounts) functional explanations are evolutionary explanations and functional ascriptions have some close relation to them. Were Mayr's distinction the focus of this paper I would try to develop a better set of terms, but it is not so I'll only warn against confusion.

9. There is still considerable terminological controversy among biologists and philosophers of biology. Should the word "teleology" with all its Aristotelian connotations be used or should we drop it and use a less loaded term like "teleolonom"? Mayr would even object to my calling the questions evolutionary theory answers *what-for*-questions (see his 1961 paper). These terminological disputes are neither trivial nor momentous. Later some justification for my choice of terms will be given.

10. I found "survival of the stable" in Dawkins (1976). I suspect it has been used before.

11. For a good presentation of a plausible scenario of how physical evolution led to, or if you like, evolved into, biological evolution see Dawkins (1976, chap. 2).

12. Actually both phenotypic variation and genetic variation are relevant. Natural selection acts on phenotypes not in any direct manner on genes. Natural selection would not act on a population of phenotypically identical organisms no matter how genetically diverse they were. Yet only genetically transmittable variations are of (genetic) evolutionary significance. Genetic variation is the *sine qua non* of genetic evolution. Thus variation at both levels is relevant.

13. For definition and discussion of relative adaptedness and natural selection see Brandon (1978b).

14. See Darwin (1859, p. 186). One should not confuse this process with the process of individuals 'adapting' to their environment. For example frogs of the species *Hyla versicolor* change their color to match their background. This individual adaptability is, presumably, the result of evolutionary adaptation, but it is not an instance of it.

15. There is an interesting problem concerning just what should count as a trait, but I will not deal with it here. For the purposes of this paper a trait is any describable feature of an organism. But see Lewontin (1978).

16. *E.g.* Ruse (1971), Nagel (1977) and, it seems, Munson (1971).

17. See Ayala (1974). I should point out that the theory of non-Darwinian evolution is generally concerned with molecular evolution, so with selectively neutral variants of protein molecules. If these different protein molecules are functionally equivalent (which is not the same as selectively neutral) then they make no difference at any phenotypic level above the protein level. Talk of molecularly different but functionally equivalent protein molecules as different traits may be misleading. In any case the mathematical theory of non-Darwinian evolution tells us what can happen with alternative selectively neutral "traits."

18. See my basset-shepherd case (Brandon, 1978b).

19. By what Mayr calls the *founder principle*. New species are formed from small geographically isolated founder populations. The gene pool of this population is, to a large extent, a random sample of the original gene pool. Also the small size of the founder population increases the subsequent role of random drift. See Mayr (1963).

20. This example is from Lewontin (1978, p. 228), where he discusses more fully some of the topics I'm mentioning.

21. According to the example it is also the case that *a* is better adapted than *b* in *E* if and only if (in the truth-functional sense) *a* is darker colored than *b* (in *E*). But the darker color is causally irrelevant; thus it is not the case that if we *were* to paint black an organism lacking the detoxification ability he *would* become better adapted to *E*. Clearly this explanation is nonextensional. Recent literature on the extensionality of explanation, *e.g.* Levin (1976) and Stern (1978), seems largely irrelevant to this example. I am indebted to Robert Nozick for suggesting this footnote.

22. Or its "function." I've avoided the word "function" in order to avoid needless verbal polemics. Were I to use it I would do so in a manner roughly following Wright (1976).

23. *I.e.* the dynamics of the expansion of similarity classes of genotypes relative to others (Brandon, 1978a).

24. See Scheffler (1959, pp. 269 and 270) where he discusses some learning-theoretic psychological explanations which are teleological in exactly the same sense that adaptation-explanations are.

25. For those fond of counterfactuals I am willing to claim the following: If Aristotle had been convinced of the truth of Darwinian evolutionary theory then he would have found such explanations totally adequate as answers to teleological questions.

26. For a fuller presentation of this proposed explanation see Wynne-Edwards (1962) or Dawkins (1976, pp. 118–131).

27. But see Wade (1978) for a discussion of some of the heretofore unrecognized biases of models of group selection.

References

F. Ayala, "Biological Evolution: Natural Selection or Random Walk?", *Am. Scient.* 62 (1974), 692–701.

F. Ayala and T. Dobzhansky (eds.), *Studies in the Philosophy of Biology* (London: Macmillan, 1974).

B. Baumrin (ed.), *Philosophy of Science: The Delaware Seminar* (New York: Interscience, 1963), Vol. II.

R. N. Brandon, " 'Evolution,' " *Phil. Sci.* 45 (1978a), 96–109.

R. N. Brandon, "Adaptation and Evolutionary Theory," *Stud. Hist. Phil. Sci.* 9 (1978b), 181–206.

S. Bromberger, "A Theory about the Theory of Theory and about the Theory of Theories," pp. 79–106 in Baumrin (1963).

S. Bromberger, "Why Questions," pp. 86–111 in Colodny (1966).

R. Colodny (ed.), *Mind and Cosmos: Explorations in the Philosophy of Science* (Pittsburgh: University of Pittsburgh Press, 1966).

C. Darwin, *On the Origin of the Species* (London: John Murray, 1859).

R. Dawkins, *The Selfish Gene* (New York: Oxford University Press, 1976).

C. G. Hempel, *Aspects of Scientific Explanation* (New York: The Free Press, 1965).

J. L. King and T. H. Jukes, "Non-Darwinian Evolution," *Science, N.Y.* 164 (1969), 788–798.

D. Lack, *The Natural Regulation of Animal Numbers* (Oxford: Clarendon Press, 1954).

I. Lakatos, "Falsification and the Methodology of Scientific Research Programmes," pp. 91–196 in Lakatos and Musgrave (1970).

I. Lakatos and A. Musgrave (eds.), *Criticism and the Growth of Knowledge* (Cambridge: Cambridge University Press, 1970).

L. Laudan, *Progress and its Problems* (Berkeley: University of California Press, 1977).

M. Levin, "Extensionality of Causation and Causal Explanatory Contexts," *Phil. Sci.* 43 (1976), 266–277.

R. C. Lewontin, "Adaptation," *Scient. Am.* 239 (1978), 212–230.

E. Mayr, "Cause and Effect in Biology," *Science, N.Y.* 134 (1961), 1501–1506.

E. Mayr, *Animal Species and Evolution* (Cambridge: Harvard University Press, 1963).

T. H. Morgan, *The Scientific Basis of Evolution* (New York: W. W. Norton & Co., 1932).

R. Munson, "Biological Adaptation," *Phil. Sci.* 38 (1971), 200–215.

E. Nagel, "Teleology Revisited," *J. Phil.* LXXIV (1977), 261–301.

A. C. Purton, 'Biological Function,' *Phil. Q.* 29 (1979), 10–24.

M. Ruse, "Functional Statements in Biology," *Phil. Sci.* 38 (1971), 87–95.

I. Scheffler, "Thoughts on Teleology," *Br. J. Phil. Sci.* 9 (1959), 265–284.

D. Shapere, "On the Relations between Compositional and Evolutionary Theories," in Ayala and Dobzhansky (eds.) (1974).

D. Shapere, "Scientific Theories and their Domains," pp. 518–565 in Suppe (1977).

C. Stern, "Discussion: On the Alleged Extensionality of Causal Explanatory Contexts," *Phil. Sci.* 45 (1978), 614–625.

F. Suppe, *The Structure of Scientific Theories* (Urbana: University of Illinois Press, 1977).

F. Suppe and P. Asquith (eds.), *PSA 1976*, Vol. I (East Lansing: Philosophy of Science Association, 1977).

M. Wade, "A Critical Review of the Models of Group Selection," *Q. Rev. Biol.* 53 (1978), 101–114.

M. B. Williams, "The Logical Structure of Functional Explanation in Biology," pp. 37–46 in Suppe and Asquith (1977).

W. C. Wimsatt, "Teleology and the Logical Structure of Function Statements," *Stud. Hist. Phil. Sci.* 3 (1972), 1–80.

Wright, *Teleological Explanation* (Berkeley: University of California Press, 1976).

V. C. Wynne-Edwards, *Animal Dispersion in Relation to Social Behavior* (Edinburgh: Oliver & Boyd, 1962).

II

Don't Look Back: Nonhistorical Approaches to Biological Teleology

4

The Inference of Function from Structure in Fossils

M. J. S. Rudwick

1

The concept of adaptation stands at the centre of the modern debate on the mechanisms of evolutionary change.[1] The theory at present dominant among biologists, which is generally termed "Synthetic" by its proponents and "Neo-Darwinian" by its critics, stresses the ubiquity of natural selection as the prime causal factor behind the apparently "directive" element in evolutionary change. Adaptation is considered to be as ubiquitous a feature as natural selection; for it is the "pressure" of selection that evokes the first appearance, directs the development and even enforces the extinction of the features in which adaptation is expressed. (Inadaptive "genetic drift," postulated principally on the basis of mathematical models, occupies a relatively minor position in the theory.)

Standing against this Synthetic theory, and generally lacking both its brilliant advocacy and its present massive support, is a diverse group of theories which however have one important feature in common: they attribute a significant role to inadaptive or non-adaptive structures. This applies, for example, on Schindewolf's "Typostrophic" theory, to the "ground-plan" of the first members of a new major taxon; or, on most varieties of the theory of Orthogenesis, to the "bizarre" features of the last surviving members of an evolutionary trend. "Unorthodox" theories such as these are often attacked intemperately on the grounds that they are "unscientific," "metaphysical," or even "mystical."[2] This is certainly unjust to many (though not all) who have propounded these theories: internal orthogenetic factors, typostrophes, etc., have often (though not always) been postulated as strictly scientific explanations.

As Dr Marjorie Grene has shown in her lucid analysis of the disagreement between Schindewolf and Simpson, this whole debate is "an illuminating paradigm of scientific controversy."[3] These two palaeontologists see the same evidence from widely different viewpoints, and are therefore separated by disagreement on several different planes. Each of their respective ways of looking at the evolutionary record "constitutes, for its proponent, a closed interpretive system *in* which he sees the facts. In such a case the objects of one to the arguments of the other are, from the other point of view, irrelevant."[4] The Synthetic theory and the theories standing in opposition to it do indeed show many of the marks of closed all-explanatory systems. This is reinforced by the personal attitudes of their supporters: the Synthetic theory shows a disturbing tendency to become an illiberal and intolerant orthodoxy, which is matched by the tendency of its opponents to regard themselves as an embattled minority. This situation has obscured the possibility of discovering criteria that might be accepted by both sides as being valid tests.

An example of such a test is to determine whether or not a series of fossil species existed in the order predicted by an orthogenetic interpretation of their evolution. Thus Simpson has justly argued that the many orthogenetic interpretations of the evolution of the horses and the sabre-toothed "tigers" are invalid, simply because the fossil evidence shows that the evolution was not a single directional trend at all, or because the species occur in the wrong order.[5] But in fact this is a one-way test; it involves a "risky prediction" on one side but not on the other. An orthogenetic explanation involves a prediction of the order of occurrence of certain fossils; and this prediction may be (in some instances has been) proved incorrect. But there is no comparable prediction in a Synthetic explanation. If the fossils do not form a trend, that is easily explained; but if they do, that too can be explained in terms of selection and adaptation. Even if there was a genuine trend, culminating in structures that some would judge to have been non-adaptive and therefore favouring an orthogenetic explanation, the Synthetic theory is not disturbed at all. For its supporters can always argue that these structures, however "bizarre," *might* have had an adaptive value.

It is this asymmetrical relation between the theories that I want to examine more closely in this article.

2

Adaptation is a critical issue between the Synthetic theory and its rivals. Are there, or are there not, important features of the evolutionary process which cannot be explained in terms of adaptation under the influence of selection? Is the "ground-plan" of a new major taxon (for example the tetrapod vertebrates) initially no more than a slight adaptive modification of that of its immediate ancestors; or is it genuinely novel, and not in itself adaptive, however many adaptations it may subsequently give rise to? Are the first rudiments of a new organ (for example the rudimentary horns of the earliest titanotheres) themselves of adaptive value, however slight; or are they initially no more than fortuitous inadaptive variants, even though, enlarging under the influence of some internal (perhaps genetic) factor, they may subsequently become adaptively significant? Does the extreme terminal development of such an organ (for example the antlers of the "Irish elk") reflect the ever-increasing perfection of its adaptation under the continuing pressure of selection; or does it represent a trend carried to a point at which the organ has ceased to be adaptively valuable and become disadvantageous to the organism that possesses it? Does the extinction of such an organism reflect its inability to modify a highly adaptive but narrowly specialised mode of life in the face of changing circumstances; or is it the inevitable result of a trend that has equipped the organism with a feature which has become too disadvantageous to permit survival?

With such questions as these, the argument between the Synthetic theory and its rivals can be seen to depend critically on the recognition of the adaptive value—or lack of it—of the structural features of particular organisms. If this were all, the argument could be settled—at least in principle—by determining, by observation and experiment, whether or not these features perform identifiable functions in the economy of the organisms.

But the problem is more difficult. For the organisms whose adaptive status we wish to evaluate can only be recognised as such *in retrospect*. The first members of a new major taxon, if discovered today, would not be recognised for what they were. If (on the Synthetic theory) they differed only slightly from their immediate ancestors, they would be placed taxonomically in the same major taxon, and not in a new one of their own; whereas if (on the Typostrophic theory) they showed a radically novel ground-plan, they would merely be assigned to a new "minor phylum" (as for example the Pogonophora have been in recent years), and their adaptive status would be of no great significance. On either theory it is only in retrospect, in view of the abundance and diversity of the adaptive forms that subsequently arose from them, that the earliest members of a new major group can be recognised and their special significance appreciated. Similarly, the earliest rudiments of a new organ that gradually develops into great prominence can only be recognised in retrospect, after the organ has developed: such rudiments, if discovered in an existing organism, would be regarded as minor variant features of no great importance. Again, continued directional change in such an organ can only be recognised over a long period of geological time, and not in the evolutionary "instant" in which we can study living organisms. Finally, it is only in retrospect that the extreme development of a structural feature can be seen to have been connected—by whatever cause—with the extinction of the form that possessed it.

Since the critical evidence can only be recognised in retrospect, the alleged ubiquity (or near-ubiquity) of adaptation can only be tested against the alleged existence of important non-adaptive features by referring to the evidence of palaeontology. It is on fossil material, or not at all, that we can hope to test the validity of this aspect of the Synthetic theory and its alternatives. (The reality and significance of this situation has been obscured by the tendency of many palaeontologists, with more imagination than logic, to use the present tense to describe phenomena that in the nature of the case belong to the unobservable geological past.)

That the critical evidence is limited to fossil material introduces problems which, although they are strictly technical in origin, have important theoretical implications. The fossil record is highly imperfect, not only as a record of the course of evolution, but also as a record of the adaptations of individual organisms. Unavoidably, the range of specific adaptations that we can hope to detect in fossil material is therefore limited to those that are reflected in preservable features of the morphology and environment of the organism. The evidence of environmental factors may be extremely scanty, and it is often only possible to reconstruct the external environment in the most general terms. Factors of the environment, physical or biotic, which might be essential to the understanding of an adaptation, may have left no trace. Moreover, the preservation of the organism itself if also invariably imperfect: whole organs or parts of organs, which might be essential to the evaluation of the adaptive status of the organism as a whole, may have left no trace. However, favourable modes of preservation may record at least the skeletal structures in extremely fine detail, and occasionally other parts of the anatomy also; and in some taxa an intimate relation between skeletal and non-skeletal parts may allow much of the unpreserved (non-skeletal) anatomy to be reconstructed with considerable confidence from indirect traces in the skeletal anatomy (e.g. muscle insertions and cranial anatomy in fossil vertebrates). Such reconstructions depend essentially on the detection of homologies between the fossil and some related living organism; if the fossil belongs to wholly extinct group (e.g. graptolites) the problem is far more difficult.

But even with the most favourable circumstances of preservation, we can only re-create the fossil organism, with a lesser or greater degree of confidence, into a state in which it resembles a pickled specimen of some present-day organism. Anatomically it may be more or less complete, but it is isolated from most features of its original environment, and of course it lacks the ongoing activities of life. The functions—if any—of its various parts and organs therefore cannot be determined by simple observation and experiment; we cannot (as we can, in principle, with any living organism) simply observe what effects an organ in fact achieves in natural or experimentally contrived situations.

The same difficulty often arises in practice even in the study of living organisms. For example, direct observation of the function may be impossible except by dissection of a kind that modifies or even inhibits the function; or an organism may only be known from dead specimens (a common situation, for example, in the study of deep-water marine animals). But though reasons such as these may sometimes hamper the study of the functional morphology of living organisms, the impossibility of a direct study of function becomes in palaeontology not merely incidental but normative. It is this that has led some palaeontologists to

dismiss all inferences of function from structure in fossils as purely speculative, unscientific, and useless. Many such inferences have indeed been
advanced without valid arguments to support them, but clearly this does
not warrant a wholesale rejection of them all, without an examination of
the basis on which they are or can be made.

3

When the inference of function from structure is discussed in palaeontological works, it is commonly said that the fundamental methods of
reasoning involve comparisons—either homological or analogical—with
living organisms. These however are pseudomethods, which conceal the
real nature of such inferences.

Suppose that we have discovered the fossil remains of an extinct
organism, and we wish to determine the adaptive significance (if any) of
its distinctive features. Suppose that what we have found is in fact a
pterodactyl.[6] We notice its long slender forelimbs, and the traces of a
thin membrane stretched behind them. What is the adaptive significance
of these forelimbs? We notice their resemblance to the wings of birds
(though there is no trace of feathers) and to the wings of bats (though the
skeletal framework of the membrane is significantly different). It is clear
that the pterodactyl is a tetrapod vertebrate, so that we are justified in
regarding the comparison between its forelimbs and the forelimbs of
birds and bats as being—at least in a general way—a *homological* comparison. Yet if we provisionally identify the fossil structure as a "wing,"
we are obviously not doing so by virtue of its (imperfect) homology with
the wings of birds and bats. If homology were the real criterion, we
ought to be comparing it rather with the forelimbs of some more closely
related living organism, such as a lizard; but this would lead us to identify it as a crawling organ, which we would rightly reject as highly
improbable.

Why then do we call it a "wing"? We do so, of course, because we
recognise its *analogical* similarity to the wings of birds and bats. Even if
we knew of no birds or bats, we might still see its similarity to the wings
of, say, insects, with which it has not the slightest homological relation.
Birds, bats, and insects, as we know from direct observation, use their

wings for flying; the pterodactyl forelimbs have a certain structural resemblance to these organs-for-flying; therefore, we may say, the pterodactyl also probably used its forelimbs for flying.

Yet if we stopped there, at a simple analogical comparison, we would in fact be seriously wrong. For a further study of the pterodactyl anatomy shows that it could not have possessed the powerful flight muscles which would be necessary for a full analogy with the flight of birds, bats, and insects. The analogy, like the homology, is imperfect. But this imperfection gives us a glimpse of the real method of inference which underlies reasoning that appears to be based on homological or analogical comparisons. For our final conclusion that the pterodactyl wing was not an organ for powered flight but probably an organ for gliding is based only incidentally on any comparison with living organisms.

The initial clue about its possible function may have been suggested by the observation of roughly similar structures in certain living organisms; but the development and testing of the interpretation depend fundamentally on applying a criterion of *mechanical fitness*. From our knowledge of natural and artificial aerofoils, and of the structural requirements of their successful operation, we conclude that the pterodactyl forelimb would have been physically *capable* of functioning as an aerofoil. From our knowledge of the energy requirements for powered flight and of the energy output of vertebrate muscle, we conclude that it would *not* have been capable of functioning as a flapping wing for powered flight. Given enough knowledge of the physical requirements of flight-mechanisms, we could reach the same conclusions, even if there were no homologous or analogous structures in any living organisms. Even though no living organism utilises, for example, the operational principles of the helicopter,[7] an extinct organism that did so could, if discovered, be recognised for what it was. All we need, ideally, is a knowledge of the operational principles involved in all actual or conceivable flight mechanisms possible in this universe. Consequently the range of our functional inferences about fossils is limited not by the range of adaptations that happen to be possessed by organisms at present alive, but by the range of our understanding of the problems of engineering.[8]

An analysis of fossil structures in the light of operational principles is thus, I believe, fundamental and unavoidable in any inference about their possible functions or adaptive significance. This involves the limited "teleology" that is inherent in any description of a machine as a machine.[9] However much the word "teleology" may be used as a term of abuse by polemical biologists—or philosophers—it is well known that the language of modern biologists (and especially our informal unpublished language) abounds in words such as "contrivances," "mechanisms," and "devices," which reveal our common conception of the organism as being (at least from this point of view) a complex of living machinery. Machines can only be described for what they are by referring to the way in which their *design* enables them to *function* for their intended *purpose*. Such purposive language cannot be avoided for some purely physical description of the machine, and no engineer would be so foolish as to attempt to avoid it. Organisms, when considered in their aspect as living machinery, must be similarly described: purposive language is unavoidable, and should not be concealed by a "pseudo-substitution" of apparently non-teleological terms.[10] Of course, the "purpose" implied in this is no more than the existence and survival of the individual organism, which from this point of view must be regarded as an end in itself; such a "purpose" need not be evaded or concealed for fear of implying a Paleyan metaphysic.

This fear probably accounts in part for the surprising neglect of functional inferences in palaeontology. Although purposive categories are equally necessary in any functional study of living organisms, it is easier to conceal them by "pseudo-substitutions" of various kinds; but with fossil material, where functions cannot be directly observed, an explicitly purposive analysis of the machine-like character of the organism is less easily evaded. So deep-rooted is the fear of following in Paley's footsteps, that even this limited teleology is widely regarded as "unscientific."

4

I have argued that the detection of any adaptation in a fossil organism must be based on a perception of the machine-like character of its parts, and on an appreciation of their mechanical fitness to perform some function in the presumed interest of the organism. This method of inference must now be examined in more detail.

With a problematic fossil structure before us, the first requirement is to perceive its possible fitness for some function that might conceivably have been relevant to the life of the organism. Often this first clue will be suggested by the recollection of a living organism with a somewhat silmilar structure, the function of which is already known. Alternatively we may recollect some purely artificial man-made device, which solves a functional problem similar to one that the fossil organism might possibly have encountered. But in either case we shall not perceive the similarity unless we are deliberately looking for a purposive, i.e. functional, explanation of the feature we are studying. If our primary purpose is the construction of phylogenies, or taxonomic classification, we shall be blind to the clues that might lead to successful functional inferences. Only by an explicitly functional approach shall we be able to perceive that the fossil structure on the one hand, and the living or artificial structure on the other, may possibly embody the same operational principles—despite even great differences in their physico-chemical constitution (as between the wings of a pterodactyl and those of a glider).

This perception, however, is no more than a clue. The similarity perceived might be purely fortuitous, rather than the reflection of a common function. The clue must therefore be tested. The origin of the clue, in the perception of an analogy with a living or artificial mechanism, now becomes irrelevant: only the function that it has suggested to us is important. For we can only test our tentative interpretation by assessing the mechanical fitness of the fossil structure as a potential agent of this particular function. Such an assessment demands a standard against which we can judge the potential efficacy of the structure; and this must be derived from a consideration of the function itself. How can this function possibly be fulfilled? What operational principles are involved? What basic characteristics must any such mechanism possess? What, in the most general terms, is the specification to which any structure must conform, if it is to fulfil this function effectively? Such questions lead us to a generalised structural specification for the function (for example, a general specification for aerofoils for gliding). But this will probably be too vague and imprecise to be of any great use. The specification can be made more rigorous by taking into account the limitations imposed by the properties of the materials in which it would have to be embodied—if at all—in the fossil structure (for example, in a pterodactyl, the anatomical relation between the forelimb skeleton and the wing-membrane, and the finite strength of these materials, would presumably limit the possible range of "design" of the aerofoil). Such a specification then describes *the structure that would be capable of fulfilling the function with the maximal efficiency attainable under the limitations imposed by the nature of the materials*. Notwithstanding a rather different usage in other contexts, I have proposed that such a specification should be termed a *paradigm* for the function.[11]

In order to determine the optimal conditions for the performance of the function, theoretical considerations may need to be supplemented by experiments with actual models, unless the physico-chemical situation is very simple. Clearly many different paradigms, of different degrees of comprehensiveness, can be specified for any given function; this is important, for with fossil material many significant properties and parameters may be known inaccurately or not at all, so that a highly rigorous paradigm may be unusable.

Our clue that the fossil structure might have been an adaptation to a particular function can now be tested by comparing the structure itself with the paradigm. The paradigm is in effect a structural prediction, which by this comparison we expose to a "risky" test. For the degree of approximation between any paradigm and an observed fossil structure is a measure of the degree of efficiency with which the structure would have been physically *capable* of fulfilling the function. Moreover, by transforming rival possible functions into their respective paradigms, rival structural predictions can be made; and these can be judged against one another by direct comparison with the observed structure of the fossil. This comparison shows which function, of all that have been suggested, could have been fulfilled most effectively by the fossil structure. Our confidence in the result of the this test will be cumulatively strengthened if we can show that it concurs with the results of similar "risky" tests on other parts of the organism; if, in other words, we can gradually build up an intelligible reconstruction of the way in which the various organs interacted in the service of the whole organism to achieve a possible mode of life.

But the ease and confidence with which we can interpret any structural feature will be proportional to its degree of resemblance to the paradigm with which we have compared it—and hence proportional to the quality of its putative adaptation. The more highly adapted it in fact was, the more closely and unambiguously will it resemble the appropriate paradigm; and hence the more readily shall we perceive the possibility of this interpretation and confirm its validity by detailed comparison. If it shows a high degree of resemblance to a particular paradigm, we may feel justified in saying that it was, very probably, an efficient adaptation to that function. If its resemblance to the paradigm is only moderately close, on the other hand, it is more likely to bear an equal resemblance to the paradigm of some other function, and we may be unable to judge between them. This will be more difficult to interpret. It might mean that the structure was rather poorly adapted to one function, and had a purely fortuitous resemblance to the paradigm of another. Alternatively, it might mean that the structure was rather poorly adapted to two different functions independently. Or again, it might mean that it had a dual function, but represented the optimal compromise between two partially incompatible specification, and was therefore "highly adapted" to that combination of functions.

But if our structure shows only a negligible resemblance to any of the paradigms with which we compare it, all we can conclude is that it could not possibly have been an effective adaptation for any of these particular functions. We are not justified in saying that it was therefore inadaptive or non-adaptive, because of course we can never exclude the possibility that it might have been highly adapted to some other function that we have failed to consider. We may have failed to perceive its similarity to the appropriate paradigm: perhaps we are not familiar with any living or artificial structures that embody the same operational principles; or perhaps the structure as fossilised is too small and incomplete a part of the whole functional mechanism with which the paradigm would ideally be compared. But in any case we can never justly conclude that the structure had no adaptive value.

Thus there is an asymmetry between the assertion that a particular fossil structure had a definite adaptive value, and the assertion that it had no adaptive value. There can be positive and cumulative evidence for the former (based on substantiation of "risky prediction"), but it is logically impossible ever to demonstrate the latter.

5

A comparison between a fossil structure and the paradigm for a postulated function shows whether the structure would have been physically capable of fulfilling that function. It cannot however establish that in fact it did fulfil that function. Even if we can show that the forelimbs of a pterodactyl were so constructed that they could have functioned as a highly effective gliding mechanism, it is still conceivable that the animal habitually kept them folded away, and that they were inadaptive or even (perhaps by getting in the way of its other movements) actually non-adaptive. However improbable this may seem, it cannot logically be dismissed altogether. If in practice we do dismiss it, and feel no qualms in doing so, it can only be because we are applying a criterion of intelligibility. Either we accept—however provisionally—an explanation that is rationally satisfying because it "makes sense" of the phenomena before us as we understand them at present; or, in rejecting that explanation, we shall leave ourselves with no explanation at all. To proceed from "This structure could have fulfilled this function" to "This structure did fulfil this function" involves a logical jump; our readiness in practice to make that jump is grounded not only in an analogy with living organisms, whose functional mechanisms we can observe in operation, but also in a conviction that extinct organisms, like those at present alive, followed their various modes of life by the use of machine-like adaptations based on rationally intelligible principles. An appreciation of the organism in these terms cannot be derived from the detailed study of its fossil remains, for it is the necessary condition for any such study. Without such an appreciation we shall fail even to perceive the evidence on which any functional reconstruction must be based.

But this is an appreciation of the organism considered as "an end in itself." Functional inference involves an analysis of adaptation only as a static phenomenon. The perception of the machine-like character of the parts of the organism is logically independent of the origin of the structures concerned. Theories of their causal origin (e.g. by natural selection or by orthogenesis) and theories of their temporal origin (i.e. by a particular ontogenetic and phylogenetic sequence) are strictly irrelevant to the detection and confirmation of the adaptation itself. The functional reconstruction of fossils is thus logically unrelated to any and all evolutionary theories.

This probably explains why the earliest effective work on the functional morphology of fossils, by Georges Cuvier and his disciples, suffered a spectacular eclipse after the general acceptance of evolutionary theory. For Cuvier's opposition to evolutionary theories arose fundamentally from his interest in the interacting functional mechanisms displayed by the organs of the body; and this is most readily allied to a static, i.e. non-evolutionary, conception of the history of the organism. It is easier to imagine that a complex of delicately co-ordinated mechanisms was "constructed" once and for all, and that once constructed it would have to remain stable, that it is to imagine an imperceptibly gradual modification, and ultimate transformation, of such mechanisms.[12]

But although logically unrelated by any evolutionary theory, the functional reconstruction of fossils would in practice be encouraged by any theory that stressed the machine-like character of organisms. Conversely, it would be impeded by any theory which tended to devalue the machine-like character of organisms. Thus for example even the Synthetic theory, despite its great *verbal* emphasis on function, tends to dissolve genuine adaptation[13] into the non-morphological concepts of gene-pool, genetical "fitness," adaptive zone, etc. Similarly the Typostrophic theory of Schindewolf, despite its great verbal emphasis on structure, tends to dissolve genuine adaptation into the non-functional concepts of type, Bauplan, etc.

In any case, however, the evaluation of theories such as these depends in part on critical tests which are unavoidably *palaeontological,* and which involve the assessment of the adaptedness, or non-adaptedness, of particular extinct organisms. The question is: are there, or are there not, important non-adaptive phenomena in the evolutionary process? the answer, insofar as it relates to these palaeontological tests, must bear the same asymmetry that we have found in the evaluation of the critical evidence. With a certain critical fossil organism before us, we may argue whether its distinctive features were or were not adaptive. But we cannot reach any conclusion about these features except by testing them as possible embodiments of conceivable operational principles. We may then be able to demonstrate that they were probably adaptations of high efficiency for particular functions; and by doing so we shall add cumulative weight to the Synthetic theory (or any other theory that stresses the ubiquity of genuine adaptation). Yet if we are unable to do so, our failure will not add corresponding weight to some other theory, for it can always be said that the features may prove to have been adaptive, if only we can think of the right function. For there is no positive criterion by which non-adaptedness can be recognised and demonstrated. At least in this respect, therefore, the Synthetic theory would seem to be as unfalsifiable as its rivals are unverifiable.

Notes

1. This essay expands and develops a brief introductory section to my paper on "The feeding mechanism of the Permian brachiopod *Prorichthofenia*" (*Palaeontology*, 1961, 3, 450–471), which describes a functional reconstruction of a highly aberrant extinct organism.

2. See, for example, the frequent pejorative use of "metaphysical" in G. G. Simpson, *The Major Features of Evolution*, New York, 1953, pp. 28, 63, 73, 134, etc.

3. M. Grene, "Two Evolutionary Theories," *British Journal for the Philosophy of Science*, 1958, 9, 110–127, 185–193

4. Ibid. p. 185

5. Simpson, op. cit. pp. 259–265, 269

6. I use this example only because the pterodactyl is a fossil that is fairly generally known: my methodological argument does not depend on the validity of this particular functional interpretation of the pterodactyl.

7. The nearest approach is in the hovering flight of humming birds: cf. R. H. J. Brown, "The Flight of Birds," *Biol. Rev.*, 1963, 38, 460–489.

8. In view of the basic limitations of working with fossil material, the mechanisms and adaptations I have chiefly in mind throughout this discussion are those that are "mechanical" in the narrower sense; but the same reasoning would apply equally, for example, to biochemical and physiological mechanisms, provided only that they were expressed in the preservable portions of the morphology and environment of the organism.

9. Cf. M. Polanyi, *Personal Knowledge*, London, 1958, pp. 328–331

10. Ibid., pp. 359–361

11. Rudwick, op. cit., p. 450. The simpler word "model" cannot be used in this context, since it is needed to describe actual models (in the everyday sense) of reconstructed fossil organisms; in any case the word "model" would fail to convey the element of maximal functional efficiency.

12. See, for example, W. Coleman, "Georges Cuvier, Biological Variation and the Fixity of Species," *Arch. int. Hist. Sci.*, 1962, pp. 315–331.

13. i.e. the demonstrable relations between the structures of organisms and their functions, as opposed to the so-called adaptation of "increase in fitness" which is attributed to "genetical Selection": cf. M. Grene, "Statistics and selection," *British Journal for the Philosophy of Science*, 1961, 12, 25–42.

5

Adaptation and the Form-Function Complex

Walter J. Bock and Gerd von Wahlert

The current revival of morphology is heralded by a flourish of studies in functional anatomy with the general result being a renewed focus of interest in the problem of organic form. Recent morphological studies are characterized by considerations of the functional properties of structure and of the interrelationships between the structure and the environment of the organism although the traditional considerations of pure morphological description and of the phylogenetic change of morphological form are not ignored. These studies have established a broader base for morphological inquiry and have permitted a far better understanding, albeit largely theoretical, of all factors influencing the observable shape of morphological features. It may be possible, in the near future, to partition these factors and to determine the influence of function, of surrounding structures, of phylogeny, and so forth in the molding of anatomical features.

The importance of functional anatomy in the recent upsurge of morphological studies is eclipsed by its basic contribution to a deeper appreciation of biological adaptation. Yet this central position of functional anatomy in the study of adaptation is really not unexpected in view of the close historical link between morphology and evolutionary biology. Anatomical studies during the first half of the 19th century set the stage for the Darwinian theory of evolution by demonstrating the unity of structure in "natural" groups and by establishing many of the basic (and still used) principles of biological comparison. During the second half of that century, anatomists enjoyed great success in showing that their morphological sequences provided the basis for outlining phylogenies of plant and animal groups, and that these morphological–phylogenetic series were also sequences of changing adaptations. These workers were able to cite many bizarre as well as commonplace adaptations and could document the sequential stages in the development of these adaptations. The studies done in this great period of classical comparative anatomy constitute one of the major contributions, to date, by morphologists to evolutionary theory.

Most of our basic concepts of biological adaptation date back to ideas developed during the years between 1859 and 1920. Many of these notions have stood the test of time and are still accepted today. Nevertheless, our understanding of biological adaptation contains certain serious limitation which may be traced directly back to the prevailing philosophy accepted by anatomists during the last century and the beginning of this century. The most pertinent element of this philosophy is the postulate that morphology should be a study of pure form divorced form function. Morphological features were treated as geometrical units that changed during ontogeny and phylogeny according to rigid and often rather "biologically abstract" mathematical laws. Structures were not regarded, as they should be, as biological features functioning together as integral parts of the whole organism. Nor were the changes in these structures during ontogeny and phylogeny regarded as modifications in response to alterations in the relationships between the form-function complex and the environment. Not all morphologists studied only pure form: some did include functional aspects of structure in their studies, as for example Dohrn (1875), who postulated the principle of functional change as a causal factor for morphological change. Unfortunately these workers had little or no influence on the overall development of morphological thought during the 19th century (see Russell, 1916, for an excellent account of the history of anatomy in the 1800s).

A morphology based upon pure form—isolated from function and from the interaction between the form-function complex and environmental factors—does not provide a broad enough foundation on which to analyze the phenomenon of biological adaptation. Yet most of our current notions of adaptation are based upon such a morphology. The inadequacy of this foundation is revealed by continued discussions, even recently, of whether the form of a feature is adapted to its function, or whether the form of a structure is adapted to environmental factors without considering its function, or whether the form or the function of a structure changes first during evolution. Similarly, much controversy exists on the question of non-adaptive structures, and on the problem of the adaptive meaning of differences between structures, and on the possibility of non-adaptive evolution without having good definitions of adaptation, either as the process or as the state of being.

The central theme of this paper is a reexamination of biological adaptation based upon the recent welding of morphology, function, and ecology by students of functional anatomy (see Böker, 1935–1937, especially chapter 1; von Wahlert, 1957, 1961a and b; and Gans, 1963; see also, Bartholomew, 1958, 1963, for a similar approach to the interrelationships between physiology, behavior, and ecology). We hope to establish an internally consistent framework of concepts pertaining to morphology, function, and environment, including the interconnections between them, and to use this framework as the basis for the elucidation of adaptation. All concepts have been formulated within the framework of the synthetic theory of evolution (Mayr, 1963; Simpson, 1953, 1958) and in agreement with the findings of population studies although we have stated all definitions and conclusions in terms of individual organisms and have not shown explicitly how these statements should be extended to populations. We believe that all of our ideas can be rephrased in terms of populations but to do so would greatly increase the length of this paper without adding significantly to it. The greatest care, even at the risk of being pedantic, will be given to the clarification and the separation of concepts, and to the definition of terms. Some of the most commonly used notions, as for example, feature, form, function, and environment, are the most difficult to define lucidly and non-circularly. Although some of these definitions may appear superfluous, we believe that precise delineation and definition of concepts is of the essence in evolutionary biology.

The Form-Function Complex

In all considerations of the form–function complex in this study, only the phenotype, both immature and adult, will be covered. Analysis of the form–function problem on the genetic level makes little sense at the

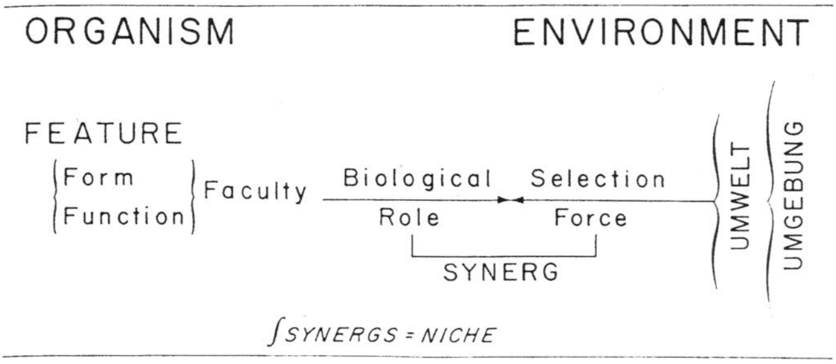

ORGANISM   ENVIRONMENT

FEATURE
{Form   }   Biological   Selection
        } Faculty   ─────────────────────────
{Function}   Role   Force
                            └──────SYNERG──────┘

∫SYNERGS = NICHE

Figure 5.1
A simplified scheme to illustrate the hierarchy and relationships between the components of the organism and of the environment that are pertinent to the understanding of biological adaptation. The bond between the organism and the environment is the synerg which is composed of the biological role and the selection force; the sum of all the synergs is the niche. The adaptation is the faculty. Every adaptation is a compromise between the demands of all the synergical relationships in which it takes part. See the text for additional explanation.

present stage of our knowledge. The two dimensions—the form and the function—of phenotypic features have usually been disassociated from one another in past evolutionary studies. In reality these dimensions constitute the two inseparable components of biological features and *must always be considered together* (see von Bertalanffy, 1960, for a discussion of form and function in his organismic conception of biology and for references to earlier works; Nagel, 1961: 425–427; and Beckner, 1959).

The interrelationships of the various components of the organism and of the environment are shown in figure 5.1.

Feature

A feature of an organism is any part, trait or character of that organism, be it a morphological feature, a physiological one, behavioral, biochemical, and so forth. Morphological features are the structures of the organism. Behavioral features are the components of display repertoires, and so forth. A clear and inclusive definition of feature is most difficult to formulate, especially if all aspects (genetical, developmental, evolutionary) and all points of view are included. We shall define feature as: *Any part or attribute of an organism will be referred to as a feature if it stands as a subject in a sentence descriptive of that organism.* Thus a feature is any subdivision of an organism according to the particular method of investigation. The limits of a feature are generally set arbitrarily. Most features are usually a part of a larger subdivision of the organism and can usually be divided into smaller subunits. The humerus, for example, is part of the skeleton (or part of the pectoral limb depending upon the study) while its head, greater tuberosity, lesser tuberosity, and so forth, are parts of the humerus. Likewise, within the greater tuberosity, sites of attachment for certain muscles and ligaments, and the arrangement of the internal bony trabeculae may be cited as even finer subdivisions of features. Moreover, the limits of a feature defined in morphological terms are almost always set in an arbitrary fashion with regard to its development, genetics, and evolution. Hence morphological structures should not necessarily be considered as real or absolute units of an organism for other biological studies. At present, our knowledge is too meager to allow us to delimit features in terms of genetics, development, or evolution, or to correlate phenotypic features with known genetical or developmental units.

The character complex or functional unit has special importance in systematic and evolutionary studies, and deserves special mention. A character complex is a feature of the organism and is comprised of a group of smaller features which act together to carry out a common biological role. No real distinction exists between individual features and character complexes: forms, functions, and biological roles may be described for each. Usually, however, the biological roles of the individual features are the same as those of the character complex: this concurrence of biological roles results from the traditional division of the major functional units of the organism into smaller but still distinguishable morphological units.

The bones, muscles, and ligaments of the jaw apparatus would be a character complex, as would be the parts of the forelimb or the hind limb. Establishing the limits of features in a character complex poses a major problem. Should, for example, a muscle-bone unit be considered as a single feature or should this unit be considered as a character complex that can be divided into smaller units—the muscle and the bones to which it attaches. Is the hind limb of a tetrapod, which everyone would agree is a character complex, a single feature or a series of smaller features? The vertebrate eye would be regarded by most workers as a single feature, but it can be divided into finer units, leaving the question of whether the eye is better considered to be a character complex. Additional difficulties exist. A single feature may participate in several character complexes, *i.e.*, the limits of character complexes may overlap. The same bone may be part of many different bone-muscle features such as the humerus and the several muscles that attach to it. Or a feature may take part in two distinct functional complexes, as for example, the pelvic girdle which is part of the hind limb and is also part of the axial skeleton, or the teeth in a carnivore which are part of the digestive system and are also part of the defensive system.

The adaptiveness of a character complex is usually dependent upon a close correspondence in the form and function of its individual features. But it should be emphasized that the interdependence between the features comprising a character complex does not mean that all of these features had to be fully evolved before the complex could have an adaptive significance (see Bock, 1963b). The composition of a character complex when it first appears can be (and usually is) quite different from the fully developed complex. New features may be added later and become, in time, indispensable parts of the complex which could no longer be adaptive in the absence of that part. In a similar fashion. new roles of existing features could arise and become an indispensable role of the entire complex. Other features which are integral components of a complex can be eliminated without affecting the basic operation of the whole unit. The iris of the vertebrate eye does not influence the basic mechanism of sight: it only permits the animal to adjust to different light intensities. If the animal lived in an environment (perhaps at the limits of light penetration in the ocean) in which the light intensity did not alter greatly, then the iris could be lost without affecting the efficiency of the eye.

Form

The form of a feature is simply its appearance, configuration, and so forth. It may be defined formally as: *In any sentence describing a feature of an organism, its form would be the class of predicates of material*

composition and the arrangement, shape or appearance of these mate-
rials, provided that these predicates do not mention any reference to the
normal environment of the organism. In morphology, the form would be
the shape of the structure. In behavior, it would be the configuration of
the display, including the involved structures, their movements, intensity,
and so forth. Commonly, structure and form are used interchangeably by
morphologists. Such usage should be discouraged because it perpetuates
a narrow and even erroneous concept of form, and confuses the distinction between the feature (= the structure) and its form.

One very interesting problem is whether a certain morphological feature has only one form. Morphologists generally accept the notion that
any structure has only a single shape. This idea is based partly upon the
sort of material studied and partly upon the methods by which the material is prepared. The study of bones, feathers, shells, cuticles, and other
hard tissues which maintain their preserved shape for many scores of
years easily leads to the impression that these are fixed permanent forms
in the animal. Moreover, because the morphologist works mainly with
preserved material that maintains a rigid shape even if it is "soft" tissue,
he assumes quite naturally that the observed shape is the only shape of
the structures. But the problem is not so simple and clear cut. Some
structures, such as the shell of mollusks, the cuticle of insects, or feathers
and hair of vertebrates, do not alter their shape once development is
completed, except for wear. And wear is, in many cases, an important
method of altering the form of a feature, as for example, maintaining a
sharp edge on the incisors of rodents, or altering the plumage color in
birds by abrasion of the tips of the feathers which is well illustrated in the
common starling (*Sturnus vulgaris*). Still, the shape of flight feathers,
especially the outermost primaries, changes constantly, because of the
change of air pressure on the feathers, as the bird flies; and the shape
of these feathers at each instant determines their exact function. Other
structures, vertebrate bone for example, have only one form at any time,
but their form can be altered over a relatively short period of time during
the life of an individual (Kummer, 1962a). Yet, still other structures,
notably muscles, blood vessels, and the digestive tract, are constantly
changing in their form over extremely short periods of time. Muscles
undergo a radical change in shape during each contractile cycle, and

these modifications in form affect drastically the functional properties of the muscle. The lens of the tetrapod eye is constantly changing its shape. Thus, for this last class of structures, a single form cannot be described, rather the worker must consider the entire spectrum of shapes that can be assumed by the structure and must ascertain the exact function at each point in this spectrum of forms. Indeed, one of the most difficult problems in morphology is the comparison of muscles, be it morphological, functional, or adaptational. It is not sufficient to compare muscles at the same stage of contraction, *e.g.*, at resting length, they must be compared over their normal range of contraction (shortening) which may not be the same for different muscles.

In general, we cannot assume that a feature has only one form. Many features possess only a single form during the life of the organism after development is completed, but this must be determined for each case. Others have only one form at any one time, but this form can alter over a period of time because of their ability of physiological adaptation. Still others are able to change their shape within certain limits over very short periods of time. This last category of features is the most difficult to anaylze adaptively because to their ever-changing form and resultant modification in function. Again muscles are a prime example of features in this category; comparative morphologists are still struggling to develop methods with which the adaptive significance of muscles can be compared.

Function

The concept of function is one of the most controversial ones in biology and one of the most difficult to define. Basically the function of a feature is its action or how it works. It may be defined formally as: *In any sentence describing a feature of an organism, its functions would be that class of predicates which include all physical and chemical properties arising from its form (i.e., its material composition and arrangement thereof) including all properties arising from increasing levels of organization, provided that these predicates do not mention any reference to the environment of the organism.* It is obvious that the function of features must be determined with reference to chosen surrounding conditions (= environment) and one may choose conditions that are similar to the natural environment of the organism, but such a choice does not affect the definition of function. Moreover it is clear that a worker may consciously or unconsciously study functions of features that never occur during the life of the organism. A muscle may never shorten more than 10 per cent of its resting length during life, but the physiologist may study its functional properties when it shortens up to 40 per cent of its resting length. These non-utilized functions cannot be ignored because we generally do not know which functions are utilized and which are not utilized by the organisms and because the non-utilized functions form an important basis of phenomenon of preadaptation.[1]

The function of a feature may be studied and described independently of the natural environment of the organism as is done in most studies of functional anatomy. The animal is placed in an experimental device which allows ascertainment of the functions of the feature with various degrees of precision. But the conditions are almost always highly artificial. In addition to these studies of pure function are investigations of biological anatomy in which the "function" (= biological role, see p. 131) of a structure is studied with the animal living freely in its natural habitat. Both types of studies are required to obtain different, but related sets of information which are prerequisites for the study of adaptation.

We wish to stress that the definition of function as given above does not involve any aspect of purpose, design, or end-directedness. Moreover, this definition of function is free, as it should be, of any form of teleology, Aristolelian or otherwise, and of any from of teleonomy (Pittendrigh, 1958: 392–394; Huxley, 1960; Mayr, 1961). The coupling of function in biology with teleology (see Nagel, 1953, 1961, pp. 23–25 and chapter 12; Beckner, 1959, chapters 6 and 7) is unnecessary although we agree that many biologists formulate functional statements in a teleological framework. A major source of ambiquity stems from the several meanings of function in biology. Function is used in the sense of the physical and chemical properties of the feature and in the sense of the role the feature has in the life of the organism. A review of the literature of functional anatomy will reveal that both meanings are employed, and indeed that many workers treat the two meanings as absolute equals. We feel that these two concepts must be separated sharply and prefer to restrict the term function to those properties of features which may be described and treated in the same way as a typical physical and chemical law (*e.g.*, Nagel, 1961: 406). The second common meaning of function can be designated as the biological role of the feature (see p. 131). Although this restriction of the term function may seem initially strange and awkward to many biologists, it has many advantages over the other alternate actions of restricting function to the meaning of the role of the feature in the life of the organism or of abandoning the term altogether and coining new terms for these two concepts.

A feature may have, and usually does have, several functions simultaneously even if it has only one form. If a feature has several forms, as for example, a muscle, or the lens of the eye then it almost always has several functions, or at least a varying function. The skin of mammals (including the dermis and epidermal derivatives) prevents the passage of many materials such as fluids, dirt, foreign organisms, reduces the flow of heat, and stops light rays although other rays may pass through the skin. Bony skeletal elements possess great strength against compression, tensile, and shearing stresses which allow their role as support of the body, points of attachment for muscles, and mechanical protection of vital organs. Moreover bony elements are a reservoir of mineral salts. Collagenous fibered tendons and ligaments possess great strength against tensile stresses and are highly non-compliant (not stretching), but they have no strength against compression or shearing forces which allows their roles as the intermediate structure between muscles and bones and the ties between individual bones. Yellow elastic fibers possess strength against tensile forces, but are compliant and hence have quite different roles than collagenous fibers. Muscles develop tension and shorten, these properties depending upon the length of the muscle relative to its resting length and upon the load on the muscle. Another function of contracting muscles is their resistance against shearing stresses, *i.e.*, the muscle becomes rigid during contraction and will support a load acting at right angles to the longitudinal axis of the muscle fibers (see Lockhart, 1960: 8–9). This non-utilized function of most muscles is basic to the role of the M. submentalis in the frog which becomes upon contraction a structural element in the tongue flipping mechanism. The M. submentalis becomes a rigid bar that serves as the pivot about which the rigid M. geniolossus swings (Gans, 1961, 1962).

The form of a feature may be studied as a function (or correlate, in a mathematical sense) of its several functions, but it must be appreciated that the correlation between the form(s) and the function(s) of a feature is not a simple one (Dullemeijer, 1958). (Most workers discussing this problem probably use the term function in the sense of biological role; however, the question of whether the form of a feature may be completely explained in terms of past and present functions is still a legitimate one.) Nor are all the functions, including all past and present ones, the only factors determining the form of a feature. It is not correct to assume that the form of a feature may be explained completely by correlating it with all past and present functions. An accidental component, one usually associated with the origin of the feature or with the major shifts in its evolution, is also involved. The basis for this accidental component stems largely from the chance factor in genetical changes such as mutations, chromosomal aberrations, and recombinations of all sorts. The accidental component forms the basis for the principle of multiple pathways of adaptation (Bock, 1959; Bock and Miller, 1959; Mayr, 1960 and 1962). Moreover, this accidental component is just as important in understanding the evolution of a single pathway of adaptive evolution as stressed by Mayr (1962).

Lastly it should be noted that our definition of function will affect the wording of many classical and still unsolved biological problems. The question of whether the form or the function of a feature changes first in evolution must be restated because under our definition, the function of a feature changes automatically with changes in its form. This question should be phrased as whether the form or the biological role changes first. The concept of functionless features becomes meaningless, if a feature has a form then it must have a function. Functionless features are ones that lack a biological role. Preadaptation is not a change of functions but a change in biological roles. And, indeed the question of whether the form of a feature is just a function of its functions as just discussed is probably considered more as whether the form of a feature is a function of its biological roles, than a function of its functions.

Faculty

The two separate dimensions of a feature are, as shown, the form and the function no matter how many of each may be described for a single feature of an organism. Each of these dimensions considered singly or both considered independently would give only part of the total picture of the feature. The two components must be combined to create the form-function complex before the feature can be understood properly. The form-function complex may be termed the faculty of the feature. Hence, *a faculty is defined as the combination of a form and a function of a feature.* It may be defined formally as: *In any sentence describing a feature of an organism, its faculties would be that class of predicates each of which includes a combination of a form (material composition and arrangement) and a function (physical and chemical properties) of the feature, provided that these predicates do not mention any reference to the natural environment of the organism.*

The faculty, comprising a form and a function of the feature, is what the feature is capable of doing in the life of the organism and is the unit that bears a relationship to the environment of the organism. The faculty is the unit acted upon by natural selection (see beyond, p. 136) and is the aspect of the feature adapted to the environment. The faculty is, therefore, the evolutionary unit of the feature. Neither the form nor the function alone can be an adaptation or can comprise an evolutionary unit because natural selection cannot act upon just the form or the function of a feature. Neither component is sufficient by itself to interact with a factor in the environment of the organism: both must be present. The form of the forelimb in a mole is not adapted for burrowing nor is the form of the webbed foot of a duck adapted for swimming. Rather it is the form plus the function of these features that is adapted to the environment of these animals.

If a feature has one form and more than one function, as is the usual case, then it would have a number of faculties each being a combination of the form and a function. The number of faculties would be the same as the number of functions in this case. Should the feature have several forms, either at separate times, as in the example of vertebrate bone, or at the same time, as in the example of muscle. then it would have a larger number of faculties resulting from the combinations of the several forms with the several sets of functions. Again it should be emphasized that some of these faculties would be non-utilized ones corresponding to the non-utilized functions. Each utilized faculty of a feature is controlled by a different set of selection forces and hence each would have a separate evolution so far as possible. Hence it is obvious that the feature cannot be perfectly adapted to all the selection forces acting upon it, but it must be a compromise between all of these selection forces. In the case of the feature having only one form, then that particular form would be a compromise so that each of its faculties satisfies as best as possible the demands of the selection force (or forces) acting upon it. A feature with one or a few utilized functions would be better adapted to each selection force as fewer faculties enter into the compromise. Features that are important "functionally" would be generally less well adapted to each selection force as more faculties are involved in the compromise, and more likely at least some of these faculties and the corresponding selection forces would conflict with one another.

A striking example of a conflict between faculties can be seen in the wing of many diving birds. Some diving birds, *e.g.,* loons, grebes, and ducks, propel themselves underwater with their feet. Others, such as auks and diving petrels, propel themselves underwater with their wings—they "fly" underwater. The high density of water compared to the low density of the air necessitates a decrease in the size of the wing to render it a more efficient swimming organ. Yet any decrease in size reduces the efficiency of the wing in flight. A compromise structure is one that is suitable, but not the best possible, for both swimming and flying. The peculiar buzzing and inefficient flight of the auks testifies to the compromise form of the wing. When flight is lost, as in the Great Auk and in the penguins, the compromise between swimming and flying disappears, and the wing can become modified into a much better swimming structure. This modification is obvious when the rigid paddle-like wing of the penguins is compared to the swimming–flying wing of the auks.

The several possible combinations of form-function complexes may be visualized in the following systems.

a.   If a single feature of an organism has one form and several functions, then:

Form 1 + Function 1 = Faculty 1
Form 1 + Function 2 = Faculty 2
Form 1 + Function 3 = Faculty 3
$\vdots$

Form 1 + Function n = Faculty n

If the feature had seveal forms, then the list would be expanded until all of the combinations of forms and functions are given.

b.   It is also possible to have combinations of the same (or very similar) functions and different forms of usually different features in the same or in different organisms to produce different faculties; hence:

Form 1 + Function 1 = Faculty 1
Form 2 + Function 1 = Faculty 2
Form 3 + Function 1 = Faculty 3
$$\vdots$$
Form n + Function 1 = Faculty n

Such a system may be seen in a series of organisms in which different forms of the same or different features may have the same function (for example, strength against a certain amount of tension), resulting in different faculties in the structures.

c.   If a series of homologous structures in different organisms are compared and found to differ, then:

Form 1 + Function 1 = Faculty 1
Form 2 + Function 2 = Faculty 2
Form 3 + Function 3 = Faculty 3
$$\vdots$$
Form n + Function n = Faculty n

A common example of this case is when a feature is changing under the action of a selection force toward a better-adapted faculty. Note that both the form and the function are changing together, although the change in the form or in the function in certain cases may appear to the observer as a greater or more profound modification than the change in the other dimension.

In some cases, a series of different forms and functions of different features in the same organism would result in different faculties having the same adaptive advantage and hence being under the control of the same selection force. An example of such a series of faculties forming a multifactorial adaptation for the same selection force is that of the different features in the mammals which prevent heat loss (Mayr, 1956). The hair insulates the body in one way, the subcutaneous layer of fat insulates in another way, decreased subcutaneous circulation (and other circulatory system structures as rete mirabile) reduces the flow of heat to the surface of the body, reduction in the size of the extremities reduces the surface area of the animal and hence the ratio of heat production to heat loss (increased size has the same result), various behavioral habits (such as nest construction), provide more suitable microclimates, physiological mechanisms (such as hibernation) reduce the temperature gradient, and so forth. Each of these features have different faculties, but all of the faculties have the same adaptive significance.

Other combinations of forms, functions, and faculties could be given, but those cited above are ample to illustrate how combinations of the two separate components of features result in different faculties. The important point is that the form(s) and the function(s) are inseparable parts of a whole unit which is the feature of the organism.

Biological Role
The concept of biological role represents a major problem in that it must be separated clearly and sharply from the concept of function. In most earlier works, no distinction was made between function and biological role, either in terminology or in conceptual thinking; both were called functions. Some workers presented a concept identical or quite similar to our notion of biological role, which they considered to be the function of the feature. The use of different names for the same concept is of very minor significance, what is important is that these workers did not outline a separate concept similar to our notion of function. A second problem is the difficulty in presenting an unequivocal definition of the biological role.

The biological role of a faculty, and hence of the feature, may be defined as the action or the use of the faculty by the organism in the course of its life history. It may be defined formally as: *In any sentence describing a feature of an organism, the biological roles would be that class of predicates which includes all actions or uses of the faculties (the form–function complexes) of the feature by the organism in the course of its life history, provided that these predicates include reference to the environment of the organism.* Essential to the description of a biological role is the observation of the organism living naturally in its environment.

The descriptive adjective "biological" stresses this fundamental property of the biological role. A biological role cannot be determined by observations made in the laboratory or under other artificial conditions. This is the basic distinction between study of functions and study of biological roles; in morphology, the former would be functional anatomy while the latter would be biological anatomy. Moreover, it must be stressed that the biological role of a feature cannot be predicted with any certainty from the study of the form and the function of the feature. For example, Bock (1961) suggested that the biological role of the large mucus glands of the gray jays was to coat the tongue with a glue-like material which would allow the bird to obtain food from crevices in the bark of trees and thus is similar to the biological role of the large mucus glands in some woodpeckers. Dow (1965) however, has shown, by observations of these birds in captivity and in the wild, that the mucus serves as a glue to cement food particles together into a food bolus which is then stuck to branches of trees. These food boli are a device to store food during the winter; the stored boli are found and eaten during stormy weather and other periods during which these jays cannot find food. Determination of biological roles is especially difficult in fossil forms or in recent organisms whose life history is not known.

While a biological role is the action of a faculty of a feature, it is not the same as the function. These two aspects of the feature have a definite relationship to one another, but they are not equivalents or interchangeable. Nor is it wise to draw analogies between the biological role and the function, such as the obvious comparison that the biological role has a similar relationship to the faculty as the function has to the form. Each utilized faculty of a feature has at least one biological role; frequently each utilized faculty has several roles. Herein lies the distinction between utilized and non-utilized faculties and functions. A utilized faculty has at least one biological role, it is used by the organism in the course of its natural history; a non-utilized faculty does not have any biological roles. As each feature generally has several faculties, the number of biological roles associated with the feature is very great. For example, the legs of a rabbit have the function of locomotion—either walking, hopping, or running—but the biological roles of this faculty may be to escape from a predator, to move toward a source of food, to move to a favorable habitat, to move about in search of a mate, and so forth. The legs of a fox also have the function of locomotion, although the details of the form and the function of the fox leg differ greatly from the rabbit leg. Yet some of the biological roles are quite different in the fox and in the rabbit. One role of the leg in the fox would be to catch its prey when it is chasing the rabbit in which the role would be to escape from its predator. This is a trite example, but it does illustrate very clearly how different biological roles may be.

Each biological role of the faculty is under the influence of a set of selection forces, and consequently the number of different selection forces acting upon a faculty would depend upon the number of biological roles. The biological role of the faculty of the feature serves, therefore, as part of the connecting link—that on the side of the organism—between the organism and its environment. Or, one could say, that the biological role establishes a specific interrelationship between the organism and its environment.

Environmental Factors

The second prerequisite to any discussion of the concept of biological adaptation is ascertaining the distinction between the different aspects of the environment and determining their relationships to the organism. These topics will be covered only briefly here; a more complete discussion, especially of the niche concept, will be presented elsewhere (Bock and von Wahlert, ms.). Environmental factors are all external factors, physical and biotic, which interact in some way with the organism. We do not distinguish between the physical and the biotic environment because there is no basic difference to the organism whether the selection force stems from the physical part of the environment or from another organism. References to the environment are always to the external environment; we doubt the existence of an internal environment distinct from the organism itself.

Umgebung
The umgebung or prospective habitat of an organism is comprised of those factors of the environment which possibly could be utilized by that

organism or which possibly could act upon that organism; it is the sum of environmental factors generally called the habitat of the organism. The umgebung may be defined without the actual presence of the organism although some knowledge of the structure and the life history of the organism is needed to judge the suitability of the habitat. In general, the umgebung is the habitat that a worker can observe in the field, and can deduce from its composition whether certain organisms might live there. The umgebung contains both the factors which are actually being used by the organism and those which are not being used by the organism, but might conceivably be used by the organism. For example, a seed-eating finch feeds upon certain seeds, but could also eat other seeds if they became available or if the need arose; all of the seeds belong to the umgebung of the finch. A fish living in the middle levels of a lake does not make use of the surface and the bottom of the lake although these environmental factors are present. These factors may not even exist for this fish; however, since they could be used, they would be included in the umgebung of the fish. But a tree in the neighboring forest or a predator living in the upland fields would not be a part of the umgebung of that fish.

Umwelt

Those factors of the environment which are actually being used by the organism or which are actually acting upon it comprise the umwelt or the species-specific habitat. The umwelt can be ascertained only in the presence of the organism living naturally in its environment. The factors of the environment comprising the umwelt are a subgroup of those factors forming the umgebung. A shift in a factor from the umgebung to the umwelt may be considered as the incorporation of that environmental factor by the organism into its sphere of influence. This incorporation is accomplished through the action of the biological roles of the faculties of the organism. The same shift from the umgebung to the umwelt occurs when a new environmental factor acts on the organism. Whenever the organism ceases to utilize some environmental factor, that factor would shift from the umwelt to the umgebung of the organism. Ecological factors are constantly shifting between the umgebung and the umwelt of the organism. Those factors which are essential to the species would remain in the umwelt at all times, while those factors that are less essential, or are peripheral or supplemental, could shift in and out of the umwelt.

The umwelt of a species may vary between different parts of its geographical range. Moreover, the umwelt may change during the development of the organism, especially when a radical metamorphosis takes place. Seasonal or other cyclic changes may occur in the umwelt with factors shifting from the umwelt to the umgebung and back again. When a bird migrates, for example, from the cool coniferous forests of the temperate regions to the hot tropical jungles, its umwelt changes drastically. The organism must be adapted to all of the changes in its umwelt which occur during its life. Nevertheless, a particular individual need not, and usually is not, adapted to the entire spectrum of umwelts existing over the whole geographic range of the species. Nor must the individual be adapted simultaneously to all of the temporal or seasonal components of its umwelt, most likely the adaptation of the individual will change as the umwelt changes. This ability of the individual to change is the basis for physiological adaptation or acclimatization observed in many organisms.

The grouping of environmental factors into the umgebung and the umwelt is almost identical, if not completely identical, to the concepts of potential environment and operational environment (Bates, 1960: 554), and to the concepts of potential niche and operational niche advocated by Parker and Turner (1961). We choose not to use these terms because we feel that the term "niche" should not be used to describe the environmental factors (see below).

The basic distinction between the concept of the umgebung and that of the umwelt of an organism is the presence of the organism. Only in the presence of the organism can an investigator ascertain whether a certain environmental factor is interacting with the organism. It would, in general, be impossible to describe the umwelt of a species in every detail. But it is also impossible to describe its morphology or physiology completely. Moreover the umwelt of a species varies geographically and temporally (even cyclically on a daily, yearly, or longer, basis). The umwelt should never be thought as a fixed rigid system; it changes as long as environmental change exists.

The concept of umgebung or prospective habitat is similar to the concept of habitat used by most workers, to the concept of the ecological zone as defined by Simpson (1953: 201) and to the concept of the niche as used by most workers (see Bock and von Wahlert, MS.). Empty or unoccupied habitats are umgebungs. Theoretical considerations of the number of possible habitats within a broad biotope or community (as defined by Slobodkin, 1963: 11) deal with umgebungs. Without the organism, only the umgebung or prospective habitat can be discussed, never the umwelt. If an organism is not found within a certain region, or did not exist, or became extinct, then its umwelt would not exist. When a species becomes extinct, its umwelt disappears and the environmental factors comprising it revert back to the umgebung. These environmental factors may be utilized by other species or that particular umgebung may remain empty and available for a new invading species.

The same environmental factor may be a part of the umwelt of more than one species. A water hole on the African veld may be used by a number of species. This common usage of the same environmental factor might lead to competition between the species sharing it, should a shortage exist. Sometimes the several species would continue to share the same environmental factor. At other times, the factor would shift from the umwelt of one species and be present exclusively in the umwelt of the other species. Generally, however, the environmental factor would remain part of the umgebung of both species.

The Organism-Environment Bond

All biologists agree that the interrelationship between organisms and their environment is the essence of life, but they do not always agree on the details. These details are of prime importance for the appreciation of adaptation which is, after all, nothing more than some sort of correlation between the organism and the environment.

Synerg

The factors of the umwelt interact with the organism and thereby exert "pressures" on the organism; these are the selection forces acting on the species. Membership in the umgebung of a species is not a sufficient criterion for an environmental factor to be a selection force because the factor may not be interacting with the organism, and, by definition, a selection force is acting on the organism. If an environmental factor is actually interacting with the organism, it must be part of the umwelt as well as the umgebung. Selection forces do not act directly upon the form or the function of a feature, but upon the faculty through its biological role. Thus the interconnection between the organism and its environment is through a couple formed by the biological role and the selection force. This couple may be called a *synerg* signifying a working together of the organism and the environment. A synerg may be defined as *a link between the organism and its umwelt formed by one selection force of the umwelt and one biological role of a faculty*. Obviously, most selection forces and most biological roles have many different interactions and hence each may take part in the formation of a number of different synergs.

The exact nature of the synerg depends upon the manner in which the organism uses the environmental factor, *i.e.*, the way in which a part of the umgebung is taken into the sphere of influence of the organism (von Wahlert, 1963). For example, a flatfish uses the ocean bottom in a number of different ways: for resting, for sleeping, for hiding, as an ambush site, and so forth (von Wahlert, 1961b). The ocean bottom remains the same, but the use of it by the flatfish differs greatly. The ecological factor, the ocean bottom, has a series of different facies which are presented to the organism; each of these facies will constitute a different selection force. Insect-eating songbirds could obtain the same insects by gleaning them from the surface of leaves and twigs or by catching them in midair when they are flying. The insects remain the same, but the means of capturing them differs greatly, and so would the anatomical and behavioral features of the birds. It is essential to consider not only the environmental factor, but how it is used by the organism. The exact nature of this use would influence the selection force and, hence, the synerg and the resulting adaptation.

The influence of the organism on the facies of the environmental factor and therefore on the selection force could be described in terms of a feedback mechanism between the organism and the synerg. An environmental factor usually does not have a single selection force, but a spectrum of possible selection forces. Determination of which one of these selection forces will act upon the organism depends upon the synerg, and the choice of the synerg may depend largely upon the organism. A choice may not be possible in plants and in many groups of lower invertebrates, these organisms would have more rigid synergs with few or no feedback systems. But in the higher animals which can select their habitat, the determination of the synerg by the animal may have an important bearing upon its future evolution. Evolutionary change of the species depends upon the selection forces acting upon it, and the final choice of the exact selection force may be by the organism (see von Wahlert, 1963). This choice of the synerg by the organism is not necessarily a conscious one and is not a completely free choice. It is influenced greatly by the evolutionary history of the organism.

The Niche

The snyerg describes the interaction between individual selection forces and individual biological roles of an organism. A complex pattern of synergs would exist for the entire organism with all of the biological roles of all the faculties of all the features interacting with all the selection forces of the multifactorial umwelt. *The niche of an organism is defined as the sum of all of its synergs. Thus the niche is the total relationship between the whole organism and its complete umwelt.* The niche can only be defined with the presence of the organism. Otherwise a relationship between the environment and the organism would not exist. The niche is not the physical and biotic factors of the habitat, but how the organism uses these factors. Because the niche is defined as an interrelationship between organism and environment, it depends upon the organism. An empty niche cannot exist any more than can a hypothetical organism. No organism means no niche. Indeed the logical and quite useful extension of this argument is that the niche is the species. After all, a species is not just the object we can see and describe, but it is a complex series of interactions of this object with its umwelt. Niches are not filled, but new ones appear gradually as new organisms evolve and occupy different habitats. When a species becomes extinct, its niche disappears along with its umwelt.

The niche, being the total of the synergs, determines the selection forces acting upon the species and hence the direction of its evolutionary change. Usually, changes in the niche imply changes in the umwelt of the species. But it is possible for the umwelt of a species to remain the same and to still have changes in the niche. This would depend upon whether the animal is using the factors of the umwelt in a different way; thereby the synergs will change. An example would be the changing use of the ocean bottom by flatfish. With any modification in the niche, either with or without a change in the umwelt, the selection forces acting on the species would alter and new adaptations could evolve. (See Bock and von Wahlert, MS., for a more detailed analysis of the niche concept.)

Adaptation

Biological adaptation, no matter what the exact definition may be, has always been interpreted as an interaction between the organism and its environment. Such a notion is not only not a consequence of any theory of organic evolution, but is quite independent of it. Indeed, the idea of a close correlation between the features of living organisms and the conditions of their environment predates by many years the general acceptance of any theory of organic evolution by biologists. Pre-evolutionary biologists understood the general notion and many of the details of this correlation between organisms and environment as well as we do today; what they lacked was a solid scientific explanation of the how and the way of adaptation. Rather than the notion of adaptation being a consequence of the acceptance of organic evolution, the search for an explanation of these observations was a major impetus in the development of a scientifically acceptable theory of organic evolution.

The adaptation problem was not solved automatically with the agreement that biological adaptation is the result of organic evolution; considerable differences of opinion continued to exist. Much of the past and current disagreement on adaptation centers about the definition of the concept and its application to particular examples; these arguments would lessen greatly if precise definitions for adaptations were available. In the following discussion, we shall first concentrate upon this problem of definitions and then inquire into the mechanism of adaptation.

A brief perusal of the writings on adaptation will show that several quite different, although related, concepts are grouped under one term. (We exclude any non-evolutionary concepts of adaptation, such as the adaptation of a sense organ to a continuous, steady stimulus.) These concepts have not always been clearly separated. The ensuing confusion has been largely responsible for the formulation of some rather erroneous ideas on adaptation. One such problem that continues to plague students of evolution is whether the entire process of the origin and the subsequent evolution of a new major taxonomic group is adaptive or whether inadaptive stages exist during this process (see Bock, 1963b).

We shall distinguish between three major concepts of adaptations: that of universal adaptation, that of physiological adaptation, and that of evolutionary adaptation. Although some lack of clarity will remain, we shall follow tradition and refer to each of these concepts as adaptation.

Universal Adaptation

Life, as we know it, originated as a result of the particular environment present on the primeval earth, and all life continues to exist because of the persistence, although with modifications, of this environment. Life cannot exist in a vacuum, separated from its environment. The bond between living organisms and their environment is an absolute property of life, a property that cannot increase or decrease. This property may be termed universal adaptation.

Universal adaptation, being an absolute property of life, is not a result of evolution; rather it is a cause of evolution. As the environment changes, organisms must change accordingly to retain their universal adaptation. Otherwise they would become extinct and life would exist no longer. If the environment did not change, then evolution would have never occurred because the first organisms would have had to remain unchanged to retain their adaptation. We do not mean to imply that evolution was an automatic consequence of a changing environment. Quite to the contrary, because any evolutionary change following environmental modification is quite fortuitous depending upon the ability of the species to alter. Environmental modifications occur; if the organisms can change accordingly, they evolve and hence survive. But if the organisms did not happen to change, they would have become extinct. No other possibility exists. It is possible that at some time after life had first originated on the earth, all organisms were not able to cope with the particular environmental changes that occurred. All life would have become extinct, and thus living organisms would have had to redevelop from non-living materials. This cycle may have happened several times. But at least once the primitive forms of life modified with the changing environment and still continue to do so. Hence, evolution took place and the universal adaptation of living organisms was preserved. It is still possible to have some environmental change to which no living organism can adapt and hence all life would cease. This has apparently never happened since the beginning of the known fossil record, but it has happened to major groups, orders and classes, of organisms as shown by the complete, and often rapid, disappearance of these groups from the fossil record.

Universal adaptation cannot be lost. If it is absent, the organism is no longer living. Species that become extinct have not lost their universal adaptation. They become extinct because the particular environment to which they are adapted has disappeared and with it, the species disappears. Extinction occurs when the species cannot change in accordance with modifications in environmental conditions. A species on the road to extinction still possesses universal adaptation, no more and no less than the most successful species. It is still adapted to a certain set of environmental conditions. Should these particular conditions reappear before all individuals of the species die out, then that species will survive. It is incorrect to assume, as does Simpson (1953: 293–303), that species lose adaptation, either slowly or rapidly, when becoming extinct. They cannot lose universal adaptation. The only correct interpretation of loss of adaptation as used by Simpson would be loss of evolutionary adaptation, *i.e.*, adaptation to a particular set of environmental factors.

Physiological Adaptation
The ability of tissues to modify phenotypically in response to environmental stimuli is generally termed physiological adaptation, somatic adaptation, or phenotypical adjustment (see Prosser, 1958, 1960; Simpson, 1953; Waddington, 1960; Bateson, 1963). We regard physiological adaptation as a special case of the general principle that the phenotype is an expression of the genotype in a particular environment. No apparent sharp distinction separates developmental lability and physiological adaptation; the two phenomena grade into one another. Physiological

adaptation proper may be restricted to events which occur after development is complete, considering either the whole organism or the particular feature in question. Developmental lability exists only while the feature is undergoing its development to the adult condition. Tissues that possess the ability to adapt physiologically have an equal amount of developmental lability; however, many features which have great developmental lability cannot undergo any physiological adaptation.

A sharp distinction does not exist between physiological adaptation and physiological processes in which there are no long-lasting modifications in the tissues. Modifications in bone shape, in muscle size, in tanning of the skin, and in changes in the number of red blood cells are all generally considered as physiological adaptations. Accommodation of the eye by modifying the opening of the iris, reduction of the subcutaneous blood flow, and the rise in the secretion of digestive enzymes are physiological processes. Although the two types of changes approach one another very closely, distinction between them is obvious in most cases. The best single criterion for separation is the nature of the stimulus on the organism. If the stimulus is, and indeed must be, steady for a long time in terms of the life-span of the organism, then the response if any, would be a physiological adaptation. If the stimulus is of short duration, changeable (especially cyclic) and quickly reversible, then the response would be a physiological process.

Although physiological adaptation is of extreme importance in many sorts of evolutionary change, we shall not discuss it further in this paper aside from expressing the opinion that physiological adaptation is clearly a special form of evolutionary adaptation. Physiological adaptation allows the tissues of an organism to respond and modify favorably to changes in the environment; hence the organism has a greater chance to survive and contribute to the gene pool of the next generation. These changes are generally of a duration shorter than the life-span of the organism. Because the environmental changes are often recurrent and cyclic, as for example, seasonal changes, physiological adaptation is generally a reversible process. If tissues of an organism did not have the ability to change in response to a fluctuating environment, then the organism would not be able to cope with the entire range of its possible habitats; it would only possess the ability to cope with a part of it. A single form and function of a feature cannot be as well adapted to a changing environment as would be a varying form and function. The ability to cope physiologically apparently evolved early in the phylogeny of animals and plants because it is a widespread, if not universal, phenomenon of most animal and plant tissues.

Evolutionary Adaptation

The most general and widespread usage of the term adaptation by evolutionists and most biologists is with the meaning of a long term, hereditary adjustment of a species to a particular set of environmental conditions. Clarity of meaning would be increased if the general term "adaptation" were restricted to evolutionary adaptation with new terms coined for universal adaptation and physiological adaptation. We hesitate to propose such a change at this time. Nevertheless we would recommend strongly that when no descriptive adjective is used before adaptation, the term should imply only evolutionary adaptation; this would be in line with past usage in which the term adaptation is usually in the meaning of evolutionary adaptation. Moreover, evolutionary adaptation denotes both a process and a state of being. Although the term for the process and the state of being are identical, it is generally very easy to determine which usage is meant by an author. In either case, evolutionary adaptation must always be considered in relationship to a particular set of environmental conditions. It is correct to say that an organism is adapted to winter temperatures of 0° C, or that the early reptiles were adapted to drier terrestrial conditions. It is not correct to say that horses are adapted or that birds were adapting during the course of their evolution from reptiles. The particular set of environmental conditions must be established first, and only then is it possible to judge whether an organism is adapted to these conditions or is becoming better adapted to them.

Given a certain set of environmental conditions, different species may be judged to be better or worse adapted to it. Judgment can be made, at least theoretically, between forms within a single phyletic lineage or between organisms of different lineages which are either coexisting at the same time or have existed at different times. The major snare in the comparisons is establishing just what is meant by "being better or poorer adapted" or "becoming better adapted." Generally, a closer correlation or adjustment between the organism and its environment, or between the feature and the selection force would constitute better adaptation. In terms of the framework developed above, the state of being an adaptation would be a good correlation between the biological role of the faculty and the selection force of the environmental factor. The process of adaptation would be the improvement of this correlation. But, with a little reflection, it is obvious that such a definition of adaptation (*i.e.*, a close correlation or a close adjustment between the organism and the environment) is either circular reasoning or simple redundant or without any real meaning; in any case, the definition is of little use.

What does it mean to say that adaptation is a close correlation or adjustment between the organism and the environment? The word "correlation" has been substituted for adaptation with the meaning of correlation being left as vague as adaptation. Not only has there been no clarification of meaning, but there is no operational basis for measuring the degree of adaptation of a particular feature of an organism or for comparing the relative adaptations of a feature in different species. To define adaptation as a good correlation begs the question of how the correlation is to the measured. The problem is to define adaptation in terms of some reasonable concrete concept that permits independent and repeatable measurement.

The definition of an evolutionary adaptation must include two separate and different elements. An adaptation is (a) a feature of the organism that has a role in the life history of that organism (*i.e.*, must perform some definite task, such as obtaining food or escaping from a predator), and (b) performs this role with a certain degree of efficiency. Hence, an evolutionary adaptation is always part of the organism (see Dobzhansky, 1942, 1956, 1960) and it is a faculty (a form–function complex) of a feature which has a biological role and interacts with some environmental factor of the umwelt of the organism. The bond between the faculty and the environmental factor is the synerg formed by the biological role interacting with the selection force. This relationship holds for single features or for entire character complexes. Adaptation of the entire organism would be to its umwelt with the bond being the niche. The problem of judging the quality of the adaptation still remains—the above statement that an adaptation is a faculty adjusted to some environmental factor is not sufficient. We believe that the quality or the degree of adaptation, both the process and the state of being, can be expressed best in terms of energy requirements.

The first fact supporting our notion for defining the degree of adaptation in terms of energy requirements is that every organism has available to it, at any stage of its life cycle, only a certain amount of energy with which it must maintain all of its life processes. This notion is simply an extension of the general idea of energy flow in populations (*e.g.*, see Slobodkin, 1963) to the rate of energy flow in each individual of that population (see also Kendeigh, 1949; Orians, 1961; Gates, 1962). In addition to the supply of stored energy, the amount of "new" energy that the organism can obtain over a certain period of time is also limited. The amount of available energy, both stored and obtainable, and the period of time over which this energy must last, will vary between individuals of the same population and even within the life-span of a single individual. The interindividual and intraindividual variation in the amount of available energy, and the finite limit of energy possessed by an organism at any one time are all basic to our concept.

The second fact is that an organism must expend energy to maintain any synerg. Every time that a synerg is operating, no matter whether the organism or the environment has set the synerg into action, some energy must be expended by the organism. The amount of energy expended may be very minute for some synergs such as breathing or sitting quietly and scanning the surroundings with its eyes, but every little bit adds up. Even a seed or a spore uses energy very slowly. Other synergs, such as escaping from a predator or maintaining body heat in arctic winters require a large percentage of the available energy. Obviously some synergs, such as photosynthesis or digestion result in a net gain of energy for the organism, but the operation of the synerg itself requires energy.

It is obvious that the organism must maintain all of its synergs—that is, it must maintain its niche—with its available energy. Because this energy is limited, it is advantageous for the organism to use the minimum amount of energy possible to operate each individual synerg. Hence we define *the degree of evolutionary adaptation the state of being, as the*

minimal amount of energy required by the organism to maintain success-
fully the synerg if a single biological role of a faculty is considered, or to
maintain successfully its niche if the whole organism is considered. The
adaptation is, of course, the faculty. If the faculty was well adapted to the
environmental factors, the operation of the synerg would require less
energy, but if it was poorly adapted it would require more energy. Survival of the organism, which is the crux of adaptation is accounted for by
specifying successful maintenance of the synerg. Unsuccessful maintenance of the synerg or of the niche would imply the elimination of that
individual. The relative factor of survival or the relative number of progeny left which is usual when comparing the adaptiveness of individuals is
accounted for by the relative nature of the term "successful." Success is
not an absolute all or none concept, but a relative one. The less energy
used, the more successfully the synerg or the niche will be maintained.

Evolutionary adaptation, the process, is defined as any evolutionary
change which reduces the amount of energy required to maintain suc-
cessfully a synerg, or the niche as the case may be, toward the minimum
possible amount.

Under this definition of adaptation, a perfect adaptation would be
when the organism can operate the synerg with the theoretically minimum amount of energy. The level of perfect adaptation is rarely if ever
achieved. Yet, this definition permits us to define good adaptation as
approaching this minimum and poor adaptation as being far removed
from this minimum. The upper limit for poor adaptation would be when
the organism is using all of its available energy to maintain a synerg. An
organism is not adapted to a particular set of environmental conditions
when it does not have sufficient energy available or cannot obtain sufficient energy to operate successfully a synerg or its niche. In this case, the
organism perishes—selection has eliminated it.

If the degree of adaptation is measured in terms of energy requirements, the absolute units of measurement must be converted into relative
values by including the size of the organism in the calculations. Otherwise it would not be possible to compare the relative adaptations of
closely related forms which differ markedly in size.

Evolution of the degree of adaptation in terms of energy has a decided
advantage to the student of evolution for it is possible to express the

degree of adaptation in a system of fixed and comparable units. It would be possible, at least theoretically although difficult practically, to compare the same or different adaptations to a particular environmental condition in two or more species. Or the relative adaptiveness of different structures in a single species to a certain selection force could be compared. Many non-climbing birds are able to cling to vertical surfaces for short times, as for example, the house sparrow (*Passer*) which can cling to the sides of brick buildings for five to 10 minutes. Yet this bird is poorly adapted for clinging to vertical surfaces. The amount of energy expended by the sparrow could be determined by measuring oxygen consumption. The degree to which the house sparrow is adapted for clinging to vertical surfaces could be ascertained by comparing this energy with the energy expended by good climbers such as woodpeckers, nuthatches, and creepers. The degree of adaptation for climbing in woodpeckers could be compared by measuring the amount of oxygen consumed by different species while clinging to vertical surfaces. This could be compared with the morphology of the birds, the stance used while clinging to vertical surfaces, and the arrangement of forces on the bird due to each stance. The contribution of the hair, the subcutaneous layer of fat, blood flow to the body surface, and so forth, to preventing heat loss in a mammal could be ascertained and, from these measurements, the adaptive significance of each feature for heat loss could be judged.

An excellent example illustrating degree of adaptation in terms of available energy was described to us by S. Charles Kendeigh for which we are most grateful. Kendeigh has frequently kept migratory songbirds, captured in the autumn, in outdoor aviaries at Urbana. Illinois. These birds were given all the food they could eat and kept under excellent conditions. The object of the study was to ascertain the degree of adaptation to cold in healthy, well-fed birds. As the weather became colder during the autumn, the birds ate more and more food to compensate for the heat loss. When the temperature dropped sufficiently, the birds could not obtain enough food and metabolize it to compensate for their heat loss, even though they spent a large part of the day eating. Consequently the birds began to use their stored energy (stored as body fat) until even that was depleted. The birds were now using all the energy they could obtain each day by eating to compensate for loss of heat that occurred that day. They had no reserves to utilize during any strenuous period and hence perished during the next cold period, usually at night when their food supply was used up and could not be replace at once. In these experiments, the adaptations possessed by the birds for surviving winter conditions were inadequate for conditions existing at Urbana. Although food was abundant, the birds simply could not obtain enough energy to maintain the synerg formed by the faculties to prevent heat loss and the selection forces resulting from low ambient temperatures; hence, these species were eliminated by selection. The dickcissel, *Spiza americana*, is an example of a bird that is not able to metabolize sufficient energy to compensate for heat loss and succumbs to the winter cold at Urbana (Zimmerman, 1963). The field sparrow, *Spizella pusilla*, is partly adapted to these cold winter temperatures (Olson, 1964). Other species of birds, normally wintering in Urbana, have better adaptations for cold weather and were able to operate this synerg with the utilitation of far less energy (Kendeigh, personal communication). They lost much less heat in low ambient temperatures and hence are able to survive the cold winters with the intake of less food. Migratory birds could become better adapted to these cold conditions if they evolved faculties which required less energy to combat cold temperatures and heat loss. The testing of the relative adaptation of different birds to cold conditions can be done very accurately by regulating the temperature of the experimental room, controlling the amount of food available, and so forth, as has been done for many years by Kendeigh and his associates. They have found that the birds wintering further north could withstand lower temperatures before perishing and that these birds required less energy to survive at a fixed temperature than birds wintering further south.

Timing of various events during the year becomes clear with the notion of energy levels for the degree of adaptations (Pitelka, 1958; Orians, 1961). Reproduction generally occurs when large amounts of energy (=food) can be obtained quickly and easily by the organism. Periods during which energy is short or difficult to obtain (winter and dry periods) are avoided by migration to a more favorable location or by a dormant period in the life cycle of the organism. Usually two activities which require large amounts of energy do not occur at the same time. Breeding, molt, and migration are three high energy demanding periods in the yearly cycle of birds. Each of these take place at different times of the year, although some overlap exists. The division of labor between sexes during reproduction or the division of labor between the various castes in social insects can be appreciated readily under this definition of adaptation.

Survival may be expressed in the same terms as adaptation. An organism can survive if the amount of energy required to maintain successfully all of its synergs is less than the energy the organism can mobilize. A coefficient of survival may be defined as the amount of required energy divided by the amount of available energy. Thus, this coefficient would vary from 0 (which is never obtainable) to one. A value higher than one is not possible as the organism cannot use more energy than is available (stored and obtainable)—it simply could not survive and would die. Probably an organism could not exist for a long time with a survial coefficient of one or even close to one. Simply, it would not have any surplus energy available for unexpected stress periods. This coefficient permits an objective comparison of the survival ability of different adaptive traits or of the entire organism. If two organisms require the same amount of energy to maintain an identical or similar synerg, then we would say that they are similar in this adaptation. But if one organism was able to mobilize twice as much energy as the other, then it would have a better chance of survival. Similarly, if one organism required more energy than another to maintain the same synerg, it would appear to be less adapted in terms of this trait. But if this organism was able to mobilize more energy, it would have a better chance of survival than a second organism that could only obtain less energy.

The advantage for the organism to use as little energy as possible to operate the necessary synergs lies in the fact that when less energy is utilized in operating the essential every day synergs, more energy will remain of the limited supply available to the organism to meet unexpected or strenuous conditions. A rabbit living at the very limits of its energy resources would have little or no energy available if it had to run away to escape a predator that happened to pass by. A bird wintering in the cold temperate regions would have less chance of surviving an unusual cold period if it had been using all of its energy to meet the normal conditions. Reduction in the amount of energy needed to maintain essential synergs would permit the organism to cope with more strenuous conditions without being forced to acquire additional energy.

Strenuous conditions would exert stronger selection forces on the organism. A better-adapted organism would be one having more energy available to meet the greater demands upon it during stress periods or one which would have to utilize less energy to meet these stress demands. In any case, selection is most strenuous because of the greater demands on the available energy of the organism. This idea agrees well with the concept that natural selection is most effective under adverse conditions (see, for example, Lewis, 1962) and that in normal environmental conditions of the species, selection may have little influence altering the features of the organism.

This notion also agrees with the well-known fact that most features of organisms (anatomical structures, energy stores, physiological process) have a margin of safety of at least two to three times that which is required by the normal environmental conditions of the organism. This safety factor, which has puzzled biologists and has been used as an argument against natural selection, evolved because of the stronger selection pressures in the abnormal years of severe conditions. The problem of "hyperadaptation" (Nicholson, 1960: 519) may be explained as adaptations to the severest conditions of the environment. Competition between organisms and the resulting selection exerted by one species upon the other would be greatest when stress conditions exist because of the low surplus energy available.

Comparison of adaptations in terms of energy balance does not remove all existing problems. Certain limitations exist on the sorts of comparisons that may be made between evolutionary adaptations of different organisms. These limitations appear to be independent of the use of energy balance or any other method of comparing the degree of adaptations. The most important limiting factor is that the environmental factor(s), against which the adaptations are judged, must be the same or at least reasonably similar. Ideally, a comparison should be made in respect to the same environment. It is not possible to compare the evolutionary adaptations of such diverse creatures as amoebae and zebras even with the notion of energy balance simply because the environments of these organisms differ too greatly. Nor is it correct to compare major life processes, such as feeding, locomotion, reproduction, and so forth, in very diverse forms to ascertain which of these organisms are the best adapted in each of these processes. It is feasible to compare the relative adaptation of reproduction in closely related forms by comparing the energy needed for reproduction against the total amount of energy available to each organism. The form using the smallest percentage of the total energy for reproduction would be the best adapted in this respect. But in diverse forms, this comparison is meaningless. An amoeba using 40 per cent of its total available energy for reproduction may be better adapted in this respect than a zebra requiring 10 per cent. It is certainly not possible to compare the total evolutionary adaptations of diverse forms (*i.e.*, comparing the adaptation of the respective niches) by comparing the amounts of energy, or even the "relative" amounts, needed to maintain all life processes. Such a comparison would be comparing the universal adaptation of these forms which cannot be done for this is an absolute, non-varying property of life.

Adaptation and Selection

The definition of selection has long remained a perturbing problem because of the different viewpoints on which a valid definition may be based. The correlation between some of these definitions is not always clear nor can each of the valid definitions be transposed easily into another of these definitions (see Olson, 1960, for a discussion of some of these problems; Waddington, 1960; Hecht, 1963).

We prefer to define selection in terms of the environment and its action upon the organism. A single selection force would be the action of a single environmental factor of the umwelt on the organism through a biological role of a faculty. Selection on the individual would be the summation of all the individual selection forces acting upon that organism; the summation would not be a simple addition because some selection forces might counteract others. Natural selection always acts upon the phenotype of the whole individual; it does not act directly upon the genotype nor does it act independently upon single features.

Selection as we define it affects the gene pool of the population by its action on the phenotype of individual members of the interbreeding population. Selection acts upon the individuals of one generation, permitting some to reproduce and preventing others from taking part in the reproduction of the next generation. The production of the next generation by interbreeding of only those individuals favored by selection and by differential reproduction of theses individuals will (usually) lead to a modification in the gene pool of the population. The individuals of the next generation would be somewhat different from the last one because of the modification in the gene pool.

It is possible for selection to operate and to allow only individuals of a certain phenotype to produce the next generation, without any resultant modification in the gene pool. An example may be when some individuals had a phenotype differing from normal arose because these individuals encountered different environmental condition during their ontogenetical development. If this phenotype was better adapted, these individuals would be favored by selection and contribute more to the next generation. However if the genotypes were the same in the normal and modified phenotypes, no genetical change would result from this selection. Exactly the same would result if all individuals possessed a new and better-adapted phenotype because of some environmental change. Selection would favor this new phenotype but no change would occur in the gene pool of the following generation. Either of these possibilities are probably so rare that they can be ignored in the formulation of general principles.

Students of genetics and especially of population genetics are most concerned with changes in the genetical frequencies of the gene pool. Consequently, they have defined selection in terms of its effect on the genetical composition of the population. Lerner, for example, defines selection (1958: 5) as: "*Selection* can be defined in terms of its observable consequences as the *non-random differential reproduction of genotypes.*" This definition is just as correct as our definition, but expresses selection in terms of its results instead of its action. This definition does not clarify what factors act upon the organism to produce the non-random differential reproduction of genotypes nor does it indicate how these factors act. Our definition is clear on these points, but does not give any clue as to the consequences of natural selection on the genetic composition of the population. The differences between these definitions are minor and do not deserve further discussion. A composite definition of selection may be offered as follows: Selection is the action of any factor of the umwelt on an organism through the appropriate synerg that results in a non-random differential reproduction of genotypes. This definition excludes only the single possibility of selection not resulting in a modification of the gene pool which is probably quite rare.

One point of considerable interest is the very simple relationship that exists between the genetical definition of selection and the definition of the degree of adaptation advanced in this study. We wish to thank Dr. Robert Bader for suggesting this relationship to us. Selection is defined, genetically, as differential reproduction of genotypes and expressed as a coefficient S that varies from 0 to 1. The coefficient of adaptation (W) is expressed in terms of selection (*i.e.*, $W = 1 - S$ or $S = 1 - W$). If selection is acting to alter a population, then degree of adaptation of this population will change. It is possible to measure change in degree of adaptation by comparing the energy requirements before and after the action of selection which is a simple comparison of the energy required to operate the synergs in the two populations. Hence, if the same adaptive trait is being compared in two populations, *i.e.*, being tested against the same selection force, then selection would be the ratio of the energy required to maintain successfully each of the two synergs [$S = 1 -$ (synerg A/synerg B)]. The direction of selection and the change in adaptation would depend upon whether population A or populations B is the ancestral population. This information must be provided separately as it cannot be incorporated easily into the equation. The synergs must be chosen so that the energy of synerg B is larger than the energy of synerg A so that the coefficient of selection (S) would be between 0 and 1. As synerg B becomes much larger than synerg A, *i.e.*, its degree of adaptation decreases, then S approaches 1 which means that selection would become greater for population A. When synerg A is equal to synerg B, then there would not be any selective difference between the two populations. This relationship holds true when the difference between an ancestral and a descendant population is compared or when the difference between two contemporaneous populations is being compared.

The same method may be employed when the adaptiveness of two populations of organisms are being compared to the same environment (Gause's principle). In this case, selection would be the ratio of the energy required to maintain successfully each of the two niches [$S = 1 -$ (niche A/niche B)]; niche B being again the larger of the two figures, *i.e.*, the one requiring the larger amount of energy.

The energy relationships between the synergs for single adaptive traits or between the niches for the entire organism are exactly the same as non-random differential reproduction of genotypes which is the genetical definition of selection.

Preadaptation

In an earlier paper, Bock (1959: 201) defined preadaptation as: "A structure is said to be preadapted for a new function if its present form which enables it to discharge its original function also enables it to assume the new function whenever need for this function arises." While this definition is basically correct, it was formulated with a loose and rather erroneous concept of function so that it is necessary to redefine preadaptation within the framework of concepts developed in this paper. We would define preadaptation as follows: *A feature is said to be preadapted when its present forms and functions (both utilized and non-utilized ones) allow one of the faculties (either currently utilized or non-utilized) to acquire a new biological role and hence establish a new synergical relationship with the umwelt whenever the need (= appearance of the selection force) for this new adaptation (the faculty) should arise.* The assumption of a new biological role or the switch of one of the formally non-utilized functions of the feature to a utilized function does not mean that the feature must lose its former biological roles. Moreover it should be noted that because no change in the form of the feature will occur with the acquisition of a new biological role, the feature will not lose any of its existing functions and faculties; modification of the functions (utilized plus non-utilized) is dependent upon a change in the form of the feature. Thus preadaptation should not be construed as a change in functions as has been expressed in earlier papers, but as a change in biological roles. With the origin of a new adaptation, a new selection force acts upon the feature which would alter the entire pattern of selection upon that feature. The result would be that the feature would acquire a new compromise condition even though it has not lost any of its existing adaptations.

Evolution of the feature to the point where it is preadapted for a new biological role is strictly fortuituous in terms of the new biological role. Up to the level of preadaptation, at which point the selection forces associated with the new biological role can operate, evolution of the feature has been directed only by the action of the selection forces associated with the already existing faculties and biological roles.

We should note that preadaptation applies to the feature and more precisely to the forms and functions of the feature considered together. Preadaptation must not be applied only to the form of the feature as it has been in many past studies. For additional details on the principle of preadaptation, the reader is referred to Bock (1959, the discussion in this earlier paper is in basic agreement with the present one although the terminology must be changed greatly with biological role substituted for function in most places), Osche (1962), and Kummer (1962b).

Many preadaptations permit the organism to enter a new niche relationship and exploit a new habitat; these preadaptations are generally termed key innovations (Miller, 1949; Bock, 1961, 1963b; von Wahlert, 1961c).

Postadaptation

Any change in a faculty, in its form and function, after it has acquired a new biological role through preadaptation is postadaptation. Post-adaptational changes result in a more perfect correlation between the biological role and the selection force—in a synerg requiring less energy.

Origin of an Adaptation

If an adaptive trait is the faculty and the faculty is composed of a form and a function, the old question of whether form or function evolves first can be reopened. The question is, in part, meaningless because of (a) failure to appreciate the inseparability of form and function, and (b) the assumption that the form or the function becomes fully developed before the other aspect has appeared. Before this problem can be examined, it must be rephrased in more accurate terms.

The proper expression of this problem is whether the form of the adaptation or the biological role appears first in evolution. We should emphasize that the functions of the feature are determined exactly, given the form. And it is the faculty which is the adaptation. Thus the question is whether the faculty or the biological role appear first in evolution. Neither of these aspects should be considered either as not present or fully developed. Each passes through the entire spectrum from rudimentary to fully developed in the evolution of the adaptation. And it is clear that neither of these aspects lags far behind the other in any evolutionary change.

New adaptations can be grouped into two classes, those involving an already present feature that is preadapted and those involving a completely new feature (see Bock, 1959; Mayr, 1960). The first group includes the large majority of new adaptations and is the easiest to analyze. Evolution of new adaptations through preadaptation of existing features has two steps. First is the evolution to the preadapted feature: the form and functions appear. The second is acquiring a new biological role and establishing a new synergical relationship with the environment. Clearly, in the evolution of a new adaptation through preadaptation, the form of the feature (or the faculty) precedes the biological role; the evolution of this form is not under the control of selection forces associated with the new adaptation. Rather the evolution of the form of the feature is determined by selection forces associated with the already existing adaptations of the feature. Needless to say, during the following period of postadaptation, both the form of the adaptation and the biological role may be modified.

The origin of new adaptations as *de novo* features is more difficult to analyze. The major problem is proving that the feature is really absent before the appearance of the new adaptation. Many *apparently* new features, such as the lung or the eye, actually had their origin in the acquisition of a new biological role of an existing feature—obtaining of oxygen in the membranes lining the pharynx in the case of the lung and perception of light rays in part of the central nervous system in the case of the eye. These examples involve preadaptation without much doubt. Mayr (1960) in his discussion of the origin of evolutionary novelties presents only one possible explanation that is clearly not preadaptation. This is the orgin of a new feature as a pleiotropic by-product, be it because of genetical or developmental linkage, of some other evolutionary change. We agree completely with this explanation and would like to point out that whenever a new adaptation appears as a pleiotropic by-product, the form would first appear and the biological role acquired later. It is obvious that in this case, the new feature has a form and functions, but it would not be an adaptation until it has a biological role and established a synergical relationship with a factor of the umwelt. The faculty would be the adaptation and, without possessing a biological role, the faculty cannot be an adaptation. And it is obvious that a biological role cannot be present in the absence of the appropriate faculty.

We would conclude that the origin of any new adaptation, no matter by which method it arose, always had the sequence of the form (and functions) of the feature appearing first followed by the biological role. In concluding that the form appears first, we wish to emphasize that we do not regard this form to be the final form of the fully developed adaptation. Many of the earlier discussions on the form or "function" first problem considered form only in its fully developed condition. The feature was envisioned as having reached its final form and then, and only then, acquiring its biological role. We would reject this. The form of a new adaptive trait is not the fully developed one for that adaptation, but a very rudimentary one. The biological role would also be poorly developed at its beginning stage. Both form and biological role would change in close accord during postadaptive evolution, neither being very far ahead of the other.

Conclusion

Life may be defined as a continuous self-perpetuating interaction of organic matter with the physicochemical environment. In accordance with this definition, all living organisms must be in proper adjustment with their particular environment, and they must obtain and expend enrgy to maintain themselves. Any definition of biological adaptation must be in harmony with these notions. Evaluation of the degree of adaptation in terms of energy requirements and balances is in accord with this concept of life, but it is not the only possible system of defining adaptation.

In our analysis of biological adaptation, we have arranged the components of the organism and of the environment and the various interactions between these components into a formal framework of concepts (summarized in Fig. 1). We do not maintain that this system is the only possible one, but we do believe that it does cover all of the pertinent factors involved in the organism-environment interaction and, most importantly, that it is an internally consistent system. We recognize that many of our terms are new or are old ones redefined in a more restricted and unfamiliar fashion so that they appear strange to many workers. We would like to emphasize that only the concepts deserve serious consideration, not the terms standing for the concept. Indeed we are not committed to the system of terms used in this study and would welcome suggestions for changes that would clarify the underlying concepts. Moreover, the discussions in this study have, by necessity, been brief and we regard this study as an abstract rather than a fully documented analysis. The reader is urged to provide additional examples from his own experience.

The hierarchical scheme of concepts proposed above not only allows an operational definition for the degree of adaptation in terms of energy balance, but will allow inquiry into additional problems of biological adaptation which have not yet been discussed in this paper. We would like to touch briefly upon some of these questions because they are fundamental to any conception of biology and of the philosophy of biology.

A definition of biological adaptation is inadequate unless it can be used as the foundation for an explanation of the mechanisms underlying the evolution of new adaptations and the adaptive origin of new major groups of organisms. These are not independent parts of evolutionary mechanisms; indeed, the solution of the latter question is completely dependent upon the solution of the former.

The evolutionary principles for the emergence of novelties may be built upon the concept of adaptation advanced in this study by a synthesis of ideas such as those advanced by Mayr (1958, 1960, 1962, and 1963) and Bock (1959, 1963a) and with the concepts of population genetics such as those summarized by Lerner (1954, 1958). At present, the greatest need is for sufficient case studies against which the general theories may be tested.

The origin of new major groups presents more difficulties for evolutionists, all of which can be condensed into the question: How do the many individual features comprising a major adaptive complex, which characterizes larger groups of organisms, evolve and become integrated into this single adaptive complex? The many features essential for flight in birds, provide an excellent example. Some workers (*e.g.*, Russell, 1962) have outlined a series of pertinent objections to the generally accepted explanation for the adaptive origin of new groups provided by the synthetic theory of evolution and have proposed alternate explanations. Because these alternate explanations raise more problems than they solve, it is reasonable to assume that the synthetic theory is adequate to explain the adaptive origin of new major groups, but that some aspects of the theory may have to be rephrased or clarified further. Thus we have undertaken studies to clarify some of the disputed sections in the synthetic theory by analyzing the sequence of events (Bock, 1963b) and the pattern of ecological changes (von Wahlert, 1963) in the evolution of new groups; many of the ideas presented in these studies were used by von Wahlert (1961b) to explain the evolution of the flatfishes, which are a classical problem group in evolutionary studies. The essential element in our analyses is the concept of mosaic or stepwise evolution as first publicized by de Beer (1954). On the basis of these studies, we feel that the evolution of new higher groups is by known adaptive mechanisms and is in full agreement with the concepts of adaptation advanced in this study.

The dichotomy of general adaptation versus special adaptation (*e.g.*, Brown, 1958) continues to evade solution. A major difficulty lies in the several concepts attributed to these terms by different biologists, often without the realization that their ideas differ. It is not always clear if an author considers general and special adaptation as *a priori* or *a posteriori* notions. No one has yet provided convincing operational definitions for distinguishing between these two kinds of adaptations. That a distinction exists between the evolutionary potentials of adaptations seems quite definite, but considerably more study is needed before the nature of this distinction is understood.

One of the basic axioms of biology is that adaptation is the core of any theory of evolutionary mechanics (*e.g.*, Simpson, 1958). Hence, any analysis of biological adaptation has a most important bearing upon such perplexing biological and philosophical topics as whether organic evolution is directional, or progressive, or purposeful, or finalistic in any sense (Simpson, 1960: 174–177; Mayr, 1961; Slobodkin, 1964). These problems can all be grouped under the general heading of whether organic evolution is a teleological, or a teleonomical, or any other end-directed process.[2] Much has been written upon this question, and we do not wish to add to this literature at this time. Little significant advance can be made in further clarification of this most difficult problem of progress in evolution before all aspects of biological adaptation are fully understood.

In conclusion, we feel that the mechanism of biological adaptation, like all other evolutionary processes, can be understood best by starting at a certain point and working forward through time. Looking backward may provide hindsight, but it also commits the worker to a particular result of evolutionary change and the necessity of explaining why that particular feature arose. A lack of hindsight is far more advantageous when formulating general principles of evolutionary change because the worker is completely uncommitted to particular results of past evolution.

Summary

1.   A feature (= character, trait, part) is any subdivision of the organism in terms of the particular method of investigation (precise definitions of these various concepts may be found in the text). Its form may be defined as its material composition and arrangement of these materials; a feature may have one fixed form or one that changes slowly (*e.g.*, a bone) or rapidly (*e.g.*, a muscle). Its function is its action or simply how the feature works—as stemming from the physical and chemical properties of the form; a feature may have several functions that operate simultaneously or at different times. A faculty is defined as the combination of a form

and a function of a feature; it is what the feature is capable of doing in the life of the organism. The biological role is the action or the use of the faculty by the organism in the course of its life history. A biological role can be ascertained only by observation of the organism living naturally in its normal environment.

2.   The environment of an organism is the external environment and consists of all external factors, physical and biotic, which interact in some way with the organism. The umgebung or prospective habitat of an organism are those factors of the environment which possibly could be utilized by that organism or which possibly could act upon that organism. The umgebung is commonly called the habitat of the organism and may be defined without the actual presence of that organism. The umwelt or the species-specific habitat are those factors of the environment which are actually being used by the organism or which are actually acting upon it; the umwelt can be established only with the presence of the organism. Environmental factors shift between the umgebung and umwelt as the organism begins and ceases to use them. The factors of the umwelt are those that interact with the organism and thereby exert "pressures" on the organism; these are the selection forces acting on the species.

3.   The interaction between the organism and its environment is through a couple formed by the biological role and the selection force. This interaction may be called a synerg which is defined as a link between the organism and its umwelt formed by one selection force of the umwelt and on biological role of a faculty of the organism. The niche of an organism is defined as the sum of all of its synergs and hence it is the total relationship between the whole organism and its complete umwelt.

4.   Biological adaptation has always been interpreted as an interaction between the organism and its environment: it is used in three different ways in evolutionary studies. Universal adaptation is the bond between living organisms and their environment and is an absolute property of life that can neither increase or decrease. Physiological adaptation is the ability of tissues to modify phenotypically in response to environmental stimuli and is a special case of the general principle that the phenotype is an expression of the genotype in a particular environment. Evolutionary adaptation is the long term, hereditary adjustment of an organism to a

particular set of environmental conditions; the particular environmental conditions must always be established first.

5.   An evolutionary adaptation is a faculty of a feature which interacts with the umwelt through a synerg formed by a biological role coupled with a selection force. Adaptation of the entire organism would be to its umwelt with the bond being the niche. The degree of evolutionary adaptation, the state of being, is defined as the minimal amount of energy required by the organism to maintain successfully the synerg if a single biological role of a faculty is considered, or to maintain successfully its niche if the whole organism is considered. Evolutionary adaptation, the process, is any evolutionary change which reduces the amount of energy required to maintain successfully a synerg, or the niche, as the case may be, toward the minimum possible amount.

6.   Several aspects of biological adaptation, such as preadaptation and selection, are discussed briefly. The general problem of evolutionary progress was mentioned but not discussed. As a general conclusion, it was emphasized that all evolutionary mechanisms can be studied best by working forward through time and without the aid of hindsight.

Acknowledgments

We have discussed all or part of this study with so many of our friends over the last several years that we do not know where to start and where to end the list of acknowledgments; we do wish to extend our deep appreciation and sincere thanks to everyone with whom we have discussed our notions of adaptation. In particular, we want to thank Ernst Mayr for reading and commenting on the manuscript, S. C. Kendeigh for describing several studies that support the concept of judging the degree of adaptation in terms of energy balance, R. S. Bader for suggesting the relationship between our definition of adaptation and the genetical definition of selection, and Gregory Bateson for his valuable suggestions on the formulation of the various definitions.

This study was possible only because of the several programs for foreign postdoctoral research. Each of us had the opportunity to visit the home country of the co-author during which time we were able to acquaint ourselves firsthand with the evolutionary thinking in each country. We wish to express our appreciation to the National Science Foundation and to the Fulbright Exchange Program for making possible our visits to each other's country and for the chance to collaborate on this study. Much of the final writing of the manuscript was done with the support of a National Science Foundation Grant (G. B. 1235).

Notes

1. Usually the biological roles of a feature are the guides to the function that are studied; however, this procedure may hinder the clarification of important functions of the feature which may be utilized in some or all cases. Ossification of tendons changes the general functions of these structures from those of collagenous fibers to those of bone. For tendons of origin, such as seen in certain jaw muscles of birds, the essential function of the ossified tendon is strength against shearing forces, *i.e.*, a rigid tendon of origin as compared to a flexible one. In the case of the long tendons of the leg of birds, the important function of the ossified tendon is not the obvious one of a rigid tendon, but it is probably the generally non-utilized function of the absence of any stretch as compared to the stretch of a few per cent in the collagenous fibered tendon. This stretch of a few per cent in the long tendons running to the toes may elongate these tendons a significant amount before the toes are moved. Hence the muscle must shorten further to achieve the desired movement of the toes which would place a great disadvantage on the short-fibered pinnate muscles that flex and extend the toes. These fibers would have to contract isotonically well into the low tension portion of the tension–length curve and hence the force produced by the muscle would be decreased considerably. Ossification of the tendon would eliminate the stretch and would reduce the amount of shortening of the muscle when it contracts isotonically; hence the muscle would remain in the high tension portion of the tension-length curve and would produce considerably more force.

2. "March 24, Easter Sunday" from John Steinbeck's "The log from the Sea of Cortez" (Chapter 14) is one of the most readable and interesting discussions of teleological versus non-teleological thinking. Steinbeck's comments are nicely summarized in the quote: "Non-teleological ideas derive through 'is' thinking associated with natural selection as Darwin seems to have understood it. They imply depth, fundamentalism, and clarity—seeing beyond traditional or personal projections. They consider events as outgrowths and expressions rather than as results; conscious acceptance as a desideratum, and certainly as an all-important prerequisite. Non-teleological thinking concerns itself primarily not with what should be, or could be, or might be. but rather what actually 'is'—attempting at most to answer the already sufficiently difficult questions *what* or *how*, instead of *why*." We heartily recommend this chapter and indeed the entire book to all interested in the philosophical aspects of adaptation and in general natural history.

Several of chapters in G. G. Simpson's new book "This view of life" (Harcourt, Brace and World, New York, 1964) have a direct bearing upon these philosophical problems, and we would strongly recommend his book to anyone concerned with philosophical aspects of evolutionary biology.

Literature Cited

Bartholomew, G. A. 1958. The role of physiology in the distribution of terrestrial vertebrates. *In* Zoogeography, C. L. Hubbs (ed.), Washington, D.C., A.A.A.S. Publ. 51, pp. 85–95.

————. 1963. Behavioral adaptations of mammals to desert environments. Proc. XVI Internat. Cong. Zool., 3: 49–52.

Bates, M. 1960. Ecology and evolution. *In* the evolution of life, S. Tax (ed.), Univ. Chicago Press, pp. 547–568.

Bateson, G. 1963. The role of somatic changes in evolution. Evolution, 17: 529–539.

Beckner, M. 1959. The biological way of thought. Columbia Univ. Press, New York, vii+200 pp.

Beer, G. De. 1954. *Archaeopteryx* and evolution. The Advancement of Sci., 11(42): 160–170.

Bertalanffy, L. von. 1960. Problems of life. Harper and Brothers, Harper Torchbook reprint, New York, viii+216 pp.

Bock, W. J. 1959. Preadaptation and multiple evolutionary pathways. Evolution, 13: 194–211.

————. 1961. Salivary glands in the gray jays (*Perisoreus*). Auk, 78: 355–365.

————. 1963a. Evolution and phylogeny in morphologically uniform groups. Amer. Nat., 97: 265–285.

————. 1963b. The role of preadaptation and adaptation in the evolution of higher levels of organization. Proc. XVI Internat. Cong. Zool., 3: 297–300 (also, in press, Syst. Zool.).

Bock. W. J., and W. de W. Miller. 1959. The scansorial foot of the woodpeckers, with comments on the evolution of perching and climbing feet in birds. Amer. Mus. Novit. 1931: 1–45.

Bock, W. J., and G. von Wahlert. ms. The niche concept in evolutionary biology.

Böker, H. 1935–7. Einführung in die vergleichende biologische Anatomie der Wirbeltiere. G. Fischer, Jena, 1: xi+228 and 2: xi+258.

Brown, W. L. 1958. General adaptation and evolution. Syst. Zool., 7: 157–168.

Dobzhansky, T. 1942. Biological adaptation. Sci. Monthly, 55: 391–402.

————. 1956. What is an adaptive trait? Amer. Nat., 90: 337–347.

————. 1960. Evolution and environment. *In* The evolution of life, S. Tax (ed.), Univ. Chicago Press, pp. 403–428.

Dohrn, A. 1875. Princip des Functionswechsels. Englemann, Leipzig.

Dow, D. D. The role of saliva in food storage by the gray jay (*Perisoreus canadensis*). Auk, 82: 139–154.

Dullemeijer, P. 1958. The mutual structural influence of the elements in a pattern. Arch. Neerl. Zool., 13 (1 suppl.): 74–88.

Gans, C. 1961. A bullfrog and its prey. Nat. Hist., 70: 26–37.

————. 1962. The tongue protrusion mechanism in *Rana catesbeiana*. Amer. Zool., 2: 524.

————. 1963. Functional analysis in a single adaptative radiation. Proc. XVI Internat. Cong. Zool., 3: 278–282.

Gates, D. M. 1962. Energy exchange in the biosphere. Harper and Row Biological Monographs, New York, viii+151 pp.

Hecht, M. 1963. The role of natural selection and evolutionary rates in the evolution of higher levels of organization. Proc. XVI Internat. Cong. Zool., 3: 305–308 (also, in press, Syst, Zool.).

Huxley, J. S. 1960. The openbill's open bill: a teleonomic enquiry. Zool. Jahrb. (Anat.) 88: 1–17.

Kendeigh, S. C. 1949. Effect of temperature and season on energy requirements of the English sparrow. Auk, 66: 113–127.

Kummer, B. 1962a. Funktioneller Bau und funktionelle Anpassung des Knockens. Anat. Anz., 110: 261–293.

————. 1962b. Funktionelle Anpassung und Präadaptation. Zool. Anz., 169: 50–67.

Lerner, I. M. 1954. Genetic homeostasis. John Wiley and Sons, New York.

————. 1958. The genetical basis of selection. John Wiley and Sons, New York.

Lewis, H. 1962. Catastrophic selection as a factor in speciation. Evolution, 16: 257–271.

Lockhart, R. D. 1960. The anatomy of muscles and their relation to movement and posture. *In* The structure and function of muscles, G. H. Bourne (ed.), Academic Press, New York and London, vol. 1, pp. 1–20.

Mayr, E. 1956. Geographic character gradients and climatic adaptations. Evolution, 10: 105–108.

————. 1958. The evolutionary significance of the systematic categories. Uppsala Univ. Arssks., 1958: 13–20.

————. 1960. The emergence of evolutionary novelties. *In* The evolution of life, S. Tax, ed., Univ. Chicago Press, pp. 349–380.

————. 1961. Cause and effect in biology. Science, 134: 1501–1506 (see also, Science, 135: 972–981, 1962).

―――. 1962. Accident or design, the paradox of evolution. *In* The evolution of living organisms, G. W. Leeper (ed.), Melbourne Univ. Press, Victoria, pp. 1–14.

―――. 1963. Animal species and their evolution. Harvard Univ. Press, xiv +797 pp.

Miller, A. H. 1949. Some ecological and morphological considerations in the evolution of higher taxonomic categories. *In* Ornithologie als biologische Wissenschaft, E. Mayr and E. Schüz (eds.), Carl Winter, Heidelberg, pp. 84–88.

Nagel, E. 1953. Teleological explanation and teleological systems. *In* Readings in the philosophy of science, H. Feigl and M. Brodbeck (eds.), Appleton-Century-Crofts, New York, pp. 537–558.

―――. 1961. The structure of science. Harcourt, Brace and World. Inc., New York, xiii +618 pp.

Nicholson, A. J. 1960. The role of population dynamics in natural selection. *In* The evolution of life, S. Tax, ed., Univ. Chicago Press, pp. 477–521.

Olson, E. C. 1960. Morphology, paleontology and evolution. *In* The evolution of life, S. Tax (ed.), Univ. Chicago Press, pp. 523–545.

Olson, J. 1964. The effect of temperature and season on the bioenergetics of the eastern field sparrow, *Spizella pusilla pusilla*. Unpublished Ph. D. thesis, Univ. Illinois.

Orians, G. H. 1961. The ecology of blackbird (*Agelaius*) social systems. Ecol. Monogr., 31: 285–312.

Osche, G. 1962. Das Praedadaptationsphänomen und seine Bedeutung für die Evolution. Zool. Anz., 169: 14–49.

Parker, B. C., and B. L. Turner. 1961. "Operational niches" and "community-interaction values" as determined from *in vitro* studies of some soil algae. Evolution, 15: 228–238.

Pitelka, F. 1958. Timing of molt in steller jays of the Queen Charlotte Islands, British Columbia, Condor, 60: 38–49.

Pittendrigh, G. S. 1958. Adaptation, natural selection, and behavior. *In* Behavior and evolution. A. Roe and G. G. Simpson (eds.), Yale Univ. Press, pp. 390–416.

Prosser, C. L., ed. 1958. Physiological adaptation. American Physiological Society, Washington, D.C.

―――. 1960. Comparative physiology in relation to evolutionary theory. *In* The evolution of life, S. Tax (ed.), Univ. Chicago Press, pp. 569–594.

Russell, E. S. 1916. Form and function. A contribution to the history of animal morphology. J. Murray, London, 383 pp.

―――. 1962. The diversity of animals. An evolutionary study. Acta Biotheoretica. 9: 1–151.

Simpson, G. G. 1953. The major features of evolution. Columbia Univ. Press, xx +434 pp.

————. 1958. The study of evolution. Methods and present status of theory. *In* Behavior and evolution, A. Roe and G. G. Simpson (eds.), Yale Univ. Press, pp. 7–26.

————. 1960. The history of life. *In* The evolution of life, S. Tax (ed.), Univ. Chicago Press, pp. 117–180.

Slobodkin, L. B. 1963. Growth and regulation of animal populations. Holt, Rinehart and Winston, New York, viii +184 pp.

————. 1964. The strategy of evolution. Amer. Sci., 52: 342–357.

Waddington, C. H. 1960. Evolutionary adaptations. *In* The evolution of life, S. Tax (ed.), Univ. Chicago Press, pp. 381–402.

Wahlert, G. von. 1957. Merkmalsverschiebung und Merkmalskopplung—ein neontologischer Beitrag zur Evolutionsforschung. Paläont. Zeits., 31: 23–31.

————. 1961a. Diskussionsbemerkungen zur Parallelbildungen und Stammesgeschichte. Zool. Anz., 166: 437–446.

————. 1961b. Die Entstehung der Plattfische durch ökologischen Funktionswechsel. Zool. Jahrb. (Syst.), 89: 1–42.

————. 1961c. Die Schüsselmerkmale der Rocken. Zool. Anz., 167: 9–12.

————. 1963. The role of ecological factors in the evolution of higher levels of organization. Proc. XVI Internat. Congr. Zool., 3: 301–304 (also, in press, Syst. Zool.).

Zimmerman, J. L. 1963. The bioenergetics of the dickcissel, *Spiza americana*. Unpublished Ph. D. thesis, Univ. Illinois.

6

Functional Analysis

Robert Cummins

I

A survey of the recent philosophical literature on the nature of functional analysis and explanation, beginning with the classic essays of Hempel in 1959 and Nagel in 1961, reveals that philosophical research on this topic has almost without exception proceeded under the following assumption.[1]

A. The point of functional characterization in science is to explain the presence of the item (organ, mechanism, process, or whatever) that is functionally characterized.

B. For something to perform its function is for it to have certain effects on a containing system, which effects contribute to the performance of some activity of, or the maintenance of some condition in, that containing system.

Putting these two assumptions together, we have: a function-ascribing statement explains the presence of the functionally characterized item i in a system s by pointing out that i is present in s because it has certain effects on s. Give or take a nicety, this fusion of (A) and (B) constitutes the core of almost every recent attempt to give an account of functional analysis and explanation. Yet these assumptions are just that: assumptions. They have never been systematically defended; generally they are not defended at all. I think there are reasons to suspect that adherence to (A) and (B) has crippled the most serious attempts to analyze functional statements and explanation. as I will argue in sections II and III below. In section IV, I will briefly develop an alternative approach to the problem. This alternative recommended largely by the fact that it emerges as the obvious approach once we take care to understand why accounts involving (A) and (B) go wrong.

II

I begin this section with a critique of Hempel and Nagel. The objections are familiar for the most part, but it will be well to have them fresh in our minds as they form the backdrop against which I stage my attack on (A) and (B).

Hempel's treatment of functional analysis and explanation is a classic example of the fusion of (A) and (B). He begins by considering the following singular function-ascribing statement.

(1)   The heartbeat in vertebrates has the function of circulating the blood through the organism.

He rejects the suggestion that "function" can *simply* be replaced by "effect" on the grounds that, although the heartbeat has the effect of producing heartsounds, this is not its function. Presuming (B) from the start, Hempel takes the problem to be how one effect—the having of which is the function of the heartbeat (circulation)—is to be distinguished from other effects of the heartbeat (e.g., heartsounds). His answer is that circulation, but not heartsounds, ensures a necessary condition for the "proper working of the organism." Thus Hempel proposes (2) as an analysis of (1).

(2)   The heartbeat in vertebrates has the effect of circulating the blood, and this ensures the satisfaction of certain conditions (supply of nutriment and removal of waste) which are necessary for the proper working of the organism.

As Hempel sees the matter, the main problem with this analysis is that functional statements so construed appear to have no explanatory force. Since he assumes (A), the problem for Hempel is to see whether (2) can be construed as a deductive nomological explanans for the presence of the heartbeat in vertebrates and, in general, to see whether statements having the form of (2) can be construed as deductive nomological explananda for the presence in a system of some trait or item that is functionally characterized.

Suppose, then, that we are interested in explaining the occurrence of a trait *i* in a system *s* (at a certain time *t*), and that the following functional analysis is offered:

(a)   At *t*, *s* functions adequately in a setting of kind *c* (characterized by specific internal and external conditions.)

(b)   *s* functions adequately in a setting of kind *c* only if a certain necessary condition, *n*, is satisfied.

(c)   If trait *i* were present in *s* then, as an effect, condition *n* would be satisfied.

(d)   Hence, at *t*, trait *i* is present in *s*.[2]

(d), of course, does not follow from (a)–(c), since some trait *i′* different from *i* might well suffice for the satisfaction of condition *n*. The argument can be patched up by changing (c) to (c′): "condition *n* would be satisfied in *s* only if trait *i* were present in *s*," but Hempel rightly rejects this avenue on the grounds that instances of the resulting schema would typically be false. It is false, for example, that the heart is a necessary condition for circulation in vertebrates, since artificial pumps can be, and are, used to maintain the flow of blood. We are thus left with a dilemma. If the original schema is correct, then functional explanation is invalid. If the schema is revised so as to ensure the validity of the explanation, the explanation will typically be unsound, having a false third premise.

Ernest Nagel offers a defense of what is substantially Hempel's schema with (c) replaced by (c′).

A teleological statement of the form, "The function of *A* in a system *S* with organization *C* is to enable *S* in the environment *E* to engage in process *P*," can be formulated more explicitly by: every system *S* with organization *C* and in environment *E* engages in process *P*; if *S* with organization *C* and in environment *E* does not have *A*, then *S* does not engage in *P*; hence, *S* with organization *C* must have *A*.[3]

Thus he suggests that (3) is to be rendered as (4):

(3)   The function of chlorophyll in plants is to enable them to perform photosynthesis.

(4)   A necessary condition of the occurrence of photosynthesis in plants is the presence of chlorophyll.

So Nagel must face the second horn of Hempel's dilemma: (3) is presumably true, while (4) may well be false. Nagel is, of course, aware of this objection. His rather curious response is that, as far as we know, chlorophyll *is* necessary for photosynthesis in the green plants.[4] This may be so, but the response will not survive a change of example, Hearts are *not* necessary for circulation, artificial pumps having actually been incorporated into the circulatory systems of vertebrates in such a way as to preserve circulation and life.

A more promising defense of Nagel might run as follows. While it is true that the presence of a working heart is not a necessary condition of circulation in vertebrates under all circumstances, still, under *normal* circumstances—most circumstances, in fact—a working heart is necessary for circulation. Thus it is perhaps true that, at the present stage of evolution, a vertebrate that has not been tampered with surgically would exhibit circulation only if it were to contain a heart. If these circumstances are specifically included in the explanans, perhaps we can avoid Hempel's dilemma. Thus instead of (4) we should have:

(4′)   At the present stage of evolution, a necessary condition for circulation in vertebrates that have not been surgically tampered with is the operation of a heart (properly incorporated into the circulatory system).

(4′), in conjunction with statements asserting that a given vertebrate exhibits circulation and has not been surgically tampered with and is at the present stage of evolution, will logically imply that that vertebrate has a heart. It seems, then, that the Hempelian objection could be overcome if it were possible, given a true function-ascribing statement like (1) or (3), to specify "normal circumstances" in such a way as to make it true that, in those circumstances, the presence of the item in question is a necessary condition for the performance of the function ascribed to it.

This defense has some plausibility as long as we stick to the usual examples drawn from biology. But if we widen our view a bit, even within biology, I think it can be shown that this defense of Nagel's position will not suffice. Consider the kidneys. The function of the kidneys is to eliminate wastes from the blood. In particular, the function of my blood. Yet the presence of my left kidney is not, in normal circumstances, a necessary condition for the removal of the relevant wastes. Only if something seriously abnormal should befall my right kidney would the operation of my left kidney become necessary, and this only on the assumption that I am not hooked up to a kidney machine.[5]

A less obvious counterexample derives from the well-attested fact of hemispherical redundancy in the brain. No doubt it is in principle possible to specify conditions under which a particular duplicated mechanism would be necessary for normal functioning of the organism, but (a) in most cases we are not in a position to actually do this, though we are in a position to make well-confirmed statements about the functions of some of these mechanisms, and (b) these circumstances are by no means the normal circumstances. Indeed, given the fact that each individual nervous system develops somewhat differently owing to differing environmental factors, the circumstances in question might well be different for each individual, or for the same individual at different times.

Apparently Nagel was pursuing the wrong strategy in attempting to analyze functional ascriptions in terms of necessary conditions. Indeed, we are still faced with the dilemma noticed by Hempel: an analysis in terms of necessary conditions yields a valid but unsound explanatory schema; analysis in terms of sufficient conditions along the lines proposed by Hempel yields a schema with true premises, but validity is sacrificed.

Something has gone wrong, and it is not too difficult to locate the problem. An attempt to explain the presence of something by appeal to what it does—its function—is bound to leave unexplained why something else that does the same thing—a functional equivalent—isn't there instead. In itself, this is not a serious matter. But the accounts we have been considering assume that explanation is a species of deductive inference, and one cannot deduce hearts from circulation. This is what underlies the dilemma we have been considering. At best, one can deduce circulators from circulation. If we make this amendment, however, we are left with a functionally tainted analysis; "the function of the heart is to circulate the blood" is rendered "a blood circulator is a (necessary/sufficient) condition of circulation, and *the heart is a blood circulator*." The expression in italics is surely as much in need of analysis as the analyzed expression. The problem, however, runs much deeper than the fact that the performance of a certain function does not determine how that function is performed. The problem is rather that to "explain" the presence of the heart in vertebrates by appeal to what the heart *does* is to "explain" its presence by appeal to factors which are causally irrelevant to its presence. Even if it were possible, as Nagel claimed, to *deduce* the presence of chlorophyll from the occurrence of photosynthesis, this would fail to explain the presence of chlorophyll in green plants in just the way deducing the presence and height of a building from the existence and length of its shadow would fail to explain why the building is there and has the height it does. This is not because all explanation is causal explanation: it is not. But to explain the presence of a naturally occurring structure or physical process—to explain why it is there, why such a thing exists in the place (system, context) it does—this does require specifying factors which causally determine the appearance of that structure or process.[6]

There is, of course, a sense in which the question "Why is x there?" is answered by giving x's function. Consider the following exchange. X asks Y, "Why is that thing there [pointing to the gnomon of a sundial]?" Y answers, "Because it casts a shadow on the dial beneath, thereby indicating the time of day." It is exchanges of this sort that most philosophers have had in mind when they speak of functional explanation. But it seems to me that, although such exchanges do represent genuine explanations, the use of functional language in this sort of explanation is quite distinct from its explanatory use in science. In section IV below I will sketch what I think *is* the central explanatory use of functional language in science. Meanwhile, if I am right, the evident propriety of exchanges like that imagined between X and Y has led to premature acceptance of (A), hence to concentration on what is, from the point of view of scientific explanation, an irrelevant use of functional language. For it seems to me that the question, "Why is x there?" can be answered by specifying x's function only if x is or is part of an artifact. Y's answer, I think, explains the presence of the gnomon because it rationalizes the action of the agent who put it there by supplying a *reason* for putting it there. In general, when we are dealing with the result of a deliberate action, we may explain the result by explaining the action, and we may explain a deliberate action by supplying the agent's reason for doing it. Thus when we look at a sundial, we assume we *know* in a general way how the gnomon came to be there: someone deliberately put it there. But we may wish to know *why* it was put there. Specifying the gnomon's function allows us to formulate what we suppose to be the unknown agent's reason for putting it there, viz., a belief that it would cast a shadow such that . . . , and so on. When we do this, we are elaborating on what we assume is the crucial causal factor in determining the gnomon's presence, namely a certain deliberate action.

If this is on the right track, then the viability of the sort of explanation in question should depend on the assumption that the thing functionally characterized is there as the result of deliberate action. If that assumption is evidently false, specifying the thing's function will not answer the question. Suppose it emerges that the sundial is not, as such, an artifact. When the ancient building was ruined, a large stone fragment fell on a kind of zodiac mosaic and embedded itself there. Since no sign of the roof remains, Y has mistakenly supposed the thing was designed as a sundial. As it happens, the local people have been using the thing to tell time for centuries, so Y is right about the function of the thing X pointed to.[7] But it is simply false that the thing is there because it casts a shadow, for there is no agent who put it there "because it casts a shadow." Again, the function of a bowl-like depression in a huge stone may be to hold holy water, but we cannot explain why it is there by appeal to its function if we know it was left there by prehistoric glacial activity.

If this is right, then (A) will lead us to focus on a type of explanation which will not apply to natural systems: chlorophyll and hearts are not "there" as the result of any deliberate action; hence the essential presupposition of the explanatory move in question is missing. Once this becomes clear, to continue to insist that there *must* be *some* sense in which specifying the function of chlorophyll explains its presence is an act of desperation born of thinking there is no other explanatory use of functional characterization in science.

Why have philosophers identified functional explanation exclusively with the appeal to something's function in explaining why it is there? One reason, I suspect, is a failure to distinguish teleological explanation from functional explanation, perhaps because functional concepts do loom large in "explanations" having a teleological form. Someone who fails to make this distinction, but who senses that there is an important and legitimate use of functional characterization in scientific explanation, will see the problem as one of finding a legitimate explanatory role for functional characterization within the teleological form. Once we leave artifacts and go to natural systems, however, this approach is doomed to failure, as critics of teleology have seen for some time.

This mistake probably would have sorted itself out in time were it not the case that we do reason from the performance of a function to the presence of certain specific processes and structures, e.g., from photosynthesis to chlorophyll, or from coordinated activity to nerve tissue. This is perfectly legitimate reasoning: it is a species of inference to the best explanation. Our best (only) explanation of photosynthesis requires chlorophyll, and our best explanation of coordinated activity requires nerve tissue. But once we see what makes this reasoning legitimate, we see immediately that inference *to* an explanation has been mistaken for an explanation itself. Once this becomes clear, it becomes equally clear that (A) has matters reversed: given that photosynthesis is occurring in a particular plant, we may legitimately infer that chlorophyll is present in that plant precisely because chlorophyll enters into our best (only) explanation of photosynthesis, and given coordinated activity on the part of some animal, we may legitimately infer that nerve tissue is present precisely because nerve tissue enters into our best explanation of coordinated activity in animals.

To attempt to explain the heart's presence in vertebrates by appealing to its function in vertebrates is to attempt to explain the occurrence of hearts in vertebrates by appealing to factors which are causally irrelevant to its presence in vertebrates. This fact has given "functional explanation" a bad name. But it is (A) that deserves the blame. Once we see (A) as an undefended philosophical hypothesis about how to construe functional explanations rather than as a statement of the philosophical problem, the correct alternative is obvious: what we can and do explain by appeal to what something does is the behavior of a containing system.[8]

A much more promising suggestion in the light of these considerations is that (1) is appealed to in explaining *circulation*. If we reject (A) and adopt this suggestion, a simple deductive-nomological explanation with circulation as the explicandum turns out to be a sound argument.

(5) a. Vertebrates incorporating a beating heart in the usual way (in the way *s* does) exhibit circulation.

  b. Vertebrate *s* incorporates a beating heart in the usual way.

  c. Hence, *s* exhibits circulation.

Though by no means flawless, (5) has several virtues, not the least of which is that it does not have biologists passing by an obvious application of evolution or genetics in favor of an invalid or unsound "functional" explanation of the presence of hearts. Also, the redundancy examples are easily handled, e.g., the removal of wastes is deduced in the kidney case.

The implausibility of (A) is obscured in examples taken from biology by the fact that there are two distinct uses of function statements in biology. Consider the following statements.

(a)   The function of the contractile vacuole in protozoans is elimination of excess water from the the organism.

(b)   The function of the neurofibrils in the ciliates is coordination of the activity of the cilia.

These statements can be understood in either of two ways. (1) They are generally used in explaining how the organism in question comes to exhibit certain characteristics or behavior. Thus (a) explains how excess water, accumulated in the organism by osmosis, is eliminated from the organism; (b) explains how it happens that the activity of the cilia in paramecium, for instance, is coordinated. (2) They may be used in explaining the continued survival of certain organisms incorporating structures of the sort in question by indicating the survival value which would accrue to such organisms in virtue of having structures of that sort. Thus, (a) allows us to infer that incorporation of a contractile vacuole makes it possible for the organism to be surrounded by a semi-permeable membrane, allowing the passage of oxygen into, and the passage of wastes out of, the organism. Relatively free osmosis of this sort is obviously advantageous, and this is made possible by a structure which solves the excess water problem. Similarly, ciliates incorporating neurofibrils will be capable of fairly efficient locomotion, the survival value of which is obvious.[9]

The second sort of use occurs as part of an account which, if we are not careful, can easily be mistaken for an explanation of the presence of the sort of item functionally characterized, and this has perhaps encouraged philosophers to accept (A). For it might seem that natural selection provides the missing causal link between what something does in a certain type of organism and its presence in that type of organism. By performing their respective functions the contractile vacuole and the neurofibrils help species incorporating them to survive, and thereby contribute to their own continued presence in organisms of those species, and this might seem to explain the presence of those structures in the organisms incorporating them.

Plausible as this sounds, it involves a subtle yet fundamental misunderstanding of evolutionary theory. A clue to the mistake is found in the fact that the contractile vacuole occurs in marine protozoans which have no excess water problem but the reverse problem. Thus the function and effect on survival of this structure is not the same in all protozoans. Yet the explanation of its presence in marine and fresh-water species is almost certainly the same. This fact reminds us that the processes actually responsible for the occurrence of contractile vacuoles in protozoans are totally insensitive to what that structure does. Failure to appreciate this point not only lends spurious plausibility to (A) as applied to biological examples, but seriously distorts our understanding of evolutionary theory. Whether an organism o incorporates s depends on whether s is "specified" by the genetic "plan" which o inherits and which, at a certain level of abstraction, is characteristic of o's species. Alterations in the plan are not the effects of the presence or exercise of the structures the plan specifies. This is most obvious when the genetic change is the result of random mutation. Though not all genetic change is due to random mutation, some certainly is, and that fact is enough to show that specifying the function of a biological structure cannot, in general, explain the presence of that structure. If a plan is altered so that it specifies s' rather than s, then the organisms inheriting this plan will incorporate s' regardless of the function or survival value of s' in those organisms. If the alteration is advantageous, the number of organisms inheriting that plan may increase, and, if it is disadvantageous, their number may decrease. But this typically has no effect on the plan, and therefore no effect on the occurrence of s' in the organisms in question.

One sometimes hears it said that natural selection is an instance of negative feedback. If this is meant to imply that the relative success or failure of organisms of a certain type can affect their inherited characteristics, it is simply a mistake: the characteristics of organisms which determine their relative success are determined by their genetic plan, and the characteristics of these plans are typically independent of the relative success of organisms having them. Of course, if s is very disadvantageous to organisms having a plan specifying s, then organisms having such plans may disappear altogether, and s will no longer occur. We could, think of natural selection as reacting on the *set* of plans generated by weeding out the bad plans: natural selection cannot alter a plan, but it can trim the set. Thus we may be able to explain why a given plan is not a failure by appeal to the functions of the structures it specifies. Perhaps this is what some writers have had in mind. But this is not to explain why, e.g., contractile vacuoles occur in certain protozoans; it is to explain why the sort of protozoan incorporating contractile vacuoles occurs. Since we cannot appeal to the relative success or failure of these organisms to explain why their genetic plan specifies contractile vacuoles, we cannot appeal to the relative success or failure of these organisms to explain why they incorporate contractile vacuoles.

Once we are clear about the explanatory role of functions in evolutionary theory, it emerges that the function of an organ or process (or whatever) is appealed to in order to explain the biological capacities of the organism containing it, and from these capacities conclusions are drawn concerning the chances of survival for organisms of that type. For instance, appeal to the function of the contractile vacuole in certain protozoans explains how these organisms are able to keep from exploding in fresh-water. Thus evolutionary biology does not provide support for (A), but for the idea instanced in (5): identifying the function of something helps to explain the capacities of a containing system.[10]

(A) misconstrues functional explanation by misidentifying what is explained. Let us abandon (A), then, in favor of the view that functions are appealed to in explaining the capacities of containing systems, and turn our attention to (B).

Whereas (A) is a thesis about functional explanation, (B) is a thesis about the analysis of function-ascribing statements. Perhaps when divorced from (A), as it is in (5), it will fare better than it does in the accounts of Hempel and Nagel.

III

In spite of the evident virtues of (5), (5a) has serious shortcomings as an analysis of (1). In fact it is subject to the same objection Hempel brings to the analysis which simply replaces "function" by "effect": vertebrates incorporating a working heart in the usual way exhibit the production of heartsounds, yet the production of heartsounds is not a function of hearts in vertebrates. The problem is that whereas the production of certain effects is essential to the heart's performing its function, there are some effects the production of which is irrelevant to the functioning of the heart. This problem is bound to infect any "selected effects" theory, i.e., any theory built on (B).

What is needed to establish a selected effects theory is a general formula which identifies the appropriate effects.[11] Both Hempel and Nagel attempt to solve this problem by identifying the function of something with just those effects which contribute to the maintenance of some special condition of, or the performance of some special activity of, some containing system. If this sort of solution is to be viable, there must be some principled way of selecting the relevant activities or conditions of containing systems. For no matter which effects of something you happen to name, there will be some activity of the containing system to which just those effects contribute, or some condition of the containing system which is maintained with the help of just those effects. Heart activity, for example, keeps the circulatory system from being entirely quiet, and the appendix keeps people vulnerable to appendicitis.[12]

Hempel suggests that, in general, the crucial feature of a containing system, contribution to which is to count as the functioning of a contained part, is that the system be maintained in "adequate, or effective, or proper working order."[13] Hempel explicitly declines to discuss what constitutes proper working order, presumably because he rightly thinks that there are more serious problems with the analysis he is discussing than those introduced by this phrase. But it seems clear that for something to be in working order is just for it to be capable of performing its functions, and for it to be in adequate or effective or proper working order is just for it to be capable of performing its functions adequately or effectively or properly. Hempel seems to realize this himself, for in setting forth a deductive schema for functional explanation, he glosses the phrase in question as "functions adequately."[14] More generally, if we identify the function of something x with those effects of x which contribute to the performance of some activity a or to the maintenance of some condition c of a containing system s, then we must be prepared to say as well that a function of s is to perform a or to maintain c. This suggests the following formulation of "selected effects" theories.

(6)   The function of an F in a G is f just in case (the capacity for) f is an effect of an F incorporated in a G in the usual way (or: in the way *this* F is incorporated in this G), and that effect contributes to the performance of a function of the containing G.

It seems that any theory based on (B)—what I have been calling "selected effects" theories—must ultimately amount to something like (6).[15] Yet (6) cannot be the whole story about functional ascriptions.

Suppose we follow (6) in rendering, "The function of the contractile vacuole in protozoans is elimination of excess water from the organism." The result is (7).

(7)   Elimination of excess water from the organism is an effect of a contractile vacuole incorporated in the usual way in a protozoan, and that effect contributes to the performance of a function of a protozoan.

In order to test (7) we should have to know a statement of the form "f is a function of a protozoan." Perhaps protozoans have no functions. If not, (7) is just a mistake. If they do, then presumably we shall have to appeal to (6) for an analysis of the statement attributing such a function and this will leave us with another unanalyzed functional ascription. Either we are launched on a regress, or the analysis breaks down at some level for lack of functions, or perhaps for lack of a plausible candidate for containing system. If we do not with to simply acquiesce in the autonomy of functional ascriptions, it must be possible to analyze at least some functional ascriptions without appealing to functions of containing systems. If (6) can be shown to be the only plausible formulation of thories based on (B), then no such theory can be the whole story.

Our question, then, is whether a thing's function can plausibly be identified with those of its effects contributing to production of some activity of, or maintenance of some condition of, a containing system, where performance of the activity in question is not a function of the containing system. Let us begin by considering Hempel's suggestion that functions are to be identified with the production of effects contributing to the proper working order of a containing system. I claimed earlier that to say something is in proper working order is just to say that it properly performs its functions. This is fairly obvious in cases of artifacts or tools. To make a decision about which sort of behavior counts as working amounts to deciding about the thing's function. To say something is working, though not behaving or disposed to behave in a way having anything to do with its function, is to be open, at the very least, to the charge of arbitrariness.

When we are dealing with a living organism, or a society of living organisms, the situation is less clear. If we say, "The function of the contractile vacuole in protozoans is elimination of excess water from the organism," we do make reference to a containing organism, but not, apparently, to its function (if any). However, since contractile vacuoles do a number of things having nothing to do with their function, there must be some implicit principle of selection at work. Hempel's suggestion is that, in this context, to be in "proper working order" is simply to be alive and healthy. This works reasonably well for certain standard examples, e.g. (1) and (3): circulation does contribute to health and survival in vertebrates, and photosynthesis does contribute to health and survival in green plants.[16] But once again, the principle will not stand a change of example, even within the life sciences. First, there are cases in which proper functioning is actually inimical to health and life: functioning of the sex organs results in the death of individuals of many species (e.g., certain salmon). Second, a certain process in an organism may have effects which contribute to health and survival but which are not to be confused with the function of that process: secretion of adrenalin speeds metabolism and thereby contributes to elimination of harmful fat deposits in overweight humans, but this is not a function of adrenalin secretion in overweight humans.

A more plausible suggestion along these lines in the special context of evolutionary biology is this:

(8)   The functions of a part or process in an organism are to be identified with those of its effects contributing to activities or conditions of the organism which sustain or increase the organism's capacity to contribute to survival of the species.

Give or take a nicety, (8) doubtless does capture a great many uses of functional language in biology. For instance, it correctly picks out elimination of excess water as the function of the contractile vacuole in fresh water protozoans only, and correctly identifies the function of sexual organs in species in which the exercise of these organs results in the death of the individual.[17]

In spite of these virtues, however, (8) is seriously misleading and extremely limited in applicability even within biology. Evidently, what contributes to an organism's capacity to maintain its species in one sort of environment may undermine that capacity in another. When this happens, we might say that the organ (or whatever) has lost its function. This is probably what we would say about the contractile vacuole if fresh-water protozoans were successfully introduced into salt water, for in this case the capacity explained would no longer be exercised. But if the capacity explained by appeal to the function of a certain structure continued to be exercised in the new environment, though now to the individual's detriment, we would not say that that structure had lost its function. If, for some reason, flying ceased to contribute to the capacity of pigeons to maintain their species, or even undermined that capacity to some extent,[18] we would still say that a function of the wings in pigeons is to enable them to fly. Only if the wings ceased to function as wings, as in the penguins or ostriches, would we cease to analyze skeletal structure and the like functionally with an eye to explaining flight. Flight is a capacity which cries out for explanation in terms of anatomical functions regardless of its contribution to the capacity to maintain the species.

What this example shows is that functional analysis can properly be carried on in biology quite independently of evolutionary considerations: a complex capacity of an organism (or one of its parts or systems) may be explained by appeal to a functional analysis regardless of how it relates to the organism's capacity to maintain the species. At best, then, (8) picks out those effects which will be called functions when what is in the offing is an application of evolutionary theory. As we shall see in the next section, (8) is misleading as well in that it is not *which* effects are explained but the style of explanation that makes it appropriate to speak of functions. (8) simply identifies effects which, as it happens, are typically explained in that style.

We have not quite exhausted the lessons to be learned from (8). The plausibility of (8) rests on the plausibility of the claim that, for certain purposes, we may assume that a function of an organism is to contribute to the survival of its species. What (8) does, in effect, is identify a function of an important class of (uncontained) containing systems without providing an analysis of the claim that a function of an organism is to contribute to the survival of its species.

Of course, an advocate of (8) might insist that it is no part of his theory to claim that maintenance of the species is a function of an organism. But then the defense of (8) would have to be simply that it describes actual usage, i.e., that it is in fact effects contributing to an organism's capacity to maintain its species which evolutionary biologists single out as functions. Construed in this way, (8) would, at most, tell us *which* effects are picked out as functions; it would provide no hint as to *why* these effects are picked out *as functions*. We know why evolutionary biologists are interested in effects contributing to an organism's capacity to maintain its species, but why call them functions? This is precisely the sort of question a philosophical account of function-ascribing statements should answer. Either (8) is defended as an instance of (6)—maintenance of the species is declared a function of organism—or it is defended as descriptive of usage. In neither case is any philosophical analysis provided. For in the first case (8) relies on an unanalyzed (and undefended) function-ascribing statement, and in the second it fails to give any hint as to the point of identifying certain effects as functions.

The failings of (8) are I think bound to cripple any theory which identifies a thing's functions with effects contributing to some antecedently specified type of condition or behavior of a containing system. If the theory is an instance of (6), it launches a regress or terminates in an unanalyzed functional ascription; if it is not an instance of (6), then it is bound to leave open the very question at issue, viz., why are the selected effects seen as functions?

IV

In this section I will sketch briefly an account of functional explanation which takes seriously the intuition that it is a genuinely distinctive style of explanation. The assumptions (A) and (B) form the core of approaches which seek to minimize the differences between functional explanations and explanations not formulated in functional terms. Such approaches have not given much attention to the characterization of the special explanatory strategy science employs in using functional language, for the problem as it was conceived in such approaches was to show that functional explanation is not really different in essentials from other kinds of scientific explanation. Once the problem is conceived in this way, one is almost certain to miss the distinctive features of functional explanation, and hence to miss the point of functional description. The account of this section reverses this tendency by placing primary emphasis on the kind of problem which is solved by appeal to functions.

Functions and Dispositions

Something may be capable of pumping even though it does not function as a pump (ever) and even though pumping is not its function. On the other hand, if something functions as a pump in a system s, or if the function of something in a system s, is to pump, then it must be capable of pumping in s.[19] Thus function-ascribing statements imply disposition statements; to attribute a function to something is, in part, to attribute a disposition to it. If the function of x in s is to ϕ, then x has a disposition to ϕ in s. For instance, if the function of the contractile vacuole in freshwater protozoans is to eliminate excess water from the organism, then there must be circumstances under which the contractile vacuole would actually manifest a disposition to eliminate excess water from the protozoan which incorporates it.

To attribute a disposition d to an object a is to assert that the behavior of a is subject to (exhibits or would exhibit) a certain law-like regularity: to say a has d is to say that a would manifest d (shatter, dissolve) were any of a certain range of events to occur (a is put in water, a is struck sharply). The regularity associated with a disposition—call it the dispositional regularity—is a regularity which is special to the behavior of a certain kind of object and obtains in virtue of some special facts(s) about that kind of object. Not everything is water-soluble: such things behave in a special way in virtue of certain (structural) features special to water-soluble things. Thus it is that dispositions require explanation: if x has d, then x is subject to a regularity in behavior special to things having d, and such a fact needs to be explained.

To explain a dispositional regularity is to explain how manifestations of the disposition are brought about given the requisite precipitating conditions. In what follows I will describe two distinct strategies for accomplishing this. It is my contention that the appropriateness of function-ascribing statements corresponds to the appropriateness of the second of these two strategies. This, I think, explains the intuition that functional explanation is a special *kind* of explanation.

Two Explanatory Strategies[20]

The Instantiation Strategy   Since dispositions are properties, not events, to explain a disposition requires explaining how it is instantiated. To explain an event, we cite its cause, and to explain an event type requires a recipe (law) for constructing causal explanations of its tokens. But dispositions, being properties, not events, are not explicable as effects. The *acquisition* of a property is an event, but explaining the acquisition of a property is quite distinct from explaining the property itself. One can explain why/how a thing became fragile without thereby explaining fragility, and one can explain why/how something changed properties— e.g., why something changed temperature—without thereby explaining the property that changed. To explain a property one must show how that property is instantiated in the things that have it.

Simple dispositions are explained by exhibiting their instantiations: water solubility is instantiated as a certain kind of molecular structure, temperature as (average) kinetic energy of molecules, flammability as a kind of subatomic structure (allowing for bonding with oxygen at relatively low temperatures). When we understand how a disposition is instantiated, we are in a position to understand why the dispositional regularity holds of the disposed objects.

Brian O'Shaughnessy has provided an example that allows a particularly simple illustration of this strategy.[21] Consider the disposition he calls elevancy: the tendency of an object to rise in water of its own accord. To explain elevancy, we must explain why freeing a submerged elevant object causes it to rise.[22] This we may do as follows. In every case, the ratio of an elevant object's mass to its nonpermeable volume is less than the density (mass per unit volume) of water: that is how elevancy is instantiated. Once we know this, we may apply Archimedes' Principle, which tells us that water exerts an upward force on a submerged object equal to the weight of the water displaced. In the case of an elevant object, this force evidently exceeds the weight of the object by some amount f. Freeing the object changes the net force on it from zero to a net force of magnitude f in the direction of the surface, and the object rises accordingly. Here we subsume the connection between freeings and risings under a general law connecting changes in net force with changes in motion by citing a feature of elevant objects which allows us (via Archimedes' Principle) to represent freeing them under water as an instance of introducing a net force in the direction of the surface.

The Analytical Strategy   Rather than deriving the dispositional regularity that specifies d (in a) from the facts of d's instantiation (in a), the analytical strategy proceeds by analyzing a disposition d of a into a number of other dispositions $d_1, \ldots, d_n$ had by a or components of a such that programmed manifestation of the d_i results in or amounts to a manifestation of d.[23] The two strategies will fit together into a unified account if the analyzing dispositions (the d_i) can be made to yield to the instantiation strategy.

When the analytical strategy is in the offing one is apt to speak of capacities (or abilities) rather than of dispositions. This shift in terminology will put a more familiar face on the analytical strategy,[24] for we often explain capacities by analyzing them. Assembly-line production provides a transparent example of what I mean. Production is broken down into a number of distinct tasks. Each point on the line is responsible for a certain task, and it is the function of the components at that point to complete that task. If the line has the capacity to produce the product, it has it in virtue of the fact that the components have the capacities to perform their designated tasks, and in virtue of the fact that when these tasks are performed in a certain organized way—according to a certain program—the finished product results. Here we can explain the line's capacity to produce the product—i.e., explain how it is able to produce the product—by appeal to certain capacities of the components and their organization into an assembly line. Against this background we may pick out a certain capacity of an individual component the exercise of which is its function on the line. Of the many things it does and can do, its function on the line is doing whatever it is that we appeal to in explaining the capacity of the line as a whole. If the line produces several products—i.e., if it has several capacities—then, although a certain capacity c of a component is irrelevant to one capacity of the line, exercise of c by that component may be its function with respect to another capacity of the line as a whole.

Schematic diagrams in electronics provide another obvious illustration. Since each symbol represents any physical object whatever having a certain capacity, a schematic diagram of a complex device constitutes an analysis of the electronic capacities of the device as a whole into the capacities of its components. Such an analysis allows us to explain how the device as a whole exercises the analyzed capacity, for it allows us to see exercises of the analyzed capacity as programmed exercise of the analyzing capacities. In this case the "program" is given by the lines indicating how the components are hooked up. (Of course, the lines are themselves function-symbols.)

Functional analysis in biology is essentially similar. The biologically significant capacities of an entire organism are explained by analyzing the organism into a number of "systems"—the circulatory system, the digestive system, the nervous system, etc. each of which has its characteristic capacities.[25] These capacities are in turn analyzed into capacities of component organs and structures, Ideally, this strategy is pressed until physiology takes over—i.e., until the analyzing capacities are amenable to the instantiation strategy. We can easily imagine biologists expressing their analyses in a form analogous to the schematic diagrams of electrical engineering, with special symbols for pumps, pipes, filters, and so on. Indeed, analyses of even simple cognitive capacities are typically expressed in flow-charts or programs, forms designed specifically to represent analyses of information-processing capabilities generally.

Perhaps the most extensive use of the analytical strategy in science occurs in psychology, for a large part of the psychologist's job is to explain how the complex behavioral capacities of organisms are acquired and how they are exercised. Both goals are greatly facilitated by analysis of the capacities in question, for then acquisition of the analyzed capacity resolves itself into acquisition of the analyzing capacities and the requisite organization, and the problem of performance resolves itself into the problem of how the analyzing capacities are exercised. This sort of strategy has dominated psychology ever since Watson attempted to explain such complex capacities as the ability to run a maze by analyzing the performance into a series of conditioned responses, the stimulus for each response being the previous response, or something encountered as the result of the previous response.[26] Acquisition of the complex capacity is resolved into a number of distinct cases of simple conditioning—i.e., the ability to learn the maze is resolved into the capacity for stimulus substitution, and the capacity to run the maze is resolved into abilities to respond in certain simple ways to simple stimuli. Watson's analysis proved to be of limited value, but the analytic strategy remains the dominant mode of explanation in behavioral psychology.[27]

Functions and Functional Analysis
In the context of an application of the analytical strategy, exercise of an analyzing capacity emerges as a function: it will be appropriate to say that x functions as a ϕ in s, or that the function of x in s is ϕ-ing, when we are speaking against the background of an analytical explanation of some capacity of s which appeals to the fact that x has a capacity to ϕ is s. It is appropriate to say that the heart functions as a pump against the background of an analysis of the circulatory system's capacity to transport food, oxygen, wastes, and so on, which appeals to the fact that the heart is capable of pumping. Since this is the usual background, it goes without saying, and this accounts for that "the heart functions as a pump" sounds right, and "the heart functions as a noise-maker" sounds wrong, in some context-free sense. This effect is strengthened by the

absence of any actual application of the analytical strategy which makes use of the fact that the heart makes noise.[28]

We can capture this implicit dependence on an analytical context by entering an explicit relativization in our regimented reconstruction of function ascribing statements.

(9) x functions as a ϕ in s (or: the function of x in s is to ϕ) relative to an analytical account A of s's capacity to ψ just in case x is capable of ϕ-ing in s and A appropriately and adequately accounts for s's capacity to ψ by, in part, appealing to the capacity of x to ϕ in s.

Sometimes we explain a capacity of s by analyzing it into other capacities of s, as when we explain how someone ignorant of cookery is able to bake cakes by pointing out that he or she followed a recipe each instruction of which requires no special capacities for its execution. Here we don't speak of, e.g., stirring as a function of the cook, but rather of the function of stirring. Since stirring has different functions in different recipes, and at different points in the same recipe, a statement like, "The function of stirring the mixture is to keep it from burning to the bottom of the pan," is implicitly relativized to a certain (perhaps somewhat vague) recipe. To take account of this sort of case, we need a slightly different schema: where e is an activity or behavior of a system s (as a whole), the function of e in s is to ϕ relative to an analytical account A of s's capacity to ψ just in case A appropriately and adequately accounts for s's capacity to ψ by, in part, appealing to s's capacity to engage in e.

(9) explains the intuition behind the regress-ridden (6): functional ascriptions do require relativization to a "functional fact" about a containing system—i.e., to the fact that a sertain capacity of a containing system is appropriately explained by appeal to a certain functional analysis. And, like (6), (9) makes no provision for speaking of the function of an organism except against a background analysis of a containing system (the hive, the corporation, the eco-system). Once we see that functions are appealed to in explaining the capacities of containing systems, and indeed that it is the applicability of a certain strategy for explaining these capacities that makes talk of functions appropriate, we see immediately why we do not speak of the functions of uncontained containers. What (6) fails to capture is the fact that uncontained containers can be func-

tionally analyzed, and the way in which function-analytical explanation mediates the connection between functional ascriptions (*x* functions as a ϕ, the function of *x* is to ϕ) and the capacities of the containers.

Function-Analytical Explanation

If the account I have been sketching is to draw any distinctions, the availability and appropriateness of analytical explanations must be a nontrivial matter.[29] So let us examine an obviously trivial application of the analytical strategy with an eye to determining whether it can be dismissed on principled grounds.

(10)   Each part of the mammalian circulatory system makes its own distinctive sound, and makes it continuously. These sounds combine to form the "circulatory noise" characteristic of all mammals. The mammalian circulatory system is capable of producing this sound at various volumes and various tempos. The heartbeat is responsible for the throbbing character of the sound, and it is the capacity of the heart to beat at various rates that explains the capacity of the circulatory system to produce a variously tempoed sound.

Everything in (10) is, presumably, true. The question is whether it allows us to say that the function of the heart is to produce a variously tempoed throbbing sound.[30] To answer this question we must, I think, get clear about the motivation for applying the applytical strategy. For my contention will be that the analytical strategy is most significantly applied in cases very unlike that envisaged in (10).

The explanatory interest of an analytical account is roughly proportional to (i) the extent to which the analyzing capacities are less sophisticated than the analyzed capacities, (ii) the extent to which the analyzing capacities are different in type from the analyzed capacities, and (iii) the relative sophistication of the program appealed to, i.e., the relative complexity of the organization of component parts/processes which is attributed the system. (iii) is correlative with (i) and (ii): the greater the gap in sophistication and type between analyzing capacities and analyzed capacities, the more sophisticated the program must be to close the gap.

It is precisely the width of these gaps which, for instance, makes automata theory so interesting in its application to psychology. Automata theory supplies us with extremely powerful techniques for constructing diverse analyses of very sophisticated tasks into very unsophisticated tasks. This allows us to see how, in principle, a mechanism such as the brain, consisting of physiologically unsophisticated components (relatively speaking), can acquire very sophisticated capacities. It is the prospect of promoting the capacity to store ones and zeros into the capacity to solve problems of logic and recognize patterns that makes the analytical strategy so appealing in cognitive psychology.

As the program absorbs more and more of the explanatory burden, the physical facts underlying the analyzing capacities become less an less special to the analyzed system. This is why it is plausible to suppose that the capacity of a person and a machine to solve a certain problem might have substantially the same explanation, while it is not plausible to suppose that the capacities of a synthesizer and a bell to make similar sounds have substantially similar explanations. There is no work for a sophisticated hypothesis about the organization of various capacities to do in the case of the bell. Conversely, the less weight borne by the program, the less point to analysis. At this end of the scale we have cases like (10) in which the analyzed and analyzing capacities differ little if at all in type and sophistication. Here we could apply the instantiation strategy without significant loss, and thus talk of functions is comparatively strained and pointless. It must be admitted, however, that there is no black-white distinction here, but a case of more-or-less. As the role of organization becomes less and less significant, the analytical strategy becomes less and less appropriate, and talk of functions makes less and less sense. This may be philosophically disappointing, but there is no help for it.

Conclusion

Almost without exception, philosophical accounts of function-ascribing statements and of functional explanation have been crippled by adoption of assumptions (A) and (B). Though there has been widespread agreement that extant accounts are not satisfactory, (A) and (B) have escaped critical scrutiny, perhaps because they were thought of as somehow setting the problem rather than as part of proffered solutions. Once the problem is properly diagnosed, however, it becomes possible to give a more satisfactory and more illuminating account in terms of the explanatory strategy which provides the motivation and forms the context of function-ascribing statements. To ascribe a function to something is to ascribe a capacity to it which is singled out by its role in an analysis of some capacity of a containing system. When a capacity of a containing system is appropriately explained by analyzing it into a number of other capacities whose programmed exercise yields a manifestation of the analyzed capacity, the analyzing capacities emerge as functions. Since the appropriateness of this sort of explanatory strategy is a matter of degree, so is the appropriateness of function-ascribing statements.

Notes

1. Cf. Carl Hempel, The logic of functional analysis, in *Aspects of Scientific Explanation*, New York, Free Press, 1965, reprinted from Llewellyn Gross, ed., *Symposium on Sociological Theory*, New York, Harper and Row, 1959; and Ernest Nagel, *The Structure of Science*, New York, Harcourt, Brace and World, 1961, chapter 12, section 1. The assumptions, of course, predate Hempel's 1959 essay. See, for instance, Richard Braithwaite, *Scientific Explanation*, Cambridge, Cambridge University Press, 1955, chapter X; and Israel Scheffler, Thoughts on teleology, *British Journal for the Philosophy of Science*, 11 (1958). More recent examples include Francisco Ayala, Teleological explanations in evolutionary biology, *Philosophy of Science*, 37 (1970); Hugh Lehman, Functional explanations in biology, *Philosophy of Science*, 32 (1965); Richard Sorabji, Function, *Philosophical Quarterly*, 14 (1964); and Larry Wright, Functions, *Philosophical Review*, 82 (1973).

2. Hempel, p. 310.

3. Nagel, p. 403.

4. Ibid., p. 404.

5. It might be objected here that although it is the function of the kidneys to eliminate waste, that is not the function of a particular kidney unless operation of that kidney *is* necessary for removal of wates. But suppose scientists had initially been aware of the existence of the left kidney only. Then, on the account being considered, anything they had said about the function of that organ would have been false, since, on that account, *it has no function in organisms having two kidneys!*

6. Even in the case of a designed artifact, it is at most the designer's *belief* that x will perform f in s which is causally relevant to x's presence in s, not x's actually performing f in s. The nearest I can come to describing a situation in which x performing f in s is causally relevant to x's presence in s is this: the designer of s notices a thing like x performing f in a system like s, and this leads to belief that x will perform f in s, and this in turn leads the designer to put x in s.

7. *Is* casting a shadow the function of this fragment? Standard use may confer a function on something: if I standardly use a certain stone to sharpen knives, then that is its function, or if I standardly use a certain block of wood as a door stop, then the function of the block is to hold my door open. If nonartifacts *ever* have functions, appeals to those functions cannot explain their presence. The things functionally characterized in science are typically not artifacts.

8. A confused perception of this fact no doubt underlies (B), but the fact that (B) is nearly inseparable from (A) in the literature shows how confused this perception is.

9. Notice that the second use is parasitic on the first. It is only because the neurofibrils explain the coordinated activity of the cilia that we can assign a survival value to neurofibrils: the survival value of a structure *s* hangs on what capacities of the organism, if any, are explicable by appeal to the functioning of *s*.

10. In addition to the misunderstanding about evolutionary theory just discussed, biological examples have probably suggested (A) because biology was the *locus classicus* of teleological explanation. This has perhaps encouraged a confusion between the teleological *form* of explanation, incorporated in (A), with the explanatory role of functional ascriptions. Function-ascribing statements do occur in explanations having a teleological form, and when they do, their interest is vitiated by the incoherence of that form of explanation. It is the legitimate use of function-ascribing statements that needs examination, i.e., their contribution to nonteleological theories such as the theory of evolution.

11. Larry Wright (op. cit.) is aware of this problem but does not, to my mind, make much progress with it. Wright's analysis rules out "the function of the heart is to produce heartsounds," on the ground that the heart is not there because it produces heartsounds. I agree. But neither is it there because it pumps blood. Or if, as Wright maintains, there is a sense of "because" in which the heart *is* there because it pumps blood and not because it produces heartsounds, then this sense of "because" is as much in need of analysis as "function." Wright does not attempt to provide such an analysis, but depends on the fact that, in many cases, we are able to use the word in the required way. But we are also able to use "function" correctly in a variety of cases. Indeed, if Wright is right, the words are simply interchangeable with a little grammatical maneuvering. The problem is to make the conditions of correct use explicit. Failure to do this means that Wright's analysis provides no insight into the problem of how functional theories are confirmed, or whence they derive their explanatory force.

12. Surprisingly, when Nagel comes to formulate his general schema of functional attribution, he simply ignores this problem and thus leaves himself open to the trivialization just suggested. Cf. Nagel, p. 403.

13. Hempel, p. 306.

14. Ibid., p. 310.

15. Hugh Lehman (op. cit.) gives an analysis that appears to be essentially like (6).

16. Even these applications have their problems. Frankfurt and Poole, Functional explanations in biology, *British Journal for the Philosophy of Science*, 17 (1966), point out that heartsounds contribute to health and survival via their usefulness in diagnosis.

17. Michael Ruse has argued for a formulation like (8). See his Function statements in biology, *Philosophy of Science*, 38 (1971), and *The Philosophy of Biology*, London, Hutchinson, 1973.

18. Perhaps, in the absence of serious predators, with a readily available food supply, and with no need to migrate, flying simply wastes energy.

19. Throughout this section I am discounting appeals to the intentions of designers or users. *x* may be intended to prevent accidents without actually being capable of doing so. With reference to this intention, it *would* be proper in certain contexts to say, "*x*'s function is to prevent accidents, though it is not actually capable of doing so."

There can be no doubt that a thing's function is often identified with what it is typically or "standardly" used to do, or with what it was designed to do. But the sorts of things for which it is an important scientific problem to provide functional analyses—brains, organisms, societies, social institutions—either do not have designers or standard or regular uses at all, or it would be inappropriate to appeal to these in constructing and defending a scientific theory because the designer or use is not known—brains, devices dug up by archaeologists—or because there is some likelihood that real and intended functions diverge—social institutions, complex computers. Functional talk may have originated in contexts in which reference to intentions and purposes loomed large, but reference to intentions and purposes does not figure at all in the sort of functional analysis favored by contemporary natural scientists.

20. For a detailed discussion of the two explanatory strategies sketched here, see Cummins, *The Nature of Psychological Explanation*, Bradford Books/MIT Press, Cambridge, 1983. In the original version of this paper, I called the two strategies the Subsumption Strategy and the Analytical Strategy. I have retained the latter term, but the former I have replaced. What I was calling the subsumption strategy in 1975 was simply a confusion, a conflation of causal subsumption of events, and the nomic derivation of a property via the facts of its instantiation. Since functions are dispositions and dispositions are properties, only the latter is relevant here.

21. Brian O'Shaughnessy, The powerlessness of disposition, *Analysis*, October (1970). See also my discussion of this example in Dispositions, states and causes, *Analysis*, June (1974).

22. Also, we must explain why submerging a free elevant object causes it to rise, and why a free submerged object's becoming elevant causes it to rise. One of the convenient features of elevancy is that the same considerations dispose of all these cases. This does not hold generally: gentle rubbing, a sharp blow, or a sudden change in temperature may each cause a glass to manifest a disposition to shatter, but the explanations in each case are significantly different.

23. By "programmed" I simply mean organized in a way that could be specified in a program or flow chart: each instruction (box) specifies manifestation of one of the d_i such that if the program is executed (the chart followed), a manifests d.

24. Some might want to distinguish between dispositions and capacities, and argue that to ascribe a function to x is in part to ascribe a *capacity* to x, not a disposition as I have claimed. Certainly (1) is strained in a way (2) is not.

(1)   Hearts are disposed to pump. Hearts have a disposition to pump. Sugar is capable of dissolving. Sugar has a capacity to dissolve.

(2)   Hearts are capable of pumping. Hearts have a capacity to pump. Sugar is disposed to dissolve. Sugar has a disposition to dissolve.

25. Indeed, what makes something part of, e.g., the nervous system is that its capacities figure in an analysis of the capacity to respond to external stimuli, co-ordinate movement, etc. Thus there is no question that the glial cells are part of the brain, but there is some question whether they are part of the nervous system or merely auxiliary to it.

26. John B. Watson, *Behaviorism*, New York, W. W. Norton, 1930, chapters IX and XI.

27. Writers on the philosophy of psychology, especially Jerry Fodor, have grasped the connection between functional characterization and the analytical strategy in psychological theorizing but have not applied the lesson to the problem of functional explanation generally. The clearest statement occurs in J. A. Fodor, The appeal to tacit knowledge in psychological explanation, *Journal of Philosophy*, 65 (1968), 627–640.

28. It is sometimes suggested that heartsounds do have a psychological function. In the context of an analysis of a psychological disposition appealing to the heart's noise-making capacity, "The heart functions as a noise-maker" (e.g., as a producer of regular thumps) would not even *sound* odd.

29. Of course, it might be that only arbitrary distinctions are to be drawn. Perhaps (9) describes usage, and usage is arbitrary, but I am unable to take this possibility seriously.

30. The issue is not whether (10) forces us, via (9), to say something false. Relative to *some* analytical explanation, it may be true that the of the heart is to produce a variously tempoed throbbing. But the availability of (10) should not support such a claim.

7

Teleology Revisited

Ernest Nagel

Goal-Directed Processes in Biology

"Naturalism" is the label for a number of distinct though related philosophical doctrines; but this is not the occasion for listing their varieties or identifying what is common to them. It does seem appropriate, however, to begin these lectures with a reminder of the sense in which John Dewey characterized his logical theory (or theory of inquiry) as "naturalistic." He so described his theory, because in it "there is no break of continuity between operations of inquiry and biological operations" as well as physical ones, and because he believed he had succeeded in showing that the "rational operations [of inquiry] ... grow out of organic activities, without being identical with that from which they emerge."[1] This summary account of Dewey's conception of naturalism is not information about the nature of the "continuity" he sought not to breach. However, it does make evident that for him biological processes and biological theory constitute a substantial part of the matrix in which his theory of logic has its roots. Dewey also maintained that a touchstone of whether Darwinian ideas (as he understood them) concerning logical method have been properly understood and assimilated, is the way the much-debated problem of "design versus chance" is treated.[2] The conspicuous place this problem and its treatment occupied in his assessment of the significance of Darwinian theory for philosophy, testifies to the importance Dewey attached to the roles teleological notions play in biological inquiry.

Dewey's interest in the issues raised by the use of teleological conceptions, is *one* reason for my choice of teleology as the theme of these

Dewey Lectures. However, the debt owed to piety is not the *only* reason. Although the ideas associated with the term "teleology" have had a long and checked history, and teleological language continues to be viewed with suspicion by many scientists and philosophers, its use in biology as well as in the psychological and social sciences is widespread. Indeed, many biologists believe that such language is indispensable for describing and explaining a large variety of important biological phenomena. But despite the voluminous literature devoted to analyzing teleological concepts that has appeared in recent years at an accelerated rate, there is much disagreement both over the meaning of teleological statements and concerning the kinds of events and processes about which it is appropriate to assert them. These disagreements are not easily resolved for a variety of reasons. But whether or not they can be resolved at all, it can be instructive to reexamine some of them in the light of fresh evidence and new perspectives on the issues at stake; and I want in these lectures to do so, though restricting myself mainly to questions that are germane to biology. The recent literature sometimes breaks fresh ground; but much of it consists of modified (and sometimes much improved) versions of well-known views, presentations of difficulties in older analyses, or challenges to basic assumptions underlying customary approaches. My discussion will, in consequence, inevitably revisit much familiar territory, though with some different objectives than on previous journeys.

I

A commonly recognized but loosely delimited trait of biological organisms, a trait that is often said to distinguish living from inanimate things, is the apparently purposive character of living organisms. Teleological language reflects this distinction. Although there is no exhaustive set of criteria for distinguishing teleological locutions from those which are not, the occurrence of certain expressions in statements is usually a fairly reliable criterion that the statements are teleological. For example, the statements "The purpose of the liver is to secrete bile," "the function of the kidneys is to eliminate waste products from the blood," and "Peacocks spread their tail feathers in order to attract peahens," are teleological; and the occurrence in them of the expressions "the purpose of," "the function of," and "in order to" is a sure indication that this is so. However, these are not the only expressions whose occurrence in a statement marks the latter as teleological. There are also statements that are generally acknowledged to be teleological, even though they contain no typically teleological phrase—for example, the statement 'The hare sought refuge in its burrow from the packs of pursuing hounds." But despite the absence of a formal criterion, there is little disagreement in actual fact as to whether a given statement is teleological or not.

Teleological statements are not all of the same kind, and they can be classified in a number of ways. One distinction that will be useful in what follows is between "goal ascriptions" and "function ascriptions." The former state some outcome or goal toward which certain activities of an organism or its parts are directed. For example, each of the following is a goal ascription: "The goal of the pecking of woodpeckers is to find larvae of insects," and "'The goal of the activities in various animals of the sympathico-adrenal apparatus as well as of certain cells in the pancreas, is to keep the concentration of blood sugar within relatively narrow limits." On the other hand, function ascriptions state what are some of the effects of a given item or of its activities in an organism. For example, the following statement is a function ascription: "The function of the valves in the heart of a vertebrate is to give direction to the circulation of the blood."

Some biologists use the words "goal" and "function" interchangeably, ignoring distinctions that linguistic usage recognizes—they do so possibly because the distinctions are not relevant to the tasks on which they are engaged.[3] But however this may be, *seeing* is customarily said to be a function of eyes, rather than their goal; and *escape from a predator* is said to be the goal of a hare's flight, rather than its function. Moreover, although the end products of certain goal-directed processes have a function—for example, one function of the homeostasis of sugar concentration in the blood is prevention of convulsions—this is not always the case—for example, survival may be the goal of a hare's flight from a hound, but survival itself does not appear to have any function. Also, the lachrymal glands have a function, namely, to lubricate the exposed surface of the eyeballs, but do not have a goal. But the main reason for retaining the distinction is twofold: the analysis of what constitutes goal-directed behavior is different from the analysis of what counts as a function; and secondly, the structure of functional explanations in biology differs from the structure of explanations of goal-directed behavior. The present lecture will be devoted to the discussion of goal-directed processes in biology, and the second one to functional explanations.

II

There are a number of alternative (but not necessarily incompatible) explications of the notions of goal and goal-directed processes, and I will examine three of them. Undoubtedly the most familiar account of these notions takes the primary (or "core") meaning of these terms to be determined by their use in connection with purposive human behavior (and perhaps also in connection with the supposedly purposive behavior of the higher animals). Accordingly, the "goal" G of an action or process is said to be some state of affairs *intended* by a human agent; the *intention* itself is an "internal mental state" which, coupled with the internal state of "wanting" G together with "believing" that an action A would contribute to the realization of G, is allegedly a causal determinant of the ensuing action A. "Goal-directed" behavior is then the action A undertaken by the agent for the sake of achieving the goal. For convenience of reference I will call this account of goals and goal-directed processes the "intentional" view.

Several features of this account should be noted. In the first place, though the *occurrence* of the action A can be explained teleologically, the explanation is ostensibly a species of *causal* explanation. For, by hypothesis, the action is initiated because the agent *desires* a certain goal and also *believes* that the action will contribute to its production.[4] The causal explanation cannot be rightly charged with the difficulty often raised against teleological explanations, that the causal explanation assumes that a future state of affairs can be causally efficacious in bringing about its own realization. For according to that explanation, it is not the *goal* that brings about the action. It is rather the agent's *wanting* the goal, together with his *belief* that the action would contribute to the realization of the goal, that does so.

In the second place, this account is fully compatible with the usual characterizations of goal-directed behavior. For example, human beings are said to be goal-directed toward some goal, even when they fail to achieve it; and this is congruent with the intentional view, according to which an action is goal-directed if it is undertaken for the sake of some *intended* goal, whether or not the goal is reached. Again, goal-directed behavior is said to be "plastic," in the sense that the goal can be reached by alternative routes and from different initial positions, the route or action that is actually adopted depending on what local circumstances prevail. But such plasticity is obviously compatible with the intentional view of goal-directed behavior; for the alternative actions that are recognized as possible roads to the goal, as well as the action that is actually adopted, will depend on the beliefs the agent has in the situation in which he finds himself concerning the available means for reaching the desired end.

On the other hand, if the intentional view of goals is taken literally, an organism can be described as goal-directed with respect to a determinate goal *only if* it is legitimate to ascribe intentions, desires, and beliefs to the organism. In consequence, goals and goal-directed behavior can be correctly predicated only of human beings, and possibly of some higher animals. It is therefore entirely inappropriate to use such language in connection with organisms such as protozoa and plants which are incapable of having intentions and beliefs; in connection with subsystems of organisms, such as the complex of glands and other parts of the human body, that are involved in the homeostasis of the blood temperature; or in connection with inanimate systems, such as a steam engine provided with a governor, and other servomechanisms that are sometimes described as goal-directed.

To permit the ascription of goals in some of these cases as well, some proponents of the intentional view have broadened the notions of goal and goal-directed behavior. According to Dr. Andrew Woodfield, for example, this more inclusive notion of having a goal is an *extension* of the "core concept" of having a goal—that is, the concept of an intentional object of desire—to systems possessing "internal states" *analogous* to the internal state of wanting a goal (*ibid.*, 164). The goal-directedness of servomechanisms, he maintains, "consists in the fact that they behave as if they had desires and beliefs in virtue of the fact that they are feedback systems.... [T]he 'desired' end-state is encoded in their internal structure" (193). If an internal state of a machine is to be described as having the goal G, the internal state must not only cause the machine to behave in a manner that leads to G, but it must also "represent" G. However, Woodfield continues, "[t]here are no clear rules for deciding when an internal state is sufficiently similar to a desire to count as the state of 'having a goal'" (195). Nor does he attempt to state necessary and sufficient conditions for an internal state to count as a representation of the goal. He does not think, for example, that an electronic feedback-subsystem, which maintains a steady output of current in the main system, is an internal state of the latter sufficiently similar to a desire to count as the state "having a goal." Since there appears to be no significant material difference between such a system and either physiological homeostats or artificial thermostats, he excludes both of those from the class of goal-directed systems (196).

It is a plausible claim that the primary meaning of the phrase "being goal-directed" is stated by the intentional view; and it is not unreasonable to suppose that the current more inclusive sense of the phrase is a "metaphorical extension" of its initial meaning. In any case, since biological inquiry deals for the most part with matters in which desires and beliefs do not occur, it is obvious that were the phrase used strictly in the sense specified by the intentional view, it would have no application in most parts of biology. However, if the notion of having a goal that is not intended by anyone is held to be an impossibility, the sole consequence that this definitional resolution would most likely have is that a new word would be coined for the steady states achieved by physiological homeostats. Moreover, it seems to me dubious that there must be close analogies, as Woodfield maintains, between the items involved in the primary sense of the term "goal" and those involved in its more inclusive current usage. Nor does his requirement that there be such analogies, help to explicate the sense of goal ascriptions in biology. Consider his claim that the inner state representing a goal in an ostensibly goal-directed biological process—for example, the process in which a tadpole develops into an adult frog—must "resemble" the inner state representing a goal that is pursued in some purposive human behavior, for example, the process in which a person seeks to recover a coin that fell into a crevice. In the latter example, the inner state according to Woodfield is a complex mental state involving an intention, a desire, and a belief. But in what way does the inner state representing the goal in the former example—an inner state which is perhaps a complex subsystem of genetic materials in the tadpole—resemble (or is it analogous to) a desire or a belief? It is not clear that the question really makes sense, and in any event it is difficult to know how to begin making the relevant comparison. Moreover, even if this hurdle is jumped, and it is assumed that the goal-directed development of the tadpole might be analogous to the goal-directed efforts of human beings, the analogy could surely not be established by comparing the *inner* states themselves. What might be established is that there is an analogy between certain *behavioral* features of the two goal-directed processes—for example, it might be shown that both processes are plastic, and that both persist in the face of obstacles provided that these are not too great. In short, whatever may be the merits of the intentional view as an analysis of goal-directed behavior of purposive human beings, it contributes little to the clarification of the concepts as it is used in biology.

III

The second account of goal-directed processes I want to examine is sometimes formulated in the language of contemporary information theory, and more specifically in terms of the notion of a "coded program." I will therefore refer to it as the "program view" of goal-directed processes.

As is well known, the materials physically transmitted from parents to progeny in biological inheritance are chiefly the genes located in the chromosomes found in the nuclei of cells. The genes themselves consist of various kinds of very large molecules, among others nucleic acid molecules called DNA. The DNA molecules have a ladderlike structure, and contain four kinds of nitrogen bases in addition to other chemical groupings. Triplets of these nitrogen bases are sometimes said to be "the letters of the genetic alphabet." Provided that no mutations have occurred in the inherited genes, the sequential order of these "letters" in the long DNA molecules is then described as the "inherited information" (sometimes as the inherited "message" or "instruction") that controls the replication of the molecules, the development of the fertilized eggs, as well as the patterns of activity of the deveolped organism. Another way of stating this important conclusion of molecular biology is to say that the sequential order of those triplets is the inherited "program" that controls or places limits upon, but without determining exhaustively, the development and the numerous activities of the organism possessing it.

It is this notion of program that has been recently used to explicate the concept of goal-directed behavior; and I will examine the proposal of Professor Ernst Mayr to define it. He first distinguishes between two sorts of "end-directed" processes. Those he describes as "teleomatic" are processes that are said to be "regulated by external forces and conditions," and to achieve their end-state "in a passive, automatic way."[5] The motion of a stone dropped from a tower is an example of a teleomatic process, since the motion is governed by the external gravitational force, and its end state is reached in an "automatic" way when the stone comes to rest on the ground.

The second sort of "end-directed" processes are called "teleonomic"— the word Mayr adopts to replace the more familiar adjective "teleological," because the latter has been used to cover a miscellaneous assortment of phenomena, and has connotations many of which he wishes to exclude. He defines a "teleonomic" process or behavior as "one which owes its goal-directedness to the operation of a program" (98); and a "program" is tentatively defined as "coded or prearranged information that controls a process (or behavior) leading it toward a given end" (102). Accordingly, a process that has no programmed end cannot properly be designated as teleonomic. However, a program must not only provide "instructions" for achieving a goal; it must also "prescribe" physiological homeostats for dealing with internal and external disruptions of processes leading to the goal (99). The examples Mayr cites of teleonomic processes in biology include (either explicitly or by implication) the migration of birds, the development of zygotes into adult organisms, and the meiotic process that reduces the number of chromosomes in the sex cells of organisms to the haploid number found in the gametes.

Two additional features of Mayr's account of the program view need to be noted. In the first place, his definition of "program" is admittedly so constructed that no chasm separates ostensibly goal-directed behaviors in organisms from those of man-made machines. For example, he counts as programmed procedures both the repeated tossing of dice loaded so as to favor a given pair of sides turning uppermost, and the behavior of a clock made so as to chime on the hour (103). In the second place, he counts as programs both the *closed* programs controlling teleonomic processes in organisms that are fully contained in the DNA molecules, and the *open* programs controlling teleonomic behavior, especially in the higher organisms, that permit the incorporation of additional information acquired through learning and other experiences. For example, the *capacity* for song of the white-crowned sparrow, as well as the species-specific *skeleton* of the song, are controlled by the inherited program of these birds; but the population-specific *regional "dialect"* of the song is learned, so that in respect to the distinctive regional features of the song the program is an open one.[6] How various portions of a program come into being—whether transmitted genetically or acquired by learning—has therefore no bearing on the question whether the program leads to a "predictable goal."[7]

As has been already noted, the concept of programmed biological processes is based on some remarkable findings of molecular biology, and it introduces a unifying perspective on important features of goal-directed behavior. However, the relevant question in the present context is: What does the concept of programmed action contribute to the clarification of the notions of being a goal and being goal-directed as these are used in biology?

(a)   In the first place, it is obvious that in general we do not ascertain whether a process is goal-directed by examining the *program* that controls it. The coded information that constitutes the program is a complex physicochemical structure in the DNA molecules located in the genes of the organism under discussion; and at present, at any rate, we do not know just what is the specific physicochemical structure that corresponds to a given process. We do not know, for example, just what is the sequence of nitrogen bases in the DNA molecules that corresponds to, and is the program for, the meiotic process in cells or the singing of sparrows. Moreover, even if we did know know these correspondences, it would be quite difficult to ascertain for a given organism what is contained in its coded program. Accordingly, there must be some *other* ways of finding out than by consulting its coded program, whether a given process is goal-directed, and if it is just what is its goal. There are indeed such other ways. It is at least a plausible conjecture that an examination of these other ways is more likely to provide effective criteria for identifying goal-directed processes, than would an inspection of coded programs.

(b)   In the second place, the face that a process is controlled by a program does not suffice to make the process a goal-directed one, unless it is made so by definition. But if such a definitional maneuver is not adopted, a program will correspond to a goal-directed process only if the program is of a *special kind*. For example, the knee-jerk reflex action in normal human beings, which is manifested when the tendon below the kneecap is tapped, is presumably controlled by an inherited program. However, this is not a process that is commonly regarded as goal-directed, chiefly because (as I think) it lacks *persistence*—that is, should some internal disturbance prevent the knee jerk from taking place, even though the patellar tendon is tapped, the human body is not equipped with any mechanism for making appropriate compensatory responses to those disturbances. It is thus evident that it is possible to state at least some of the conditions a process must satisfy if it is to be a goal-directed one, *in terms of the manifest features* of the process. But no one has yet succeeded in stating, *in terms of the components and structures of DNA molecules*, what requirements a program must satisfy if it to count as one that controls a goal-directed process. Accordingly, if we do not have some idea of what goal directed behavior is—an idea that, however vague it may be, is acquired and understood independently of Mayr's formal definition of teleonomic processes—the formal definition will neither convey the idea, nor provide instructions for applying it to biological phenomena, nor clarify the notion of being goal-directed.

(c)   In the third place, although Mayr's distinction between closed and open programs is an important one, it is debatable whether he is correct in claiming that the origin of a program has no bearing on the question whether the program leads to a "predictable goal." For it is an empirical matter that cannot be settled a priori, whether predicting the goal of a process on the assumption that the process is open, is as reliable as the prediction of a goal when it is known that the process is closed. I do not know whether a serious study of this question has ever been made. It is my impression, however, that the available evidence favors a negative answer to this question. On the other hand, it is at present difficult if not impossible to ascertain by actually examining a *program*, as distinct from the over *process* it controls, whether the program is closed or open. The distinction is therefore of no help at present for deciding whether a given process has a predictable goal.

(d)   Finally, it would be pointless to explicate the notion of being goal-directed, if all processes whatever were just of that sort—that is, if processes said to be goal-directed did not differ in some identifiable respect from processes not so characterized. But although Mayr sees this clearly, it is not clear that the distinction he draws between teleomatic and teleonomic processes with the intent of explicating that difference, attains his objective.

It will be recalled that teleomatic proceses are said by Mayr to be "the simple consequences of natural laws" and are "regulated by external forces." However, if the earmark of a teleomatic process is its regulation by "external forces," then the process in which a radioactive substance such as uranium radiates energy is not teleomatic. For the process is not controlled by conditions external to the substance. Moreover, if it can be said that this process is "programmed"—despite the fact that the word would then be used, following Mayr's own practice, in a sense far more inclusive than that associated with the word in molecular genetics—the process must therefore count as teleonomic. But this is contrary to what I think is Mayr's intent, as well as to what would be generally maintained. One of his examples of a teleomatic process is the behavior of a rock dropped into a well; and one of his examples of a "programmed" (and presumably teleonomic) process is the behavior of a clock built to strike on the hour. Now the rock's behavior is surely teleomatic and not goal-directed, for it arrives at its end state "automatically." However, since the clock's behavior is also the consequence of relevant laws of nature conjoined with a number of boundary and initial conditions, it is difficult to see why this behavior should not also be described as teleomatic, and as reaching its recurrent end states automatically. I do not know how to escape the conclusion that the manner in which teleomatic and teleonomic processes are defined, does not provide an effective way of distinguishing between processes in biology that are goal-directed from those which are not. In consequence, though the program view notes some important features of goal-directed processes, it is not an adequate explication of the concept.

IV

The third account of goal-directed behavior I wish to examine makes use of some of the ideas, though for the most part not the language, of cybernetics, systems theory, and what is known as the "organismic" standpoint in the philosophy of biology. For reasons that will soon be apparent, I will refer to it as the "system-property" view of goal-directed processes.

Partial anticipations and variant expressions of some aspects of this view can be found in a number of writers; but until the publication at mid-century of a somewhat neglected book by Gerd Sommerhoff, it had no precise systematic formulation.[8] Sommerhoff's principal aim was to state the general conditions a process must satisfy if it is to be classified as goal-directed, irrespective of the particular nature of the systems involved —that is, irrespective of whether the goal is pursued by purposive human agents, by living systems incapable of having intentions, or by inanimate systems. An important and difficult part of this aim was to produce an objective and reasonably precise characterization of what it is that distinguishes processes in biology that supposedly are goal-directed from those which are not. It is easy to see that this complex aim requires that these conditions, as well as the characterization, be formulated in behavioral terms and in a highly abstract manner. In consequence, I can only sketch the central ideas of the system-property view.

Let me first recall two somewhat vaguely described behavioral features that have already been mentioned, concerning which there is good agreement that goal-directed processes have them. One feature is the *plasticity* of such processes—that is, the goal of such processes can generally be reached by the system following alternative paths or starting from different initial positions. The second feature is the *persistence* of such processes—that is, the system is maintained in its goal-directed behavior as a result of changes occurring in the system that compensate for any disturbances taking place (provided these are not too great) either within or eternal to the system, disturbances which, were there no compensating changes elsewhere, would prevent the realization of the goal. These features can be regarded as identifying marks for ascertaining whether a process does indeed have a goal, and if so what it is. At all events, it seems to me unlikely that a proposed explication of the concept of being goal-directed would be judge as adequate, if it did not incorporate these features into its analyses.

Let us now consider how the system-property view does incorporate them, by analyzing a simple example. In a normal human being the water content of the blood is about 90 per cent. This concentration remains fairly steady, despite the constant subjection of the blood to various influences that would alter that percentage were there not compensating changes in the body to prevent it. Call the maintenance of this constant concentration of water in the blood the goal G of the process. The mechanisms that achieve this homeostasis are quite complex; but the general idea of how they work can be readily understood if the simplifying assumption is made that just two sets of organs in the body are involved in that process. One set is the kidneys, which can reduced the concentration of water in the blood; and let K be a variable whose values indicate the quantity of water they remove from the blood. The second set consists of numerous muscles and the skin, all of which can release stored water into the bloodstream; and let M be a variable, whose values indicate the amount of water so released. It should be noted that these variables are independent of (or "orthogonal" to) each other, in the sense that within certain limits the value of either variable at a given moment is compatible with *any* value of the other variable at that *same* moment.[9] As will be seen presently, such orthogonality of variables is an important requirement. If now a person drinks much water, his blood does not become diluted, because the kidneys become more active (or are "adapted," in respect to the goal G, to the momentary fluctuations in the blood's water content caused by the drinking), and the value of K increases. On the other hand, if the body loses much water (for example, by sweating), water is released from the muscles and skin into the bloodstream, and M increases.[10]

However, for the process to count as goal-directed on the system-property view, it is not sufficient that on some given occasion (or even on several occasions) the kidneys "just happen" to eliminate excess water from the blood and so "happen" to keep constant the concentration of water in it. An analogous observation has to be made about the muscles and skin. To be goal-directed, the process must satisfy the much stronger requirement that *were* the blood inundated with water to a greater or lesser extent than was actually the case, the activity of the kidneys or of the muscles and skin *would* have been appropriately modified. What this amounts to is that for each member of a sequence of possible values (within certain limits) of the water content of the blood, there is a member of the sequence of possible values of K, such that for each pair of these corresponding values the goal G *would* be achieved—that is, for each such pair, the water content of the blood would be 90 per cent. And similarly for the sequence of possible values of the water content of the blood, and the sequence of possible values of the variable M. It is evident that the process is both plastic and persistent; and it could also be shown that the relevant variables are orthogonal. The system is therefore goal-directed. It is also evident that being goal-directed is a property of a *system*, in virtue of the organization of its parts.

Although in this example the goal is the product of a homeostatic mechanism, and only two variables were assumed to be relevant to the realization of the goal, the analysis can easily be generalized to cover other types of goal-directed processes involving any number of variables. But to do so would only complicate the discussion, without adding anything of importance to what has already been said. Moreover, since the system-property view is intended to cover goal-directed behavior whether it is biological or inanimate, it is not relevant in an exposition of the view to discuss either the various particular goals that may be pursued in those processes, or the mechanisms operative in them, or the origins of the mechanisms. Goal-directed processes in living systems are patently programmed, containing "instructions" for the development (among other things) of "feed-back" subsystems; and the origins of the programs are left to be explained by evolutionary theory.

A number of difficulties have allegedly been found in the system-property account of goal-directed systems. But the only really serious difficulty in that view with which I am familiar concerns the requirement that the variables relevant to the realization of the goal of a process must be orthogonal, in the sense already stated; and I want to examine this difficulty. Before I do so, however, I must briefly explain why this requirement is important: in short, it is *this* requirement that serves as a formal criterion for distinguishing processes that are goal-directed from those which are commonly held not to be such.

An example will help to clarify this claim. When a ball at rest inside a hemispherical bowl is displaced from its equilibrium position, restoring forces come into play that in the end bring the ball to rest at its initial position. Is this a goal-directed process, whose goal is the restoration of equilibrium? Were the process so classified, *every* process in which some equilibrium state is restored would also have to be designated as goal-directed; and in consequence the designation would be applicable to well nigh all processes, so that the concept of being goal-directed would not be differentiating, and would therefore be superfluous. On purely "intuitive" grounds, however, the answer to the question just raised is negative—an answer which is also in accordance with the orthogonality requirement. For the controlling variables of the ball's motion are *not* independent of each other, since the restoring force is proportional to the magnitude of the displacement force, though oppositely directed.[11] It seems, therefore, that the question whether a process is goal-directed can be decided on the "objective" grounds stated in the requirement, rather than on the basis of "subjective" intuitions that often vary from person to person.

But how are we to understand the requirement that the value of a controlling variable at a given moment is independent of the value of any other controlling variable at the *same* moment? On pain of contradiction, this surely *cannot* mean that the variables must be orthogonal in those very circumstances in which the system *actually is* in the goal-directed state relative to a given goal G. As the example of the homeostasis of the water content of blood makes plain, a system can be in a goal-directed state *only if* there are determinate relations between the relevant variables. It is just because such relations hold that the variables cannot be completely independent of each other.[12] What the requirement *does* mean is that, apart from those situations in which determinate relations hold between the variables because of their role in goal-directed processes, the known (or assumed) "laws of nature" impose no restrictions on the simultaneous values of the variables. For example, when the so-called Watts governor of a steam engine is not hitched up to the engine, any speed of the engine is compatible with any spread of the arms of the governor; for there are no known laws of nature according to which, in the assumed circumstances, the spread of the arms depends on the engine speed.

However, this presupposes that there are uncontroversial criteria for distinguishing laws of nature from laws that hold only for various specialized structures; and in any case, since the answer to the question whether variables are orthogonal depends on what laws of nature are assumed to be known, the notion of being goal-directed is also relative to that assumption. Thus, on the system property view, the system consisting of an engine with a governor is goal-directed with respect to the goal of a certain rotation speed of the engine's driving wheel. But if the relations holding between the behavior of the governor and the engine speed were included among the laws of nature, that system could no longer be so characterized. Moreover, while the behavior of the simple pendulum is *not* goal-directed relative to the assumptions of Newtonian mechanics and gravitations theory, before Newton's time that behavior might very well have counted as goal-directed.

Is this relativization a fatal flaw in the system-property view? I do not think it is. Although the answer to the question whether a given process is goal-directed is admittedly relative, in most cases it is relative to *factual assumptions* which can, at least in principle, be tested empirically. The answer is therefore not necessarily subjective or unfounded. Indeed, it is frequently possible to decide beyond reasonable doubt whether the variables relevant to a given goal are orthogonal, at any rate within the limits of certainty required in a given inquiry. For example, the two variables relevant to the analysis of a woodpecker's search for insect grubs on a tree limb, are the position of the bird's bill and the position of the insect grubs; and these variables are clearly independent in the required sense. Moreover, it does not seem to me to be a defect in an account of goal-directed behavior that it recognizes, whether explicitly or by implication, that assumptions of goal-directedness may be mistaken and need to be corrected.

V

To provide explanations for the existence of various kinds of organic structures and for the occurrence of different sorts of vital processes, is an ongoing task of biology. I want therefore to conclude this lecture with a discussion of some broad issues concerning the nature of explanations that are often proposed for, or in connection with, goal-directed processes in biology.

(a)   On the assumption that the system-property view is correct, if only in broad outline, the concept of being goal-directed can be explicated without employing in the analysis any specifically biological notions, and in particular without using any expressions that have a teleological connotation. This remark should not be misunderstood. To say that a robin is hunting for worms in order to feed its fledglings, is indeed a teleological explanation of the robin's behavior. However, the content of that statement can also be rendered by the assertion that the robin's behavior is goal-directed with respect to the goal of feeding its young; and *this* assertion, since it does not contain such typically teleological expression as "in order to," neither is a teleological *explanation* nor is it *formally* a teleological *statement*. Moreover, again on the assumption that the system-property view is sound, the assertion can be explicated in such a way (whether or not the explication counts as a correct "translation' of the assertion) that terms like "goal" and "goal-directed," which have strong teleological overtones, do not occur in the explication.

This outcome is hardly surprising. For a major objective in developing the system-property view was to analyze the concepts of being a goal and being goal-directed so that the analytic versions of these concepts would contain no undefined teleological notions. But on the hypothesis that this objective has been attained, explanations of goal-directed processes in biology are in principle possible, whose structure is like the structure of explanations in the physical sciences in which teleological notions have no place.

(b)   In the second place, explanations that have been proposed in connection with goal-directed processes, account for the presence of various items in two different ways. One such way is the explanation of how the goal (or outcome) of some goal-directed process or system is realized, in terms of assumed capacities of the system's various organs, the organization of the system's component parts, and a number of laws concerning the effects produced by the activities of those parts. For example, the homeostasis of the blood temperature in normal human beings would be explained in just this way, if the statement that this temperature is approximately constant were shown to be a consequence of assumptions which assert, among other matters, that the body has glands whose activity increases or decreases the body's metabolic rate; that moisture is evaporated in respiration and sweating; and that there are determinate relations, expressible as general laws, between these activities and changes in the blood's temperature.

Explanations of this sort are often said to be "causal." They resemble in structure, though not in specific content, typical explanations in the physical sciences. As the example just considered suggests, they are like the latter in accounting for the occurrence of some phenomenon by deriving the statement of its occurrence from assumed laws (or general hypotheses), when these are conjoined with statements of relevant initial conditions. Putting all this briefly, one sort of explanation is in terms of antecedent conditions and causal laws; and goal-directed processes, among other things, can in principle be explained in this way. Explanations of this type are not distinctive of the life sciences, they are found in all branches of inquiry, and there is nothing teleological about them.

A second sort of explanation *is* characteristic of biology and other sciences that deal with purposive behavior. These explanations do not account for a phenomenon in terms of antecedent conditions and the mechanisms that produce it. On the contrary, they account, or seem to account, for the occurrence of a process or of some other item in terms of certain *effects* these things have on the system of which they are members, or upon some other components of the system. For example, if it is taken for granted that deer pursued by predators often survive by fleeing, explanations of this second sort seem to account for the existence of various items *upon* which the survival is contingent—such as the anatomical features of deer's limbs that make swift flight possible, or the keen sense of hearing of the animal—in terms of what those items *contribute* to the survival of deer. Unlike explanations of the first type, those of this second type are often said to answer the question *why* some process or organic structure exists, or *why* it exists at just the place and time it occupies—in the sense of the particle "why" that requires mention of some purpose in the answer—by stating certain *consequences* of the process or structure. Such explanations have traditionally been called "teleological"; and it is beyond serious doubt that many biologists as well as philosophers construe them in the way just indicated.[13]

Teleological explanations are the main concern of the second lecture, and I am postponing until then further discussion of their structure. However, it has been recently argued by Professor Robert Cummins[14] that it is a misconception to suppose that the *only* way teleological explanations can be construed is as inferences from effects to causes—that is, as explanations of the *existence* of some entity in terms of certain *effects* the entity has in the system of which it is a component. He rejects this customary interpretation of teleological explanations and proposes an alternative to it. By way of constructing a bridge between the discussion of goal-directed processes in the present lecture and the issues to be examined in the second one, I will conclude with some comments on the first part of his essay.

Cummins believes that if teleological explanations are interpreted in the customary manner, they face insuperable difficulties. What these difficulties are will be ignored for the present, though they will occupy me in the second lecture. Nevertheless, it is because of that belief that he presents a different reading of such explanations. The heart of Cummins's argument is that while he thinks it is "legitimate" to reason from effects to causes—for example, from the evaporation of water to its atomic constitution, or from the homeostasis of blood sugar to the secretion of insulin in the pancreas—the reasoning is said to be "a species of inference to the best explanation." For example, "our best explanation" of the constant water content of blood can be said to "require" kidneys and muscles, so that presumably we can assert the existence of these organs.

"But once we see what makes the reasoning legitimate," Cummins continues, "we see immediately that inference *to* an explanation has been mistaken for an explanation itself. Once this becomes clear, it becomes equally clear that [the assumption that to ascribe a goal to a process is to explain the presence of the item characterized as having that goal] has matters reversed: given that [homeostasis of the water content of blood] is occurring in a particular [organism], we may legitimately infer that [kidneys are] present in that [organism] precisely because [kidneys enter] into our best (only) explanation of [that homeostasis]" (748).

Cummins is thus replacing an interpretation of teleological explanations that takes them to be explanations of the *second* type, by an interpretation that construes them to be explanations of the *first* type— though what is *explained* in the former case becomes a factor in what *does* the explaining in the latter. But how different is the outcome when the replacement is made, from the outcome prior to the replacement? On the customary reading of a teleological explanation, some effect such as the homeostasis of the blood temperature (call it E) is explained by assuming the operation of some case such as the activity of various glands (call it C). On the reading Cummins proposes, the effect E is explained by assuming some explanatory hypothesis (call it H) which is said to "require" the cause G (that is, the activity of various glands) that presumably must be supposed to exist if the hypothesis is accepted as correct. In both cases, the existence of some item C is postulated (on Cummins's proposal via the hypothesis H), whose activity is assumed to be at least a partial cause of the effect. But if this is so, the difficulties allegedly facing the customary reading of teleological explanations (according to which causes are held to be inferable from their effects) are matched by difficulties no less serious that face Cummins's proposed interpretation (according to which explanatory hypotheses are supposedly inferable from effects deducible from them).

However reasonable Cummins's proposal may be, it does not show that the difficulties associated with the customary reading of teleological explanations can be outflanked, by adopting a plausible interpretation for them that is free of comparable problems. A more extended examination of teleological explanations and their difficulties is therefore unavoidable. This is one of the tasks to be attempted in the second lecture.

Functional Explanations in Biology

As I mentioned in the first lecture, this second one will deal mainly with teleological explanations in biology. However, since the structure of explanations for goal-directed processes has already been described, the task that remains is to examine explanations in that subclass of teleological explanations usually designated as functional. I must therefore make clear what I understand by functional explanations.

Functional *explanations* are most easily described by distinguishing them from functional *statements* on the basis of differences in grammatical structure and with the help of some examples. A functional statement ascribes a *function* to some object or process. as in assertions like "The function of gills in fish is respiration," or "One function of white blood cells (leucocytes) in human bodies is to defend the body against invading microorganisms." On the other hand, a functional explanation does not *explicitly* ascribe a function to anything—indeed, it does not contain the word "function"—but accounts for the presence of some item in a system (or states *why* the item is there) in terms of the contributions the item makes to, or in terms of certain effects the item produces in, the system of which it is a component—as in the assertions "Fish have gills in order to obtain oxygen" or "Human blood contains leucocytes for the sake of defending the body against invading bacteria."

There are a number of divergent analyses of functional explanations, the differences being often attributable to different conceptions of what it is to be a function in biology. In consequence, an examination of proposed accounts of functional explanations is bound to go hand in hand with an assessment of different explications of the notion of function. I what to discuss several types of such explications that have received some attention in the recent literature.

I

The first type is a teleologically neutral definition of the notion of function—I will refer to it as the "neutral" view—such as the one proposed by Professors Walter Bock and Gerd von Wahlert. According to them, the function of a given item in an organism is the set of all the manifest as well as dispositional properties (including the physicochemical ones) the item exhibits in diverse circumstances, properties that the item possesses in virtue of its components and their arrangement.[15] It is evident that the attribution of a function to an entity, in this sense of "function," like the ascription of mass or velocity to an object in physics, has no teleological connotations. Explanations of the presence of functions so defined, will therefore have the same structure as explanations in the physical sciences, and raise no issues that are distinctive of biological inquiry.

However. Bock and von Wahlert themselves note that their definition of "function" does not express what is perhaps the generally accepted and most wide-spread sense of the word. In fact, they graft that custom-ary meaning of "function" on the term "biological role," so that in may contexts the word "function" in its familiar meaning can be used inter-changeably with the phrase "biological role," For they define the bio-logical role of a faculty in an organism as "the action or use of the faculty by the organism in the course of its life history." For example, the bio-logical role of the large mucus glands of gray jays is said to be the use of those glands "as a glue to cement food particles together into a food bolus which is then stuck to the branches of trees" (278). Accordingly, questions that may be generated in attempts to clarify the nature of explanations of biological roles—for example, under what conditions an item can properly be said to have a biological role, or what is the objec-tive of explanations of biological roles—appear to be quite similar to questions raised in the analysis of functional explanations (in the familiar sense of "function").

II

A second type of explication of the notion of biological function is based on the assumption that the primary meaning of any teleological term is the one it has when used in statements about actions that are directed by purposive agents toward achieving selected ends. Although this view resembles in some respects the account of goal-directed processes that was designated as the intentional view, the view to be examined is suffi-ciently different from the intentional view to require independent dis-cussion. I will refer to it as the "selective agency" view. According to it, teleological characterizations of nonhuman behavior constitute a "metaphorical extension" of anthropomorphic concepts. In consequence, when the term "functions" is employed in contexts in which human intentions are irrelevant (as when the *natural* function of the heartbeat in vertebrates is said to be the circulation of the blood), there must be strong *analogies* with uses of the word in contexts in which some item has been deliberately instituted or selected to behave in some specified manner (as when the governor on a steam engine is said to have the *conscious* function of regulating the speed with which the engine works). The natural function of an item is therefore not *any* effect of the item's presence; it is that *particular* effect for the production of which the item had been *selected*.

I will waive the question as not strictly relevant to the present discussion, whether it is in fact the case that the use of teleological language in nonhuman contexts is a metaphorical extension of its alleged primary use in connection with the pursuit of conscious ends. The question that *is* relevant is whether an account of functional explanations in biology is sound, when the account rests squarely on supposition that there must be strong analogies between teleological characterizations of human and nonhuman behaviors. A number of such analyses have been proposed, but there is time for discussing only one of them.

In a recent book,[16] Professor Larry Wright offers an account of functional explanations that is admittedly based on the assumption that there is such an analogy. He points out that in the case of so-called "conscious functions"—that is, functions assigned to artifacts by their conscious makers—some entity i (which may be an object, property, or process) is "introduced" into, or is identified in, a system S, *for the sake of* the effects or consequences F that i produces in S. Thus, a governor is introduced into, or made part of, an engine in such a way that the spreading and retraction of the arms of the governor affect the speed with which the driving wheel rotates. More generally, the function of an item i in system S is an effect F that i produces. However, the function of i is not *any* effect; it is *that* effect for the sake of which the item i was selected and placed in the position it actually does have. The governor of an engine produces a variety of effects; it adds to the weight of the engine, it reflects light that would otherwise have traveled in some other direction, it makes sounds when it is spinning rapidly, and it regulates the speed of the engine. Not everyone of these effects is the function of the governor;

its function is *that* effect for the sake of which the governor was constructed and made part of the engine in just the position it actually has.

In consonance with his view on the primary context of teleological language, Wright believes that explanations of natural functions have the same pattern as do explanations of conscious ones. According to him, therefore, to say that the function of the heartbeat in vertebrates is to circulate the blood, it to say two things: first, that the circulation of the blood is an effect or consequence of the heartbeat taking place in the organism; and second, that the heart is present in the animal and engaged in the activity of beating, *just because* it circulates the blood by beating. However, beating hearts do not exist in vertebrates because they were placed there by some *human* agent who selected them for their ability to circulate the blood. If we exclude the possibility of divine intervention, they occupy the place they do in vertebrates because of the operation of *natural selection.*

Wright's analysis of functional explanations therefore yields the following abstract pattern: A functional explanation of the form "The item i is in system S in order to do F" [whose content is the same as that of the functional statement "The function (or a function) of item i in system S is F"] is equivalent to the conjunction of two statements having the forms: "F is a consequence of i's presence in S" and "The item i is in S just because F is a consequence of i's presence in S." This analysis is said to be adequate for both natural and conscious functions, with the understanding that in the case of natural functions, *natural selection* takes the place of conscious choice.

Wright thus proposes what he takes to be a "causal" analysis of functional explanations. It is said to be a causal analysis, because the *causal relation* between the item i and its effect F allegedly "plays a role" in bringing about the presence of i in the system S (*ibid.*, 22). However, Wright is very careful to point out that when a function is ascribed to a *particular* item (e.g., to the heartbeat of President Ford's heart at noon on Election Day in 1976), it is not the existence of *this particular item at the stated time* that is explained by the fact that the presence of item i in S on *that* occasion produces F. Or stating this point in terms of the example, it is not the existence of President Ford's heart at noon on Election Day that is explained by the fact that the beating of his heart at *that* time produced the circulation of his blood on *that* occasion. It would be patently false if not absurd to assert the contrary. Wright's claim seems to be that the existence of hearts in general (that is, the existence of that *type* of organ) is explained by the fact that organisms with hearts that circulate the blood are more likely to survive and reproduce than organisms with hearts that do not do so—in short, that organisms with beating hearts that circulate the blood have an advantage in the evolutionary process. Or again stating this in terms of the example, on the construal just presented of Wright's claim, President Ford has a beating heart in his body, because his ancestors had beating hearts that circulated their blood—a circumstance that gave them an advantage for survival. The outcome of Wright's analysis is that when we ascribe a function to an item, we are at the same time explaining why that item is present in the system and occupies the place in the system that it actually does.

However, Wright's proposed explication raises some problems.

(i) His analysis requires us to say that F is a function of an item i if and only if the item had been selected in some way to be present in the organism just because F is an effect of i's presence. It therefore follows that F can be *asserted* to be a function of i, if and only if it is *known* (or there are good reasons for *believing*) both that F is an effect of i and that i had been selected to be present in the organism just because F is an effect of the item i. In fact, however, biologists commonly do state (often on the basis of experimental findings) that a function of some item i in organism S is F, but without knowing or believing that one *causal determinant* of i's presence in S is that F is an effect of i. For example, when William Harvey showed that one function of the heartbeat is to circulate the blood, he was unfamiliar with the etiology of the heart's formation, or with any causal determinants of the heart's presence in vertebrates. Nor is there any evidence to show that when Walter Cannon ascribed to the adrenal medula, on experimental grounds, the function of accelerating the heart's action in circumstances of emergency for the organism, he believed that the gland having this effect is a causal determinant of the gland's presence in the organism. Such examples can be easily multiplied, and are grounds for skepticism concerning the adequacy of Wright's analysis of functional explanations.

(ii) But the main reason for doubting the soundness of his analysis is the questionable validity of Wright's central claim that a functionally characterized item is "where it is," *because* the item has that function. In the first place, the claim is mistaken even in the case of *conscious* functions. Consider, for example, the functional statement that the function of the main spring in a watch is to provide power for rotating various cogwheels in the watch. On Wright's analysis, this is equivalent to saying that the spring does have this effect, and also that the spring is where it is *because* the spring has that effect. However, the second clause of the allegedly equivalent statement is surely an error. The spring was placed where it is by the manufacturer *not* because the spring is able to rotate wheels (as is required by Wright's analysis), but because the manufacturer *knew or believed* that this was so. For springs of the required sort possess that capacity whether or not anyone knows or believes this; and a spring does not appear in a watch simply as a consequence of its possessing that capacity. It is rather the manufacturer's knowledge that a spring does have the capacity which accounts in part for the spring being "where it is."

In the second place, the claim is incorrect when made for natural functions. To be correct, the presence of beating hearts in vertebrates must be accounted for by assuming that vertebrates whose hearts circulate the blood have been "selected" for survival by natural selection *just because* their hearts circulate the blood. However, in the present context the claim can be understood in two ways, each of which requires attention. On one interpretation, the claim is about the causal conditions for the existence of a *particular* heart at a *stated* time in a *given* organism. We do not at present know in sufficient detail just how the heart is formed in the development of a specified organism from a fertilized egg. But we do know that the organism has a heart because the zygote from which it developed was "programmed" to grow one—that is, we know that the zygote has a definite physicochemical composition such that, in consequence, and under normal environmental conditions, numerous (but still largely unknown) physicochemical processes take place whose outcome is the heart. It is therefore evident that neither conscious nor natural selection plays a role in the genesis of a particular heart. Nor is there any reason for believing that the heart's causal role in circulating the blood needs to be invoked in accounting for the presence of a particular heart in a given body.

On the second interpretation of the claim under discussion (namely, that vertebrates with hearts exist because hearts produce the circulation of blood), the claim is about the causal conditions for the existence of hearts in the vertebrates that have them. Wright believes that the analysis of explanations in this case closely parallels the analysis of explanations in the case of conscious functions. As he puts it, "just as conscious functions provide a consequence-etiology [that is, behavior that occurs because it brings about some specified end] by virtue of conscious selection, natural functions provide the very same sort of etiology as a result of natural selection" (38 and 84). The claim is thus made to rest on a reading of the theory of evolution according to which vertebrates with beating hearts exist, because only those vertebrates have been selected to survive by natural selection whose beating hearts circulate the blood. Wright therefore concludes that functional explanations, whether the functions are conscious or natural, all have the same structure. Moreover, since in either case consequence etiologies are asserted, functional explanations are said to be of a distinctive kind, and are not translatable into explanations of the sort customary in the physical sciences.

But is it really the case that natural selection operates, as Wright apparently believes, so as to generate consequence-etiologies—that is, to produce organs of a particular kind just *because* the presence of such organs in the organisms in which they are components gives rise to certain effects? As I see it, to suppose that this is so is to do violence to the currently accepted neo-Darwinian theory of evolution. At the risk of carrying coals to Newcastle, it may be helpful to spell out my reasons for this assertion. According to that theory, *which* heritable traits sexually reproducing organisms possess, depends on the genes organisms carry— genes that either are inherited from parent organisms or are mutant forms of inherited genes. *Which* of its genes an organism transmits to its progeny, is determined by random processes that take place during the meiosis and fertilization of the sex cells. It is *not* determined by the *effects* that the genes produce in either the parent or daughter organisms. Moreover, mutation of genes—the ultimate source of evolutionary novelty— also occurs randomly. Genes do not mutate in response to the needs an organism may acquire because of environmental changes; and which genes mutate, as well as what alterations take place in a gene when it mutates, are independent of the effects a gene-mutation may produce in the *next* generation of organisms. Furthermore, natural selection "operates" on individual organisms, not upon the genes they carry; and whether an organism survives to reproduce itself, does not depend on whether it has traits that would be advantageous to it in some *future* environment. Natural selection is not literally an "agent" that *does* anything. It is a complicated process, in which organisms possessing one assortment of genetic materials may contribute more, in their *current* environment, to the gene pool of its species than is contributed by other members of the species with different genotypes.

In short, natural selection is "selection" in a Pickwickian sense of the word. There is nothing analogous to "foresight" in its "operation": it does not account for the occurrence of organisms with novel genotypes; it does not control environmental changes that may affect the chances organisms have of reproducing their kind; and it does not preserve organisms that have traits which are *disadvantageous* to the organisms in their *present* environment, but which may be advantageous to them in *different* environments. The term 'natural selection' is thus not a name for some individual object. It is a label for continuing sequences of environmental and genetic changes in which, partly because of the genetically determined traits organisms possess, one group of organisms is more successful, in a given environment, than are other groups of organisms of the species in reproducing their kind and in contributing to the gene pool of the species. For these reasons, Wright's analysis of functional explanations in biology seems to me to be untenable.

Before leaving the subject of natural selection, it is only fair to add that the supposition that natural selection is more than a negative "sifting" process, and is "creative" in "directing" genetic changes, is endorsed by distinguished students of organic evolution. For example, the late Professor Theodosius Dobzhansky disagreed strongly with those biologists who "doubted that natural selection can be the guiding agent in evolution because selection, allegedly, produces nothing new and merely removes from the population degenerate variants and malformations."[17] He maintained that although the raw materials of evolution are the genotypes that arise by gene and chromosome mutation and recombination, nevertheless "it is selection which gives order and shape to the genetic variability, and directs it into adaptive channels" (131). Indeed, he went on to say that "It is fair to say that selection *produces* new genotypes, even though we know that the immediate causes of the origin of all genotypes are mutation and reproduction.... In the long run, selection is the directing agent because it determines which genotypes are available for new mutations to occur in."[18]

A more moderate presentation of the "agent" conception of natural selection is given by Professor Ernst Mayr. "Evolution is not an all-or-none process," he writes "Genetic variation is enormous and does *not* consist merely in the production of a few new types; it 'selects' them precisely in the same way in which a breeder 'selects' the founder individuals for the next generation of breeding. This is a thoroughly positive process; inferior zygotes are simply lost. We do not hesitate to call a sculptor creative, even though he discards chips of marble. As soon as selection is defined as differential reproduction, its creative aspects become evident. Characters are the developmental products of an intricate interaction of genes, and since it is selection that 'supervises' the bringing together of these genes, one is justified in asserting that selection creates superior new gene combinations."[19]

It is little wonder that readers of such passages are persuaded that natural selection is an agent operating in a manner closely similar to the action of a conscious agent, such as the action of an animal breeder. But it is fairly clear that not everything in the quoted passages is intended to be taken literally, as the placing of quotation marks around some of the words in them strongly suggests. It is surely not natural selection that literally produces genotypes, for this is done by the mechanisms involved in cell division and reproduction. Nor is it natural selection that does any directing, but it is the environment together with the genotype of an organism which determine whether the organism can survive long enough to reproduce its kind. And it is certainly not the case that natural selection "selects" individuals for survival "precisely in the same way in which a breeder 'selects' the founder individuals for the next generation of breeding." For a breeder deliberately selects the animals he will mate, on the basis of what he knows of the likelihood that the traits for which he is breeding will appear in the *next* or other *future* generation of the mated animals. But natural selection has no eye to the future; and if any zygotes are eliminated by natural selection, it is because they are not adapted to their *present* environment.

III

I must now turn briefly to an account of the function concept which in one respect is like the view just discussed, but which differs from it in construing functional ascriptions as having a purely methodological or heuristic role. I will refer to it as the "heuristic view" of functional ascriptions.

This view resembles the Kantian interpretation of teleological attributions; but it has contemporary adherents as well, and its distinctive thesis can be stated independently of Kant's elaborate conceptual machinery. Kant was an heir to the thought of Descartes and Newton, and subscribed to the principle that all material processes of nature must be explained by "merely mechanical laws." However, the apparently purposive character of the organization and behavior of living things seemed incapable of being understood in terms of "the mere mechanical faculty of motion."[20] and they had to be viewed *as if* they had been produced by design. Kant therefore formulated a second principle, that *some* events cannot be explained on the basis of purely mechanical laws. But these two principles appear to be incompatible; and according to Kant, they would indeed be contradictory *if* they were assertions about the *objective constitution* of nature. He resolved the antinomy by construing the principles to be "maxims" or regulative principles for guiding inquiry. According to him, the first principle does not demand that the events of nature be investigated *only* within the framework of mechanical laws. On the contrary, accepting that maxim "does not prevent us, if occasion offers, from following out the second maxim in the case of certain natural forms ... in order to reflect upon them according to the principle of final causes." On the other hand, though we may follow the second maxim in dealing with biological phenomena, this does not exclude the possibility that after all living things have been produced in accordance with purely mechanical laws. Indeed, in investigating biological organisms *as parts of nature*, we must go as far as we can in our efforts to understand them in terms of mechanical laws; for unless we do so, "there can be no proper knowledge of nature at all" (295). The conclusion to be drawn from all this is that since we cannot really understand how final causes operate except in the case of our own actions, ascriptions of goals and functions to nonhuman organisms and their parts cannot be taken literally, as objective assertions about nature. They must be construed as statements that have only a heuristic value in guiding inquiry into the mechanisms of living organisms (279–281).

Something like Kant's views on this subject is also present in C. D. Broad's definition of teleology. "Suppose," he wrote, "than a system is composed of such parts arranged in such ways as might have been expected *if* it had been constructed by an intelligent being to fulfil a certain purpose which he had in mind. And suppose that, when we investigate the system more carefully under the guidance of this hypothesis, we discover hitherto unnoticed parts or hitherto unnoticed relations between the parts, and that these are still found to accord with the hypothesis. Then I should call this system 'teleological,'"[21] Broad believed that living organisms are teleological systems in the sense of his definition. However, he also maintained that the "intelligent being" needed to design and produce the complex systems that constitute living organisms, would have to possess powers of intellect far beyond anything displayed by minds with which we are familiar. It is therefore unclear just what is assumed in the hypothesis that enters into Broad's definition of teleological systems—that is, it seems impossible for minds like our own to know what *would* be the arrangement of parts of organisms that such a superhuman intelligence instituted to achieve his purposes. It is in consequence also unclear how Broad could have known that living organisms really are teleological systems.

Both Kant and Broad were therefore agnostics about the literal truth of any functional ascription. In both cases, the agnosticism has its source in the assumptions that a process cannot properly be characterized as purposive, if it can be explained on the basis of physicochemical laws, and that an effect on organic process can count as one of its biological functions only if that process was *intended* or *designed* to produce the stated effect. Agnosticism concerning the truth of function ascriptions seems to be the price that must be paid for explicating the notion of biological function in terms of conscious intent. But if the notion of being a function can be explicated, as I believe it can be, with no reference to the intentions or choices of conscious organisms, the validity of ascribing a function to an entity can be decided by empirical inquiry, without any need for the elaborate make-believe that seems to be an inseparable accompaniment of the heuristic view.

IV

An explication of the notion of biological function that seems far more plausible than those discussed thus far, rests on the assumption that functions contribute to the "welfare" (in a sense to be specified) of either individual organisms, or populations of organisms, or the species to which an organism belongs. I will call it the "welfare" view of biological functions. There are several varieties of this view; and commenting on two of them will enable me to state my own views on the subject more clearly.

(a)   Perhaps the best known and most carefully articulated critique of functional explanations is that of Professor Carl Hempel. Since his evaluation of them has been highly influential, it is appropriate to examine his analysis once more. His analysis is quite familiar to students of the subject; a brief summary of the salient points in his account will therefore suffice.

According to Hempel, the cognitive content of the functional statement "The function of the heartbeat in vertebrates is the circulation of the blood" is stated more explicitly in the following: "The heartbeat has the effect of circulating the blood, which ensures the satisfaction of certain conditions (e.g., supply of nutriment and removal of wastes) that are necessary for the proper working of the organism."[22] Just what is to be understood by "the proper working of an organism" and by the "certain conditions" that are allegedly required for such working, are important questions; but they can be waived for the present. The basic pattern of a functional *statement* is therefore "The function of item i occurring in organism (or system) S during period t and in environmental setting C, is to do e"; and the schematic form of the import of such statements is "Item i in system S during period t and in environment C, has the effects e

that satisfy the conditions n which are necessary for the proper working of S" (306).

On the other hand, on the assumption that a functional *explanation* must account for the presence of item i in S that is, that it must explain why it is that during t and within C, item i is present in S—Hempel proposes the following schematic form for the explanatory premises: (i) During period t and in environment C, S is in proper working order. (ii) If S is in proper working order, then condition n must be satisfied. And (iii) if i is present in S, then the effect e of i's presence in S satisfies the required condition n. [Using Hempel's initial example, the explanatory premises for the presence of a heart in a given human body are as follows: (i) During a certain period and under normal environmental conditions, a certain human being is in proper working order. (ii) If that person is in proper working order, then various conditions, such as his having nourishment, must be satisfied. And (iii) if the person has a beating heart in his body, the circulation of his blood, which is an effect of the heart's beating, satisfies the required conditions.] However, if this is a correct analysis of functional explanations, the assumed premises do *not* entail the desired conclusion, and so the presence of item i in organism S is *not* explained.

Hempel points out that the flaw in the argument would be removed if the third premise were replaced by its converse (iii')—that is, by a statement of the form: The condition n is satisfied *only if* item i is present in S. He believes, however, that in general there is no warrant for doing this, Indeed, he thinks that "it might well be that the occurrence of any one of a number of alternatives would suffice no less than the occurrence of i to satisfy n" (310). But if this is so, the explanatory premises fail to explain why it is item i, rather than one of its possible alternatives, that is present in S. Hempel therefore concludes that while functional characterizations may have considerable heuristic merit, they have little if any explanatory or predictive value (313). The main burden of Hempel's devastating critique of functional explanations is that the presence of some specified item in an organism, which is to be explained in terms of its function, is in general not a *necessary condition* (or is not known to be a necessary condition) for the performance of that function. Accordingly, the presence of a heart in vertebrates is *not* explained by the fact that it has the effect of circulating the blood, because the heart is not the *sole* thing that can do this; for example, artificial pumps when properly connected with the blood vessels could also perform this function.

It is of course beyond dispute that *if* the structure of functional explanations is as Hempel describes it, these explanations fail to explain their ostensible explananda. However, it is questionable whether functional explanations, at least in biology, do in general have the form he indicates. For example, a convincing case can be made for the claim that in normal human beings—that is, in human bodies having the organs for which they at present genetically programmed—the heart *is* necessary for circulating blood; for in normal human beings there are in fact no alternative mechanisms for effecting the blood's circulation. For physiologists seeking to explain how the blood is circulated in normal human bodies have discovered that human bodies have no organs other than the heart for performing that function. The observation that it *may* be (or actually *is*) physically possible to circulate blood by means of other mechanisms is doubtfully relevant to those investigations of how the blood is circulated in normal human beings, upon which physiologists were once embarked.

The denial of the claim that the heart is necessary for circulating the blood appears to derive part of its plausibility from the imprecise way in which the expression "human body"—and more generally the expression "the system *S*"—is usually specified. Is the normal human body to be counted as the same system as the body whose blood is being circulated by a mechanical pump? The issue Hempel's analysis raises is one that has long been discussed in connection with the "doctrine of the plurality of causes." For example, since death can result from drowning, from gun wounds, from poisons, and so on, it is sometimes said that death has a plurality of causes. But, as has been often noted, the doctrine is plausible in this example, only because the causes of death have been analyzed more precisely and into a larger number of types than has their effect. However, if the state of a body whose death has been caused by drowning is compared with a body whose death was the result of gun wounds, it is clear that each of these causes has its own distinctive effect; and it no longer seems so evident that death has many causes. Something like this point seems to be involved in the denial that the heart is necessary for circulation, except that it is the loose way in which the expression "human body" (or "the system *S*") is used that may underlie the denial.

It must of course be recognized that there are systems in which several items have a common function, so that no particular member of the set of such items is necessary for performing that function. Human beings normally have two ears, either of which, by itself, is necessary for hearing. Nevertheless, in the normal body it is still necessary that one or the other or both ears be present if the organism is to hear. In such cases. functional explanations account for the presence in the system under discussion of a *set* of items.

It seems to me that Hempel should agree, at least in principle, with this defense of premises in functional explanations which assert that some item in a given system is necessary for the performance of a stated function. Despite his doubts that such premises can be validly asserted, he does assume that there are *necessary conditions* (such as the elimination of wastes) for the proper working of organisms. He must therefore have had in mind organisms which are actually found in nature in determinate environments and which must satisfy certain conditions if they are to flourish. For if we are free to exercise our imagination and deal with were possibilities, with no limitations placed on the kinds of organisms that may be considered, organisms can be imagined that produce no waste materials and have, in consequence, no need for eliminating them. However, if necessary conditions can be discovered for the "proper working" of organisms in their natural state, what reasons are there for doubting, on general principle, that certain organs and other parts of organisms may be necessary for the performance of the functions that are associated with those organs and parts? In point of fact, examination of standard treatises on the physiology of the human body, shows that the great majority of its organs and parts *are* necessary for the performance of their several functions.

Let me return briefly to Hempel's analysis of functional explanations. As has already been noted, it was not part of the task he set for himself to provide an adequate account of what is to be understood by "the proper working" of an organism, or how the "necessary conditions" for such "proper working" are ascertained. Why then did he place a limitation, involving these notions, on the effects of an item that are to count as the item's function? The main reason for his doing this was to exclude as incorrect function ascriptions which, despite the fact that the alleged functions are the effects of stated items, are strongly counterintuitive. Hempel's example of such a counterintuitive statement is: "A function of the heartbeat in human beings is to produce heart sounds."

However, the validity of the claim that this statement is a mistaken attribution of function—as all function ascriptions are alleged to be which identify a function of an item with *any* of its effects—depends on whether, *in a given environment*, the production of heart sounds satisfies a necessary condition for the proper working of organisms possessing hearts. It is therefore arguable that although in other environments and other periods heart sounds contributed nothing to the proper working of human beings, this is not true in the present environments of human beings, since heart sounds have a diagnostic value for modern physicians. But be this as it may, the requirement Hempel places on effects if they are to count as functions, does enable him to achieve his objective (in principle and at least in part) of excluding as incorrect many counterintuitive attributions of function—provided, of course, that saying that an organism is in proper working order is not just another way of saying that the various parts of the organism are performing their functions. To be sure, without a detailed account of what the proper working of an organism is—an account that is bound to require considerable biological, medical, and perhaps even psychological knowledge—it is difficult to apply Hempel's criterion for distinguishing functions from mere effects. On the other hand, just because he leaves largely unspecified what is to be understood by "the proper working" of organisms, his analysis of functional statements has a generality that many other analyses lack. In consequence, many of these other analyses can be subsumed under, and regarded as special cases of, Hempel's formulation of the structure of functional explanations.

(b)   I want now to examine the version of the welfare view of biological function proposed by Professor Michael Ruse.[23] He presents his analysis in the context of a critique of some views of my own. An examination of his statement of the issues will enable me to assess both his own analysis and his criticism of mine.

(i) In presenting my views on biological functions which Ruse criticizes, I argued that the statement: "The function of chlorophyll in plants is to enable plants to perform photosynthesis" is equivalent to another one that no longer contains any functional terms: "When a plant is provided with water, carbon dioxide, and sunlight, it manufactures starch only if the plant contains chlorophyll." And I added the proviso that functional ascriptions presuppose that, and are appropriate only if, the system under consideration (in the example it is a plant) is "directly organized" or "goal-directed." I also explained that the term "goal-directed" was to be understood in the sense of the system-property view of the notion of being goal-directed.[24]

Ruse believes that this account of functional statements is "fundamentally misconceived," and states two objections to it. The first is directed against the assumptions that the presence of chlorophyll in plants is a necessary condition for the performance of photosynthesis. Neither Ruse nor I think this is a weighty objection, and I have already stated my reasons for believing this. On the other hand, I also think that Ruse may be correct in saying that it is "somewhat unfortunate" I placed so much emphasis on the *necessity* of an item's presence for the performance of a stated function. For the emphasis can be construed as an oversight of cases in which an organism is known to have more than one organ for the performance of a function (as in the case already discussed of organisms having two ears for hearing), as well as an oversight of cases concerning which it may be unknown whether more than one organ is present to perform the function. In such cases, it is not the presence of some *single* item that is explained, but rather the presence of one or more members of a *set* of items. An explanandum of the latter sort is in a sense weaker than an explanandum consisting of a single item; but though it is weaker, it is not necessarily a trivial one.

(ii) Ruse's second objection is that the requirement mentioned in my proviso, according to which the use of functional statements presupposes that the systems to which they are applied are goal-directed, is "altogether inappropriate." He offers the following example in support of this judgment. Suppose it were true that long hair on dogs harbors fleas, and also that dogs are goal-directed toward survival. Ruse believes that the conditions stated in my account for something to be a function are satisfied in that case, so that I am committed to saying that a function of long hair on dogs is to harbor fleas. But he thinks this functional statement is strongly counterintuitive, and that no one is likely to assent to it.

On the other hand, he believes that we would be very much inclined to accept that functional statement *if* it were the case that dogs with more fleas receive more fleabites, and that fleabites provide dogs with immunity from a certain parasite whose presence in dogs without fleabites lowers their life span. The reason we would be inclined to accept the statement in this case, Ruse thinks, is that, on the assumed evidence, harboring fleas contributes significantly to the survival and reproduction of long-haired dogs. In short, on the stated hypothesis, harboring fleas is what biologists call an"adaptation," which confers an adaptive advantage on long-haired dogs. Generalizing from this example, Ruse maintains, that to say that the function of item *i* in organism *S* is to do *F*, is to say two things: first, the organism *S* does *F* by using item *i*; and second, *F* is an adaptation—that is, *F* contributes to the survival and reproductive activity of *S*. He therefore concludes that it is a mistake to say, as I did, that a functional statement presupposes that the system about which it is made is goal-directed.

Let me first comment on this second objection. In my opinion, this criticism rests partly on a misunderstanding for which I am largely responsible. I did indeed say that a functional statement "presupposes" that the system under consideration is directively organized or goal-directed (421). (I did *not* say that a functional statement *implies* that the system is goal-directed, which is the way Ruse states my view. Although the distinction between implying and presupposing is important, I do not think that much hangs on this distinction in this portion of Ruse's criticism. I will therefore ignore those parts of his objection that are based on a conflation of the two terms of the distinction.) In any case, what I did say appears to warrant Ruse's claim that, to be consistent, I must accept as correct his allegedly counterintuitive statement that the function of long hairs on dogs is to harbor fleas.

However, I also said, at other places in my analysis of functional notions, that functional statements not only presuppose that the systems under discussion are goal-directed, but also that the function ascribed to an item *contributes* to the realization or maintenance of *some goal* for which the system is directively organized (408, 422). I failed to stress this second part of my requirement, and was therefore insufficiently clear in presenting my analysis. However this may be, I want now to make explicit in a semi-formal manner what I think is a presupposition in the application of functional statements. A functional statement of the form: a function of item *i* in system *S* and environment *E* is *F*, presupposes (though it may not imply) that *S* is goal-directed to *some* goal *G*, to the realization or maintenance of which *F* contributes. I will call this account the "goal-supporting" view of biological functions. Since in Ruse's initial example harboring fleas apparently does not contribute to the maintenance of any goal for which dogs are goal-directed, I do not believe that I am committed to holding that the function of long hair on dogs is to harbor fleas. There may be difficulties in my account of functional statements, but the one Ruse mentions does not seem to be one of them.

One difficulty I do recognize is that *every* effect of an item will have to count as one of its functions, *if* it should turn out that *each* effect contributes to the maintenance of *some goal or other*. Although I have no reason for thinking that this is the actual situation for any organism, I do not know how to eliminate this possibility. However, even if it should be the case that each effect of an item contributes to some goal, it would not follow that the notion of being a function would not be differentiating— that is, it would not follow that every effect of an item would be a function *simpliciter*. For on the goal-supporting view, being a function is relative to *some goal*, but not necessarily relative to the *same* goal.

The goal-supporting view of functions seems to me to be compatible with, but to be more general than, the accounts of both Hempel and Ruse. Ruse appears to deny that this is so. On the strength of the familiar Fregean distinction between meaning and reference, he maintains that the statement that an organism has an adaptive trait *does not imply* that the organism is goal-directed, even if, *as a matter of fact*, all actual organisms were goal-directed for survival. However, since the notion of what is an organism has not been precisely defined in the present discussion (or for that matter in any other discussion), it is endlessly debatable, though perhaps impossible to decide, whether or not there is that implication. In any event, it is not essential that there be that implication, for the claim to be correct that the goal-supporting view of functions is more general than those proposed by either Hempel or Ruse. It is sufficient, to establish that claim, that in point of fact all biological organisms are directively organized with respect to *some* goal.

(iii)   I must comment briefly on Ruse's own thesis that what are called biological functions of various items in organisms are adaptations. It is certainly true that many effects of items that are generally called functions are adaptations. But this does not seem to be so invariably. Whether a designated feature of an organism is an adaptive trait, depends on the environment in which the organism lives. In consequence, if the environment is changed, a feature that was adaptive in the earlier environment may not be adapted in the altered one. For example, the fur of polar bears helps prevent heat loss in the animal, so that in arctic regions possession of heavy fur has an adaptive value for the animal. But what if the environment of polar bears were changed, whether because of long-lasting climatic changes in the polar regions or because of a migration of polar bears to other climes? In that eventuality, possession of heavy fur may no longer contribute to the survival and reproduction of the bears, although it might still be maintained that one function of the fur is the prevention of heat loss in those animals.

Moreover, it is not incompatible with currently accepted evolutionary theory to suppose that the appearance of an adaptive trait in an organism is strongly coupled with the appearance of another trait which *is* not, or *is not known* to be adaptive, but comes to be designated nonetheless as a function of some item. More generally, biologists frequently succeed in ascertaining effects of some item which they designate as the item's functions, without being at all sure that those effects contribute anything to the survival of organisms possessing the item. For example, certain genes of the yellow onion produce the yellow color of the plant, so that the production of this color is a function of those genes. However, although yellow onions are resistant to a fungus disease while white onions are susceptible to it, the *color* of yellow onions appears to have no adaptive value in itself.[25] This objection to Ruse's thesis is perhaps not a fatal one. But the objection does suggest that making adaptedness the criterion of a trait being a function, is not always congruous with biological practice.

In conclusion, I want to state in summary manner the main outcome of these lectures that bears on the nature of functional explanations. In the first place, if the system-property account of goal-directed processes is sound, goal ascriptions can be explicated without employing any teleological notions in the explication; and goal ascriptions can be explained in a manner that is structurally identical with explanations in the natural sciences. And in the second place, if the goal-supporting view of biological functions is correct, functional statements, as well as the presuppositions of functional ascriptions, can also be rendered without using functional concepts; and functional explanations can be shown to have the same structure as explanations in the physical sciences.

However, one further question must be faced. It has been argued that explanations of both goal and function ascriptions are *structurally* similar to *causal* explanations in the physical sciences. Are the former two kinds of explanation also *causal* accounts, in one case of goal-directed activity, and in the other case of the presence of some item to which a function is attributed? The two cases require separate discussion.

(a)   It was noted earlier (214 above) that a statement of the form "The system S is goal-directed with respect to the goal G" (for example, "The normal human body is goal-directed with respect to the homeostatis of the blood temperature") is explained if it is shown to be a logical consequence of an assumed set of explanatory premises, some of which are held to be laws (for example, premises that include assumptions such as that the human body possesses adrenal glands whose activity affects the body's metabolic rate, that moisture is evaporated when the body sweats, and that these activities produce changes in the temperature of the blood.) A number of these premises are causal laws, which state just how the goal G is related to various antecedent conditions. Explanations of goal ascriptions are therefore *causal*.

(b)   Explanations of function ascriptions cannot be characterized in the same way. This will be evident from a consideration of a function ascription having the form: "During a given period t and in environment E, the function of item i in system S is to enable the system to do F"—for example, "During a period when green plants are provided with water, carbon dioxide, and sunlight, the function of chlorophyll is to enable the plants to perform photosynthesis." The explanatory premises for the assertion having the form "The item i occurs in S during a given period t

and circumstances E"—for example, "During a stated period and given circumstances, chlorophyll is present in a specified green plant"—are as follows; (i) "During a stated period, the system S is in environment E (for example, "During a stated period, a green plant is provided with water, carbon dioxide, and sunlight"); (ii) "During that period and in the stated circumstances, the system S does F" (e.g., "During the stated period, and when provided with water, carbon dioxide, and sunlight, the green plant performs photosynthesis"); (iii) "If during a given period t the system S is in environment E, then if S performs F the item i is present in S" (e.g., "If during a given period a green plant is provided with water, carbon dioxide, and sunlight, then if the plant performs photosynthesis the plant contains chlorophyll"). It is obvious that "The system S contains the item i during the stated period and in the specified circumstances"—in the example, "Chlorophyll is present in the given green plant"—follows from the premises. The first two premises are instantial statements, and the third is lawlike. However, the performance of F (photosynthesis) is not an *antecedent* condition for the occurrence of the item i (chlorophyll), and so the premise is not a causal law. Accordingly, if the example is representative of explanations of function ascriptions, such explanations are *not* causal—they do not account causally for the presence of the item to which a function is ascribed.

What then is accomplished by such explanations? They make explicit *one* effect of an item i in system S, as well as that the item must be present in S on the assumption that the item does have that effect. In short, explanations of function ascriptions make evident one role some item plays in a given system. But if this is what such explanations accomplish, would it not be intellectually more profitable, so it might be asked, to discontinue investigations of the *effects* of various items, and replace them by inquiries into the *causal* (or antecedent) conditions for the occurrence of those items? The appropriate answer, so it seems to me, is that inquiries into effects or consequences are as legitimate as inquiries into causes or antecedent conditions; that biologists as well as other students of nature have long been concerned with ascertaining the effects produced by various systems and subsystems; and that a reasonably adequate account of the scientiflc enterprise must include the examination of both kinds inquiries.

(c)   None of these conclusions concerning the character of explanations of goal and function ascriptions shows that the laws and theories of biology are reducible to those of the physical sciences, although if the conclusions really do hold water, they undermine one objection that is sometimes made to the possibility of such a reduction. What I think those conclusions do establish, is that teleological concepts and teleological explanations do *not* constitute a species of intellectual constructions that are inherently obscure and should therefore be regarded with suspicion.

Notes

1. *Logic: The Theory of Inquiry* (New York: Holt, 1938), p. 19.

2. *The Influence of Darwin on Philosophy, and Other Essays in Contemporary Thought* (New York: Holt, 1910), p. 9.

3. Cf. George C. Williams, *Adaptation and Natural Selection* (Princeton, N.J.: University Press, 1966), pp. 8/9.

4. Andrew Woodfield. *Teleology* (New York: Cambridge, 1976), p. 104.

5. "Teleological and Teleonomic: A New Analysis," in R. S. Cohen and M. W. Wartofsky, eds., *Methodological and Historical Essays in the Natural and Social Sciences*, vol. xiv of *Boston Studies in the Philosophy of Science* (Boston: Reidel, 1974), p. 98.

6. E. O. Wilson, *Sociobiology* (Cambridge, Mass.: Harvard, 1975), p. 157.

7. Mayr, *op. cit.*, p. 104.

8. *Analytical Biology* (New York: Oxford, 1950).

9. Sommerhoff, *The Logic of the Living Brain* (New York: Wiley, 1974), ch. 4.

10. Walter B. Cannon, *The Wisdom of the Body* (New York: Norton, 1939), ch. iv.

11. This example is discussed in Sommerhoff, *Analytical Biology*, p. 99. An elementary mathematical analysis of an analogous problem is given in John Cox, *Mechanics* (New York: Cambridge, 1923), pp. 210/1.

12. Woodfield, *op. cit.*, pp. 67–72.

13. Cf. Francisco Ayala, "Biology as an Autonomous Science," *American Scientist*, lvi (1968): 214.

14. "Functional Analysis," this JOURNAL, lxxii, 20 (Nov. 20, 1975): 741–765.

15. "Adaptation and the Form-Function Complex," *Evolution*, xix (1965): 274.

16. *Teleological Explanations* (Berkeley: Univ. of California Press, 1976).

17. *Evolution, Genetics, and Man* (New York: Wiley, 1955), p. 130.

18. *Ibid.*, p. 132. Cf. also F. J. Ayala, "Teleological Explanations in Evolutionary Biology," *Philosophy of Science*, xxxvii, 1 (March 1970): 1–15.

19. *Populations, Species, and Evolution* (Cambridge, Mass.: Harvard, 1963), p. 119.

20. *The Critique of Judgement*, J. H. Bernard, trans. (London: Macmillan, 1931), p. 278.

21. *The Mind and Its Place in Nature* (London: Paul, Trench, Trubner, 1925), p. 82.

22. *Aspects of Scientific Explanation* (New York: Free Press, 1965), p. 304.

23. *The Philosophy of Biology* (London: Hutchinson, 1973), ch. 9.

24. *The Structure of Science* (New York: Harcourt, Brace & World, 1961), ch. 12.

25. Verne Grant, *The Origin of Adaptations* (New York: Columbia, 1963), p. 117.

8

Functions

John Bigelow and Robert Pargetter

In describing the function of some biological character, we describe some presently existing item by reference to some future event or state of affairs. So the function of teeth at time t is to pulp food at time t', where $t' > t$. This seems to present an exact parallel to the case of the function of humanmade artifacts; for instance, the function of the nutcracker at time t is to break open nuts at time t', where $t' > t$.

In the case of biological functions and in other cases of functions where human intentions are not obviously causally active, this kind of description of function has been difficult to assimilate into our scientific view of the world. There are several reasons, but we shall here concentrate on one which arises directly from the fact that, in describing a present structure in terms of its function, we mention a future outcome of some sort. The future outcome may be, in many cases, nonexistent. A structure may never be called upon to perform that function. The function of a bee's sting, for instance, is relatively clear; yet most bees never use their stings. Likewise teeth may never pulp food, just as nutcrackers may never crack nuts.

Thus, when we describe the function of something in the present, we make reference to a future event or effect which, in some cases, will never occur. Hence, prima facie, we cannot really be describing any genuine, current property of the character.

I The Problem

Even when a character does perform its supposed function, the future events that result from it cannot play any significant "scientific" role in *explaining* the nature and existence of the character. The character has come into existence, and has the properties that it does have, as a result of prior causes. It would still have existed, with just the current properties it does have, even if it had not been followed by the events that constitute the exercise of its alleged function. Hence its existence and properties do not depend on the exercising of its function. So it is hard to see what explanatory role its functions could have. Crudely put: backwards causation can be ruled out—structures always have prior causes—hence reference to future events is explanatorily redundant. Hence functions are explanatorily redundant.

Of course, there is nothing inappropriate about describing a character *and mentioning* its future effects. But describing a character as having a function is not just mentioning that it has certain effects. Not every effect counts as part of its function. And some functions are present even when there are no relevant effects to be mentioned—as with some bees and their stings. Future events are not unmentionable; but they are explanatorily redundant in characterizing the existence and current properties of a character. Hence, what role can functions have in a purely scientific description of the world: how can they be "placed" within the framework of current science?

There are three main theories that attempt to construe functions in a way that allows them to fit smoothly into the scientific, causal order. We believe that each is nearly right or partly right. Yet they are all unsatisfactory in one crucial respect: they do not restore to functions any significant explanatory power. In particular, they deny to functions any causal efficacy. So, for instance, they will not permit us to explain the evolution of a character by saying that it evolved *because* it serves a specific function.

We will offer an account of biological function and of functions generally which, although it shares much with the most promising extant theories, is crucially different from them in that it bestows greater explanatory power upon functions.[1] But first we will briefly consider the three theories.

II Eliminativism

There are three responses that arise naturally in the face of the tension between functions and the scientific standpoint. The first is eliminativist.

It is assumed that functions, if there were any, would have to be important, *currently existing*, causally active, and explanatory properties of a character or structure. It is also assumed that functions, if there were any, would essentially involve reference to future, possibly nonexistent, events. Yet something involving essential reference to future, possibly nonexistent events could not possibly characterize currently existing, explanatory properties. It is thus concluded that there really ae no functions in nature.

To add functions to the scientific biological picture, on this view, is parallel to adding final causes to physics. Final causes have no place in the scientific account of the physical universe, and, if the psychological pressures are resisted, we find we can do without them and final causes just fade away. To the eliminativist, the same will be true of functions; as the biological sciences develop, any need for function talk will vanish, and the psychological naturalness of such talk will fade away with time and practice.

A variant on this eliminativist view adds an account of why the attributions of function seem to serve a useful purpose in everyday and scientific discourse. The eliminativist does not believe in functions as genuine, currently existing properties of a character. But the eliminativist does believe in future effects of a character. And nothing stops us from mentioning whichever future effects we take an interest in. Consequently, an eliminativist can interpret talk of "functions" as being merely the specification of effects one happens to be interested in. Which effects are deemed to relate to "functions" of a character, will depend not on the nature of the character itself, but on our interests. The function of kidneys is different for the anatomist from what it is for the chef. Insofar as function talk makes sense, it does *not* describe the current nature of a character (there are no functions in nature); rather, it relates a current character to a future outcome, in an interest-dependent, extrinsic manner.[2]

The best answer to an eliminativist theory is to come up with an adequate analysis of functions still within the scientific view. This is what we attempt to do later in this paper.

But a motive for seeking such a noneliminativist account can be cited; and this motive will also provide a less than conclusive, but nevertheless weighty, argument against eliminativism. In the biological sciences, functions are attributed to characters or structures, and these attributions are intended to play an explanatory role which cannot be squared with the eliminativist's account of function talk. For instance it is assumed that biological structures *would* have had the functions they do have even if we had not been here to take an interest in them at all. And some of the effects of structures that we take an interest in have nothing to do with their function. And some functions are of no interest to us at all. Furthermore, biology standardly treats function as a central, explanatory concept. None of this rests easily with an eliminativist theory. A powerful motive for resisting eliminativism, then, is that an adequate analysis of functions, if we can find one, will enable us to take much biological science at face value: we will be relieved of the necessity of undertaking a radical reformation of the biological sciences. The eliminativists' vision, of functions "fading away," is as yet just a pipe dream; and their explanation of the apparent usefulness of function talk fails to explain away more than a fragment of the uses functions serve n biological science.

It is not the biological sciences alone which could be cited here. Psychology could be canvassed too. And even physics has facets that raise problems for an eliminativist view. Suppose someone were to suggest that the function of water is to refract light and the function of mists to create rainbows. Presumably this is plainly false. Yet it describes something in terms of future effects in which we take an interest ... which is exactly what the eliminativist takes to be the business of function talk. So eliminativists have no good explanation of why the physicist takes such attributions of function to be plainly false. They cannot explain the manifest difference between "The function of mists is to make rainbows," and "The function of teeth is to pulp food."[3]

III   Representational Theories

There is a response to the tension between the scientific view and functions which rejects any role for future events in the characterizing of functions. The future effects of a character do not themselves play an explanatory role in characterizing that character. Yet sometimes there exists, prior to the character, a *plan*, a representation of that character *and* of its future effects. Such a representation of future effects may exist, whether or not those effects ever come to pass. And this representation exists prior to the character and so contributes to the causal processes that bring that character into being—by the usual, forward-looking, causal processes that rest comfortably within our over-all scientific image of the world. On this view, we can account for a function not by direct reference to any future event, but rather by reference to a past representation of a future event. Theories of this sort have frequently been called "goal theories"; but they are best construed as a subcategory within the class of goal theories. They are goal theories in which the identification of a goal depends on the content of prior representations.[4]

This kind of account of functions fits best with attribution of function ot artifacts. The idea of breaking open nuts seems to have played a causal role in the production of the nutcracker. It does not fit so neatly into the biological sciences. Of course, it used to provide a persuasive argument (the teleological argument) for the existence of a Creator: there *are* functions in nature; functions require prior representations; yet the creatures themselves (even when creatures are involved) have no such foresight or were not around at the right time; hence the prior representations must have been lodged in some awfully impressive being ... etc. Nowadays, however, in the clear and noncontroversial cases of functions in nature, it is taken that they can be accounted for from the standpoint of a theory of evolution by way of *natural* selection—and in such a theory there can be no room for any analysis of biological functions which rests on prior representations. Even if God foresaw the functions of biological structures, that is a matter outside biology; functions, however, are a biological and not a theological matter.

It is worth noting that, even though the representational theory seems to rest comfortably with attributions of functions to artifacts, nevertheless, some artifacts prove more problematic than might first appear. Many artifacts evolve by a process very like natural selection. Variations often occur by chance and result in improved performance. The artisan may not understand fully the reasons why one tool performs better than others. Yet, because it performs well, it may be copied, as exactly as possible. The reproduction of such tools may occur for generations. The features of the tool which make it successful and which lead it to be selected for reproduction are features that have specific functions. But they were not created with those functions in mind. They may have been produced with an over-all function in mind (say, hitting nails); but the toolmaker may not have in mind any functions for the components and features of the tool, which contribute to the over-all function. For instance, the toolmakers may copy a shape that has the function of giving balance to the tool—but they need not foresee, or plan, or represent any such function. They know only that tools like this work well at banging nails, or sawing wood, or whatever the over-all function might be. Consequently, even with artifacts, structures can serve specific functions even though there exists no prior representation of that function.[5]

There is a further reason for uneasiness about the representational theory. The theory analyzes the *apparent* forward directedness of functions by an indirect, two-step, route. The forward directedness of functions is analyzed as comprising a *backward* step to a representation, which in turn has a *forward* directedness toward a possibly nonexistent future state.

Thus the seeming forward directedness of functions is reduced to another sort of forward directedness: that of representations—plans, beliefs, intentions, and so on. And this is worrying. The worry is not just that these are "mentalistic," and just as problematic as functions—and just as hard to assimilate into the scientific picture of the world. Rather, an even greater worry for many will be that of vicious circularity. Many find it plausible that the notion of representation will turn out to be analyzable in terms that at least include functional terms. And functional terms presuppose functions. Hence the future directedness of representations may turn out to presuppose the future directedness of functions. This threatens to do more than just restrict the scope of representational theories; it undermines such theories even in their home territory, as applied to artifacts.

IV Etiological Theories

The third response to the tension between the scientific view of the world and the concept of function again involves rejecting any role for future events in the characterization of functions. Yet etiological theories also eschew reference to prior representations of future effects, as well as reference to the future effects themselves.[6]

Representational theories and etiological theories have an important feature in common. Both shun any genuine, direct reference to future effects, and refer instead only to past causes. Both construe the attribution of function as supplying information about the genesis of the character, that is, about how the character came into existence.

The difference between the two theories recalls the distinction Charles Darwin drew between artificial selection and natural selection. When animal breeders select, they represent to themselves the characters they wish to develop. Natural selection has closely analogous results, but it operates in the absence of (or at least without any need for) any conscious or unconscious representations of future effects.

The etiological theory of functions explains biological functions by reference to the process of natural selection. Roughly: a character has a certain function when it has evolved, by natural selection, *because* it has had the effects that constitute the exercise of that function.

Clearly, there is room here for an overarching, disjunctive theory, which unites the representational with the etiological. Such an overlapping theory would say: a character has a certain function when it has been selected because that character has had the relevant effects. In the case of artifacts, the selection involves conscious representations (mostly); in the case of (Darwinian) sexual selection, representations may enter the picture, but reproduction, heredity, and evolution also play a part; and in the case of natural selection, representations drop out altogether.

But, on the etiological theory, a character has a biological function only if that character has been selected for by the process of natural selection because it has had the effects that constitute the exercise of that function. This is the only kind of selection compatible with the dictates of modern biological science.

We take this etiological theory of biological function as the main alternative to the account of biological function we shall proffer. The big plus for the etiological theory is that it makes biological functions genuinely explanatory, and explanatory in a way most comfortable with the modern biological sciences. But we shall argue that this explanatory power is still not quite right: it offers explanations that are too backward-looking. The theory we offer will be more forward-looking in its explanatory nature.

But, before we turn to this matter, we should note another worry with the etiological theory, a worry that extends even to the overarching disjunctive theory of which it is a part. This worry is that there is too great a dependence of the intrinsic nature of functions on contingent matters— matters which, had they been (or even if they are) otherwise, would rob the theory of any viability as an account of functions.

The etiological theory of biological functions has such functions characterized in terms of evolution by natural selection. Most theorists take evolutionary theory to be true, but contingently so. What if the theory of evolution by natural selection were to be (or had been) false? Clearly then, on the etiological theory of biological functions, as we have specified it, there would be no biological functions. Wheather or not there are biological functions at all, on the overarching, disjunctive theory, will depend on what replaces the theory of evolution by natural selection. Suppose it is creationism. Then the representational theory would apply; for we have the representations in the mind of the creator that would have the appropriate causal role in the development of biological structures, so the representational theory would become a general theory of functions.

We noted earlier that the representational theory had problems with the functions of some artifacts: artifacts that seemed to evolve over time by a process similar to natural selection. If this analogous process is also not available, along with natural selection proper, then our representational theory will not bestow functions upon such artifacts. But perhaps this is small change.

For creationism, there would of course be an enormous epistemological problem of discovering what the functions of biological structures were; for this would depend on discovering what the creator had in mind. So we would be stuck with great difficulty in discovering whether the function of the heart is to produce the sound of a heart beat, in line with the creator's idea of a beating rhythm in nature, with the circulation of the blood as a nonfunctional effect; or whether the reverse is true. It would be much like an anthropologist discovering the nature of ancient artifacts without any presuppositions about the intentions of the earlier cultures.

But suppose creationism is not the alternative. Consider the possible world identical to this one in all matters of laws and particular matters of fact, except that it came into existence by chance (or without cause) five minutes ago. Now, on even the over-arching theory, there are no functions; for there are no biological functions on the etiological theory, and no causally active representations as required by the representational theory, and hence no functions at all.

We have the intuition that the concept of biological function, and views about what functions biological characters have, are not thus contingent upon the acceptance of the theory of evolution by natural selection and on discovering what led to the evolutionary development of particular characters. In parallel, we are also inclined to think that the representational account of the function of artifacts gives too much importance to the representations or ideas of the original planner, even in cases when there is one. As we indicated earlier, we believe a satisfactory account of functions in general, and of biological functions in particular, must be more forward-looking. We now turn to this.

V   Fitness, Function, and Looking Forward

It emerges from our discussions that the tension between functions and modern, causal science has generated, fundamentally, two stances on the nature of functions.

The first is the eliminativist stance. This has the merit of giving full weight to the forward-looking character of functions, by specifying them in terms of future and perhaps nonexistent effects; and also to the explanatory importance of functions. It is mistaken only in its despair of reconciling these two strands.

The second stance is backward-looking. This embraces theories which look back to prior representations and those which look back to a prior history of natural selection and those which look back to a history of either one sort or the other.

We will argue for a forward-looking theory. Functions can be characterized by reference to possibly nonexistent future events. Furthermore, they *should* be characterized that way, because only then will they play the explanatory role they need to play, for instance, in biology. The way to construe functions in a forward-looking manner, we suggest, is (roughly) to construe them in the manner of dispositions. The shift we recommend, in our conception of functions, has a precedent: the analysis of the evolutionary concept of fitness.

One wrongheaded, but at times common, objection to the Darwinian theory of evolution is that its central principle—roughly, "the survival of the fittest"—is an empty tautology which cannot possibly bear the explanatory weight Darwin demands of it.[7] This objection assumes that fitness can be judged only retrospectively: that it is only after we have seen which creatures survived that we can judge which were the fittest; moreover, it assumes that the fact that certain creatures have survived, whereas others did not, is what constitutes their being the fittest.

The etiological theory of biological functions rests on the same sort of misconception as that which underlies the vacuity objection to Darwin. On this theory, we can judge only retrospectively that a character has a certain function, when its having had the relevant effect has contributed to survival. Indeed, on the etiological theory, that an effect is part of the function of a character is *constituted* by the fact that having this effect has contributed to the survival of the character and of the organisms that bear it.

Consequently, the notion of function is emptied of much explanatory potential. It is no longer possible to explain why a character has persisted by saying that the character has persisted because it serves a given function. To attempt to use function in that explanatory role, would be *really* to fall into the sort of circularity often alleged (falsely) against the explanatory use of fitness in Darwinism.

This comparison with fitness serves another purpose. It has displayed why functions would lack explanatory power on the etiological theory, but it also shows how to analyze functions so as not to lose this explanatory power. Fitness is not defined retrospectively, in terms of actual survival. It is, roughly, a dispositional property of an individual (or species) in an environment, which bestows on that individual (or species) a certain survival potential or reproductive advantage. This is a subjunctive property: it specifies what will happen or what is likely to happen in the right circumstances, just as fragility is specified in terms of breaking or being likely to break in the right circumstances. And such a subjunctive property supervenes on the morphological characters of the individual (or species).[8] Hence there is no circularity involved in casting fitness in an explanatory role in the Darwinian theory of evolution. In the right circumstances fitness explains actual survival or reproductive advantage, just as in the right circumstances fragility explains actual breaking. In each case the explanation works by indicating that the individual has certain causally active properties that in such circumstances will bring about the phenomena to be explained.

What holds here of fitness holds, too, of biological functions. The etiological theory is mistaken in defining functions purely retrospectively, in terms of actual survival. Hence there need be no circularity in appealing to the functions a character serves, in explaining the survival of the character. Fitness is forward-looking. Functions should be forward-looking in the same way and, hence, are explanatory in the same way.

VI   The Propensity Theory

Here is one way to derive a "forward-looking" theory of functions.

Let us begin with the etiological theory. Consider a case in which some character has a specific effect and has been developed and sustained by natural selection because it had that effect. In such a case the etiological theory deems that it is (now) a function of the character to produce that effect.

Look more closely, then, at the past process that has "conferred" a function, according to the etiological theory. The character in question must have had the relevant effect, on a sufficient number of occasions— and in most cases, this will have been not on randomly chosen occasions, but on *appropriate* occasions, in a sense needing further clarification. (For instance, sweating will have had the effect of cooling the animal— and it will have had this effect on occasions when the animal was hot, not when it was cold.)

The history that confers a function, according to the etiological theory, will thus display a certain pattern. The effect that will eventually be deemed a function must have been occurring in appropriate contexts; that is to say, it must have been occurring in contexts in which it contributes to survival, at least in a statistically significant proportion of cases.

Further, this contribution to survival will not, in realistic cases, have been due to sheer accident. One can imagine individual incidents in which a character contributes to survival by sheer chance. Laws of probability dictate that a long run of such sheer accidents is conceivable; but it will be very unlikely—except for characters with relatively short histories. The only cases in which such long runs are likely to occur are cases in which the character confers a standing propensity upon the creature, a propensity that increases its chances of survival.

If we imagine (or find in the vast biological record) a case in which a character is sustained by a chance sequence of accidents, rather than by a standing propensity, then it would not be appropriate to describe that character as having a function. This can happen when a character is linked with another character that does bestow a propensity and where variations in the character just have not occurred to allow selection against the inoperative character. It can also happen by sheer chance, a long-run sequence of sheer flukes. Such a sequence is very improbable; but biology offers a stunningly large sample. It is very probable that many improbable events will have occurred in a sample that large.

Consequently, what confers the status of a function is not the sheer fact of survival-due-to-a-character, but rather, survival due to the propensities the character bestows upon the creature.

The etiological theory describes a character *now* as serving a function, when it *did* confer propensities that improved the chances of survival. We suggest that it is appropriate, in such a case, to say that the character *has been serving that function all along.* Even before it had contributed (in an appropriate way) to survival, it had conferred a survival-enhancing propensity on the creature. And to confer such a propensity, we suggest, is what constitutes a function. Something has a (biological) function just when it confers a survival-enhancing propensity on a creature that possesses it.

Four features of this propensity theory of biological functions should be made explicit.

First, like the corresponding account of fitness, this account of functions must be relativized to an environment. A creature may have a high degree of fitness in a specific climate—but a low degree of fitness in another climate. Likewise, a character may confer propensities which are survival-enhancing in the creature's usual habitat, but which would be lethal elsewhere. When we speak of the function of a character, therefore, we mean that the character generates propensities that are survival-enhancing in the creature's natural habitat. There may be room for disagreement about what counts as a creature's "natural habitat"; but this sort of variable parameter is a common feature of many useful scientific concepts.

Ambiguities will arise especially when there is a sudden change in the environment. At first, we will refer the creature's "natural habitat" back to the previous environment. But eventually we will transfer the term to the current environment. The threshold at which we make such a transference will be vague. The notion of natural habitat will also be ambivalent as applied to domestic animals.

In its most obvious use, the term "habitat" applies to the physical surroundings of a whole organism. But we can also extend its usage, and apply the term "habitat" to the surroundings of an organ within an organism. Or to the surroundings of a cell within an organ. In each case, the natural habitat of the item in question will be a functioning, healthy, interconnected system of organs or parts of the type usual for the species in question. When some of the organs malfunction, then other organs, which go on performing their natural functions, may no longer be contributing to survival. We still say they are performing their natural function, even though this does not enhance the chance of survival. Why? Because it would enhance survival if the other organs were performing as they do in healthy individuals.

Consequently, functions can be ascribed to components of an organism, in a descending hierarchy of complexity. We can select a subsystem of the organism, and we can ascribe a function to it when it enhances the chances of survival in the creature's natural habitat. Within this subsystem, there may be a subsubsystem. And this may be said to serve a function if it contributes to the functioning of the system that contains it—provided all other systems are functioning "normally" (that is, provided it is lodged in its own "natural habitat"). And so on.

Similar hierarchies may also occur in the opposite direction: a microscopic organism has a function in a pond, which has a function in a forest, which has a function in the biosphere, and so on.

Second, on the propensity theory, functions are truly dispositional in nature. They are specified subjunctively: they *would* give a survival-enhancing propensity to a creature in an appropriate manner, in the creature's natural habitat.[9] This is true even if the creature does not survive or is never in its natural habitat. Likewise, fragility gives a propensity to break to an object in an appropriate manner, in the right circumstances—and of course some fragile objects never break. And fitness gives a propensity to an individual or species to survive, in an appropriate manner, in a specified environment and in a struggle for existence, even if there is no struggle for existence or if the individual or species fails to survive.

Of course, when functions do lead to survival—just as when dispositions are manifested—the cause will be the morphological structural form of the creature and the relationship between this form and the environment. Functions supervene on this in the same way that dispositions supervene on their categorical bases. But the functions will be explanatory of survival, just as dispositions are explanatory of their manifestations; for they will explain survival by pointing to the existence of a character or structure in virtue of which the creature has a propensity to survive.

Third, in the long run, it will be necessary to spell out the notion of a "survival-enhancing propensity" in formal terms, employing the rigors of the probability calculus. Clearly, there will be a spectrum of theories of this general form. These theories will vary in the way they explicate the notion of "enhancement": whether they construe this as involving increasing the probability of survival above a certain threshold, or simply increasing it significantly above what it would have been, and so on. We are not attempting to find and defend *the* correct propensity theory, but only arguing that a propensity theory offers the most promising theory of functions.

Fourth, there is the question as to whether the scope of the propensity theory is limited to biological functions or whether it can be extended, in some sense, to artifacts.

Obviously, like the etiological theory, the propensity theory could be part of an overarching, disjunctive theory which analyses biological functions in terms of bestowing propensities for survival and the function of artifacts in terms of prior representations. We noted earlier some problems for a backward-looking theory, even for artifacts. Yet surely representations should have some causal role in the case of consciously produced artifacts.

We are attracted here to a general, overarching theory, but one that concentrates on the *propensity* for selection. So a character or structure has a certain function when it has a propensity for selection in virtue of that character or structure's having the relevant effects. In the case of biological functions, we have a propensity for survival in the natural habitat, and so in such a habitat natural selection is likely to be operative. In the case of artifacts, we have a selection process clearly involving representations. But the representations are those at the time of selection, at the time of bestowing the function: now, so to speak. They need not be blueprints that antedate the first appearance of the prototype. Thus we feel there is a sense in which all functions have a commonness of kind, whether they be of biological characters or of artifacts.

VII   Comparisons

We hope to have led our readers to appreciate the attractiveness of the propensity theory of biological functions (and of functions generally). We conclude by making some direct comparisons between the etiological theory and the propensity theory.

On most biological examples, the etiological theory and our propensity theory will yield identical verdicts. There are just two crucial sorts of case on which they part company.

One sort of case which distinguishes the theories is that of the *first* appearance of a character that bestows propensities conducive to survival.

On our theory, the character already has a function, and by bad luck it might not survive, but with luck it may survive, and it may survive *because* it has a function.

On the etiological theory, in contrast, the character does not yet have a function. If it survives, it does not do so because it has a function; but, after time, if it has contributed to survival, the character will have a function.

We think our theory gives a more intuitively comfortable description of such cases, at least in most instances. But there are variants on this theme, on which our theory gives less comfortable results. Suppose a structure exists already and serves no purpose at all. Suppose then that the environment changes, and, as a result, the structure confers a propensity that is conducive to survival. Our theory tells us that we should say that the structure now has a function. Over all, this seems right, but there are cases where it seems counterintuitive. Consider, for instance, the case of heartbeats—that is, the *sound* emitted when the heart beats. In this century, the heartbeat has been used widely to diagnose various ailments; so it has come to be conducive to survival. The propensity theory deems the heartbeat to have the function of alerting doctors. That sounds wrong. The etiological theory says the heartbeat has no such function because it did not evolve for that reason. That sounds plausible.

And yet, we suggest, the reason we are reluctant to grant a function to the heartbeat is not that it lacks an evolutionary past of the required kind. Other characters may lack an evolutionary past, yet may happily invite attribution of a function. Rather, our reluctance to credit the heartbeat with a function stems from the fact that the sound of the heartbeat is an automatic, unavoidable by-product of the pumping action of the heart. And that pumping action serves other purposes. Although the heartbeat does (in some countries, recently) contribute to the survival of the individual, it does not contribute to survival of the character itself. The character—heartbeat—will be present in everyone, whether or not doctors take any notice of it. Although it "contributes" to survival, it is a redundant sort of contribution if it could not fail to be present whether it was making any contribution or not.

Perhaps the propensity theory should be carefully formulated in such a way as to rule out such "automatic" contributions to survival. Nevertheless, we will note only that, although the example of heartbeats seems initially to count against a propensity theory, there are other examples, and wider theoretical considerations, which count in its favor. Further, the case introduces many complications. For these reasons, it cannot be regarded as, in any sense, conclusive.

So much for cases of *new* survival-enhancing characters. There is a dual for these cases: that of characters that were, but are no longer, survival enhancing. These cases, like the former case, serve to distinguish between the etiological and propensity theories. If a character is no longer survival-enhancing (in the natural habitat), the propensity theory deems it to have no function. The etiological theory, in contrast, deems its function to be whatever it was that is used to be, and was evolved for.

In general, we think the propensity theory gives the better verdict in such cases. Under some formulations, our judgment may be swayed in favor of the etiological theory. We may be inclined to say that the function of a character is to do such and such, but unfortunately this is harmful to the creature these days. Yet surely the crucial fact is, really, that the function *was* to do such and such. It serves no pressing purpose to insist that its function still is to do that. Especially not, once we have passed the threshold over which we redefine the creature's natural habitat. If a character is no longer survival-enhancing, because of a sudden and recent change in environment, we may continue to refer its natural habitat to the past. Consequently, our propensity theory will continue to tie functions to what would be survival-enhancing in the past habitat. In such cases, there will be no conflict between the judgments of our theory and those of the etiological theory.

The test of examples and counterexamples is important. Yet in this case, in the analysis of functions, there is a risk that it will decay into the dull thud of conflicting intuitions. Similarly with intuitions as to how unified should be the analyses of functions for biological characters and for artifacts.

For this reason, we stress the importance of theoretical grounds for preferring the propensity theory. A propensity can play an explanatory causal role, whereas the fact that something has a certain historical origin does not, by itself, play much of an explanatory, causal role. Consequently, the propensity theory has a theoretical advantage, and this gives us a motive for seeking to explain away (or even overrule) apparent counterintuitions.

In a similar way, Darwinian evolutionary theory provides strong theoretical motives for analyzing fitness in a certain way. Our intuitions—our unreflective impulses to make judgments—have a role to play, but not an overriding one.

Notes

1. Our theory will be a cousin of theories which are sometimes called "goal theories" and which have been advocated, for instance, by Christopher Boorse, "Wright on Functions," *Philosophical Review*, LXXXV, 1 (January 1976): 70–86, and "Health as a Theoretical Concept," *Philosophy of Science*, XLIV, 4 (December 1977): 542–573; and by William Wimsatt, "Teleology and the Logical Structure of Function Statements," *Studies in the History and Philosophy of Science*, III, 1 (May 1972): 1–80. Some key differences between their theories and ours will be noted later.

2. A theory of this sort can be found in Robert Cummins, "Functional Analysis," Journal of Philosophy, LXXII, 20 (Nov. 20, 1975): 741–765. We have also been influenced by a paper by Elizabeth W. Prior, "What Is Wrong with Etiological Accounts of Biological Function?" *Pacific Philosophical Quarterly*, LXVI, 3/4 (July/October 1985): 310–328.

3. This problem is treated at some length by Ernest Nagel, *The Structure of Science* (New York: Harcourt; London: Routledge, 1961), in his section on teleology, pp. 401–428. Nagel manages to blunt the force of such objections, but only by augmenting his initial theory, which differs from eliminativism only superficially, thereby generating a theory that comes close to the theories of Boorse and Wimsatt mentioned above.

4. Andrew Woodfield, *Teleology* (New York: Cambridge, 1976), argues for a view that takes the primary cases of functions to rest on a prior plan, and all other cases of (unplanned) "functions" to be mere metaphorical extensions of the primary cases.

5. There are several intriguing points made about artifacts, their reproduction, selection, survival, and so forth, by Ruth Millikan, *Language, Thought and Other Biological Categories: New Foundations for Realism* (Cambridge, Mass: MIT Press, 1984). Millikan advances a sophisticated version of the etiological theory, which we discuss below.

6. The etiological theory is widely held, but a very good exposition and defense is given by Larry Wright, "Functions," *Philosophical Review*, LXXXII, 2 (April 1973): 139–168; and *Teleological Explanations* (Los Angeles: California UP, 1976). We have also been greatly influenced by the defense of etiological theories advanced by Karen Neander, *Abnormal Psychobiology*, Ph. D. thesis, La Trobe University, 1983.

7. For those who have suggested this view, with greater or less refinement and sophistication, see J. J. C. Smart, *Philosophy and Scientific Realism* (New York: Random House, 1963), p. 59; H. G. Cannon, *The Evolution of Living Things* (Manchester: University Press, 1958); C. H. Waddington, *The Strategy of the Genes* (London: Allen & Unwin, 1957), pp. 64/5; A. O. Barker, "An Approach to the Theory of Natural Selection," *Philosophy*, XLIV, 170 (October 1969): 271–290; Robert Brandon and John Beatty, "The Propensity Interpretation of 'Fitness': No Interpretation Is No Substitute," *Philosophy of Science*, LI, 2 (June

1984): 342–347. For those who have replied to this view, see in particular Edward Manier, "'Fitness' and Some Explanatory Patterns in Biology," *Synthese*, XX, 2 (August 1969): 206–218; and Michael Ruse, "Natural Selection in the *Origin of Species*," *Studies in the History and Philosophy of Science*, 1 (February 1971): 311–351.

8. For more on dispositions and their supervenience on categorical bases, see Pargetter and Prior, "The Dispositional and the Categorical," *Pacific Philosophical Quarterly*, LXIII, 4 (October 1982): 366–370; and Prior, Pargetter, and Frank Jackson, "Three Theses about Dispositions," *American Philosophical Quarterly*, XIX, 3 (July 1982): 251–257.

9. It is this central role we give to propensities which distinguishes our theory from others, like those of Boorse and Wimsatt (mentioned above), which fall back on the notion of statistically normal activities within a class of organisms. On their theories, a character has a function for a creature when it *does* help others "of its kind" to survive, in a sufficiently high proportion of cases. On our view, frequencies and statistically normal outcomes will be important evidence for the requisite propensities. But there are many well-known and important ways in which frequencies may fail to match propensities.

9

Where's the Good in Teleology?

Mark Bedau

Contemporary analyses of teleological explanation generally attempt to "sanitize" it, usually by trying to assimilate it to some uncontroversial descriptive form of explanation. This trend is misguided. Teleological explanations are controversial, especially when applied in biology, because value plays an essential role in them. If a reference to value is largely what makes teleological explanations problematic, then it seems only natural to try to vindicate the use of teleological explanations by eliminating or at least neutralizing this offensive reference to value. This has been the predominant trend in recent writings on teleology, in which teleology is usually given some purely descriptive form of causal analysis. Over against this, my project here is to defend a causal analysis that is value-centered.[1] It is crucial to distinguish different causal roles that value might play. Once these roles are clarified, it becomes clear what the quite different forms of teleology share and why biological teleology is so controversial.

Some may remain unconvinced by the defense of a value analysis offered below, in part because the current climate in philosophy seems to be one of descriptivism, a climate in which Christopher Boorse (while discussing disease and illness, concepts which on their surface seem to involve value) could dismiss the value approach with the off-hand comment that "philosophers ... have made too much progress in giving biological function statements a descriptive analysis for [a value approach] to be very convincing."[2] Nevertheless, I hope to develop a value analysis that at least warrants more than a cursory dismissal. Since the subject of teleology is difficult and traditional analyses of teleology succumb readily to counterexamples, it is an achievement to provide even a provisional analysis that easily and naturally accommodates any new difficulties that might arise.

Sentences expressing teleological statements cover a wide range of topics, partly illustrated by the following sentences:

(1)   I am running in order to get to class on time.

(2)   The rock is on top of those papers in order to hold them down.

(3)   Automobile engines have carburetors in order to mix air with gasoline.

(4)   The robin is walking back and forth in order to find food.

(5)   The human heart pumps blood in order to circulate it throughout the body.

These statements all have the form *A Bs in order to C*, and I will call them *in-order-to statements*.[3] My analysis of teleology will be couched as an analysis of in-order-to statements. Since in-order-to statements typically express teleological explanations, my analysis will amount to a theory of teleological explanation.[4] Other forms of teleological explanations, such as *A Bs for the sake of Cing*, can be given a similar analysis.

The bulk of this paper is devoted to the development of a specific value analysis, but first we will briefly examine the main alternative approaches and note that in each case problems arise, the natural response to which is to introduce a consideration concerning value.

1   Alternative Approaches

The Mental Approach. Some philosophers, such as C. J. Ducasse and (within limits) Andrew Woodfield, assert that all genuine teleology can be traced in one way or another to minds and their designs; this is the mental approach to teleology.[5] The main problem with the mental approach is that it is too narrow, because it cannot account for the apparent teleology in non-mental organisms and in organs.[6] My heart pumps blood in order to circulate it, for example, but my heart does not *desire* to circulate the blood, and it does not *believe* that pumping will bring about circulating. Still, my heart is structured and behaves *as if* it were made by a mind, so one might hope to revive mentalism by proposing the following, as C. D. Broad once did:

S is a teleological system *iff* S is structured *as if* it had been designed by a rational mind for certain purposes.[7]

This mental-analogue analysis accommodates the biological counter-examples to mentalism, but at the price of being too broad. Almost everything is structured as if it had been designed by a rational agent for some purpose, provided only that the designer is sufficiently ignorant, sufficiently inept at designing things, or is attempting to achieve sufficiently bizarre purposes. For example, both my chair and the dried leaf outside my window are structured as if they were designed as paperweights by a designer who is both ignorant about what makes for good paperweights and inept at implementing his paperweight designs.

The natural next step for a mentalist would be to require that the designer also be well informed about the world, be good at designing things, and have sufficiently normal purposes. As will become clear below, there would still be problems with this revised mental-analogue view. However, for the moment, it is enough to notice that the revised analysis now contains recognizable value notions, such as the requirement that the designer be *good* at designing things.[8] In fact, the evaluative component can replace the hypothetical designer. Another way to achieve the effect of requiring that the item be structured as if it had been designed by a well-informed effective designer, i.e., by a mind that makes things that are *well suited* to their purposes, would be to talk simply of items that are well suited to normal purposes, as follows:

S is a teleological system *iff* S is structured in such a way that it is well suited to achieve a sufficiently normal purpose.

The hypothetical designer—the mental agent, the sole remaining vestige of mentalism—has become superfluous. Its only role was to guarantee that any items it hypothetically designed were well suited to their purposes, and we can replace it with the value that it serves to guarantee. Thus, trying to work out the mental approach reorients one in the direction of a value approach.

The Systems Approach. Some philosophers have thought that the excessive narrowness of the mental approach could be avoided by using cybernetics and general system theory to isolate the features shared by all teleological systems. This systems approach to teleology has been advocated by Morton Beckner, Christopher Boorse, R. B. Braithwaite, and Ernest Nagel, among others.[9] The cornerstone of the systems approach is the notion of a goal-directed system. Roughly, a system is goal-directed if it tends to maintain a certain state (called the goal state) in the face of external and internal perturbations, and goal-directed behavior is simply behavior that contributes to the maintenance of some goal of a goal-directed system.

The basic problem with the systems approach is excessive breadth. To find a counterexample, all you have to do is find a system that tends to maintain some state or other, and then simply call that state the goal state. A standard counterexample is a marble at the bottom of a bowl. Advocates of the systems approach attempt to fix this problem by adding new systemtheoretic conditions. However, in each case, new counter-examples arise, as the following consideration show.[10] Out in an uninhabited desert there could be a physical object that is causally structured just like a bridge, but which is a result of the purely impersonal and accidental forces of nature. (So-called natural bridges are somewhat like this.) Even though such a structure would "bridge" some gap, river, etc., it would lack the function of a bridge because nothing and no one uses it or designed it as a bridge. Likewise, the random forces of nature might accidentally produce a physical structure that happens to possess exactly the sort of causal structure that a modified systems analysis requires. Nevertheless, this structure would no more be a goal-directed system than our "bridge" has the function of a bridge.

We can diagnose what is left out of the systems approach be asking what must be added to a marble-plus-bowl system in order to make it genuinely goal directed. Notice that a creature could have a proprioceptive organ that is just like a marble-plus-bowl structure, so that the creature maintains its balance by keeping the "marble" at the bottom of the "bowl." Since maintaining its balance is useful for the creature, the behavior of this marble-plus-bowl structure benefits it. Notice also that someone could measure the stability of surfaces with an instrument made out of a marble-plus-bowl, by placing it on a surface and checking whether in time the marble remains at the bottom of the bowl. So, the behavior of this marble in the bowl is beneficial to those who use the device to tell whether surfaces are stable. Whereas the systems approach concerns only intrinsic causal dynamics, these marble-plus-bowl systems show that a system's intrinsic causal dynamics do not by themselves determine whether the system is genuinely goal directed. It is necessary to ask whether any good comes from the behavior of the system. Thus, by trying to work out the systems approach, one gets reoriented in the direction of a value approach.

The Etiological Approach. Some philosophers have hoped to steer between the excessive narrowness of the mental approach and the excessive breadth of the systems approach by concentrating on certain details of the causal histories of teleological phenomena. Roughly, their idea is that something happens in order to achieve a goal when the tendency to bring about the goal explains the process in question. This is the guiding idea behind the etiological approach, different versions of which have been advocated by Jonathan Bennett, G. Cohen, Charles Taylor, and Larry Wright, for example.[11] The gist of Wright's version, which is probably the best known, is this:

A Bs in order to C *iff* A Bs because A's Bing contributes to Cing.

To get a feel for this analysis, consider the heart. The heart pumps blood in order to circulate it. According to the etiological analysis, this means that the heart pumps blood because doing so contributes to circulating it. How are we to understand this? Well, the heart pumps blood only because the creature possessing it is alive, and the creature is alive only because blood is being circulated throughout its body. If the heart stopped circulating the blood, the creature would die and its heart would stop pumping. In this way, part of the explanation of why the heart pumps at all is that the heart's pumping contributes to circulating the blood.

There is much to be said for Wright's analysis; in fact, the value analysis defended later in this paper can be viewed as a modification of Wright's analysis to which a value condition has been added. But Wright's analysis contains a fatal flaw as it stands, because something that is not teleological might nevertheless have an etiology like the heart's. Consider a stick floating down a stream that brushes against a rock and comes to be pinned there by the backwash it creates.[12] The stick is creating the backwash because of a number of considerations, including the flow of the water, the shape and mass of the stick, etc., but part of the explanation of why it creates the backwash is that the stick is pinned in a certain way on the rock by the water. Why is it pinned in that way? The stick originally became pinned there accidentally, and it remained pinned there because that way of being pinned is self-perpetuating. Therefore, once pinned, part of the reason why the stick creates the backwash is that the backwash keeps it pinned there and being pinned there causes the backwash. In this case, the stick meets the conditions of Wright's analysis: the stick creates the backwash because doing so contributes to keeping it pinned on the rock. However, clearly the stick does not create the backwash *in order to* keep itself pinned on the rock. The stick's behavior needs no teleological explanation; it is not an instance of teleology.

Given that the heart example and the stick example involve a similar sort of etiology, why is only the heart teleological? One clear difference between the two cases is that the heart's pumping is good for the creature, but the stick's backwash is not good for anything.[13] This suggests that only when the causal history identified by the etiological analysis involves a value-centered system can teleology be present. Once again, in the process of trying to defend the etiological approach, one gets reoriented in the direction of a value approach.

2   Three Grades of Evaluative Involvement

This brief survey of alternative approaches suggests that value plays some role in the analysis of teleology. But what might that role be? It turns out to be necessary to distinguish three degrees to which someone attributing teleology would be committed to attributing value, which I will call three grades of evaluative involvement in teleology.

Most recent supporters (and critics) of a value approach focus on the first grade of evaluative involvement, which one might call the *good-consequences* approach. Versions of the good-consequences approach differ on various details, but the following rough form reflects its main components.[14] Where A is an agent who does something or has some property, B, in order to bring about some end, C:

(G1)   A Bs in order to C *iff* A Bs and A's Bing contributes to Cing and Cing is good for A.

(Where A is an organ, the beneficiary is the organism containing A, and where A is an artifact, the beneficiary is the person using A.) In a nutshell, the idea behind the good-consequences approach is that things that contribute to good consequences happen in order to bring about those good consequences.

There is an obvious problem with the good-consequences approach. Consider Ralph, who swims every day. Swimming keeps Ralph fit, which is good for him, so Ralph meets the good-consequence condition for swimming in order to stay fit. But Ralph does *not* swim in order to stay fit. He swims simply because he thrills at the feeling of floating in the water and gliding through it. In general, an action can have good consequences but not be done for their sake, if the agent does not care about them. This flaw with the good-consequences approach should not be blamed only on the mind, as the classic example of the beating-heart sound shows.[15] The human heart makes a beating sound when it pumps, and (in the current state of medical practice) this sound is turned to good advantage for us because physicians use this sound in medical diagnoses. So, the heart's pumping contributes to an effect that confers a good on us, but it does not pump in order to confer this good consequence.

Ralph's swimming and the heart's beating are not completely devoid of teleology. Although it is false that Ralph swims in order to keep himself fit and that the human heart pumps in order to make sounds used in medical diagnosis, it is *true* that Ralph's swimming *performs the function*, or *serves the function*, of keeping him fit, and it is true that the beating of the human heart *performs the function* of making sounds used in medical diagnosis, or *functions as* a device for making such sounds. Having a teleological explanation is more than merely performing a function, serving a function, or functioning as something, and Ralph's swimming and the heart's beating have only the weaker form of teleology. (Expressing this point requires careful choice of teleological terminology, especially since certain telic terminology is used differently by different people. For some, saying that Ralph's swimming *has* a certain function is the same as saying that it is *performing* that function; for others, it is the same as saying that Ralph swims *in order to* achieve that function, i.e., that Ralph's swimming has a teleological explanation. Thus, I purposefully avoid saying that Ralph's swimming *has* the function of keeping him fit, or that the heart's beating *has* the (biological) function of making diagnostically useful sounds.)

Staying fit is merely an accidental benefit of Ralph's swimming, and contributing to diagnoses is merely an incidental or accidental benefit of the heart's pumping. A value analysis of in-order-to statements needs a way to screen off accidentally beneficial consequences. Borrowing the central insight of the etiological approach, one can require that the good consequences of A's Bing are part of the *explanation* of A's Bing, for the contribution that Ralph's swimming makes to his fitness does not explain his swimming, and the aid to medical diagnosis provided by the heart sound does not explain the heart sound. This is a natural way to screen off accidental benefits, since in-order-to statements are inherently explan-atory.[16] Ralph's swimming is not explained by saying that it *performs the function* of keeping him fit, but it is explained by saying that he swims *in order to* enjoy the feel of the water.

There are different kinds of explanatory roles that good consequences can play, and it is useful to distinguish the involvement of evaluation in teleology in three increasingly strong grades or degrees.[17] (For the moment, we will focus only on those aspects of the three grades that are relevant to the distinctions between them.) The first and weakest grade of evaluation involved is simply the good-consequences analysis of in-order-to statements, (G1) above. In this grade one teleology, A's Bing is merely conjoined with its contribution to a good consequence. Our earlier diag-nosis of the limitations of (G1) suggest that this conjunction should be replaced with an explanatory connection, resulting in something of this form:

(*)   A Bs in order to C *iff* A Bs *because* A's Bing contributes to Cing and Cing is good.

But (*) is ambiguous. Is the scope of the "because" wide or narrow? That is, is the goodness of Cing part of the explanation or not? These two interpretations are different ways of ruling out accidentally beneficial consequences, and constitute the second and third grades of evaluative involvement in teleology. The weaker interpretation (in which the "because" has narrow scope) requires the good consequences to be explanatory but does not require that their *goodness* itself be explanatory. This is the second grade of evaluative involvement. In grade two teleology the consequences cannot be accidental, but the benefit they provide can. Thus the grade two scheme is that A Bs in order to C *iff* Cing is good and A Bs because A's Bing contributes to Cing, or, equivalently:

(G2)   A Bs in order to C *iff* [A Bs because A's Bing contributes to Cing] and Cing is good.

According to this schema, the explanation of A's Bing depends on the fact that A's Bing contributes to Cing, and while the explanation of A's Bing does not depend on the goodness of Cing, Cing still happens to be good. Thus, the second grade of evaluative involvement captures the idea of explanation by reference to consequences that happen to be good. A stronger way to rule out accidentally beneficial consequences is to give "because" wide scope in schema (*), yielding the third grade of evaluation involved.[18]

(G3)   A Bs in order to C *iff* A Bs because [A's Bing contributes to Cing and Cing is good].

Grade three requires that the good consequences and their goodness both figure in the explanation; that is, neither the consequence(s) nor the benefit provided can be accidental. A Bs specifically because A's Bing contributes to something *good*. Thus, the third grade captures the idea of explanations by reference to the goodness of consequences.

Does either the second or third grade of teleology succeed as an analysis of in-order-to statements? Recall the stick that got stuck against a rock by the backwash it caused by being stuck there. Since the backwash is not good for the stick (or for anything else), the stuck stick does not meet the conditions in either the second or the third grades. Could the stick example be modified so as to include an evaluative element but still not be teleological, thus perhaps providing a counterexample to the value-centered etiological analyses? Since a stick is not the kind of thing that can be a beneficiary, consider an unconscious man being swept downstream, who becomes lodged against some rock by the backwash created by his being stuck there, and is thus saved from being swept over some perilous falls. This situation is a counterexample to the sufficiency of the second grade. The man is stuck on the rocks because being stuck on the rock contributes to creating the appropriate backwash, and creating that backwash is good for him. However, the man is not stuck on the rocks *in order to* create the backwash. Thus, the second grade of evaluative involvement is not sufficient for an analysis of in-order-to statements. (Of course, if we changed the example, it would be possible for the man to become stuck on the rocks in order to create the backwash, if, for example, he were conscious and maneuvered himself into a location in the water correctly calculated to get himself stuck on the rocks.)

The unconscious man is not a counterexample to the third grade, however, for the benefit provided the man by the backwash plays no role in the explanation of why the man is stuck on the rocks. He still would have been stuck there, even if it provided no benefit or were harmful, just as a sack of rags (having roughly the man's size and weight) might get stuck there. Thus, although being stuck benefits the man, the benefit plays no part in the explanation of what happens to him, and only grade two is involved.

The three grades of evaluative involvement in teleology correspond to three grades of telic phenomena. The first grade is a familiar central member of the telic family, to wit, the idea of performing a function. As far as I can tell, there is no brief expression in ordinary English that typically expresses the second grade of teleology. Grade two is merely the conjunction of the etiological insight and a value condition. In general, the second grade is a member of the telic family only because it is a cousin of a central family member—grade three teleological explanation —for which grade two is easily mistaken if a scope ambiguity is missed or ignored; however, we will see in section eight below that grade two, when underwritten by natural selection, produces the form of teleology found in biology. Only the third grade, which gives value an essential role in the explanation, serves as an analysis of full-blooded teleological explanations.[19]

3 The Evaluative Element in Teleology

One way to clarify the three grades of teleology is to clarify the value notion in the analyses. Although I have no *theory* of value to propose, I can specify the scope of the notion and address some of the worries it might provoke.

First, the goodness of Cing need not consist in any moral goodness. In some cases Cing is good only in so far as Cing is useful or beneficial. In other cases, most of us may find Cing unsavory or even loathsome. The evil pleasure enjoyed by someone who taints Halloween treats and the self-destructive satisfaction of a masochist count as goods in my weak sense. The sorts of good involved are thus quite diverse. Some proponents of value analyses allow only certain restricted kinds of goods, e.g., usefulness or reproductive fitness,[20] but this is inappropriate in a fully general account of teleology. Furthermore, the goodness of Cing implies merely that Cing confers *a* good, not that Cing is best overall. The aquatic pleasure for the sake of which Ralph swims might be relatively minor compared with other goods that would come his way if he were to stop swimming and do something else. In fact, experiencing this aquatic pleasure might put an undue strain on his heart and so be bad. Nevertheless, Ralph could still swim for the sake of the pleasure. So the value analysis requires, not that Cing confer an important or the best good, but only that Cing confer *some* good (what we might call a "weak' rather than "dominant" good). The good might be merely instrumental, in that it is merely a means to something else that is good. In general, the good that Cing confers may be moral or non-moral or even immoral, important or insignificant, and intrinsic or merely instrumental; it must simply be good for something.

Although the notion of the good in the three grade analysis admits wide latitude, it is important to respect its limits. In particular, the statement that Cing is good for A does *not* mean merely that Cing tends to cause A to obtain or persist, or that Cing is instrumental to the maintenance of some disposition or capacity of A. Failure to respect this constraint would reopen the door to the stick-on-the-rock counterexample, for the stick creates the backwash because doing so contributes to keeping itself pinned on the rock and because remaining pinned is instrumental to the maintenance of the stick's disposition to create the backwash. Thus, although C*ing* might be merely an instrumental good for A, it must nevertheless be a means to some genuine good for A. Creating a back-wash is not a genuine good for the stick. In fact, it would seem that sticks, like stones and specks of dust, are simply not the kind of thing for which *anything* could be good (or bad); they do not have interests that can be promoted (or thwarted).

Reference to beneficiaries was suppressed in (G1)–(G3), because it was not relevant to the distinctions among the three grades. But the beneficiaries play a central role in the value analysis, so we should be more precise about them. When A Bs in order to C, the beneficiary of the C*ing* depends on what *kind* of thing A is, and generally follows the principle that the beneficiary's behavior must have an essential role in the explanation of A's behavior. If A is an agent or an organism, then typically the beneficiary is A. If A is an organ in an organism, then typically the beneficiary is the organism containing A. If A is an artifact, then typically the beneficiary is the person using A. Certain kinds of teleology may well involve other kinds of beneficiaries, such as a school of fish, or an agent's ancestors, or an organism's whole species. Again, what matters is whether a benefit to these entities can figure in the explanation of something that provides that benefit.

The beneficiary principle leaves obscure many cases on the borderline between those kids of things that can be the beneficiaries and those kinds that cannot. Is self-reproduction good for a virus, for example, or for DNA-like polymers simmering in a chemical soup? Is growth good for crystals? Many of us are unsure how to answer such questions. Because of this unclarity about what things can be beneficiaries, one might judge the value analysis to be incomplete or unhelpful. For example, using the value analysis, it is difficult to tell whether there could be teleology in borderline cases like viruses and crystals. I would be delighted, or course, to have a criterion of good that settled such controversies, but it is excessive to demand that the value analysis provide such criteria. Even without them, the value analysis is interesting and helpful. In fact, one kind of evidence in favor of the value analysis is that we are unsure whether teleological notions apply in roughly the same cases as those in which we are unsure whether value notions apply. For example, many of us vacillate about whether it really makes sense to think that something is good or bad for viruses; we also vacillate about whether it really makes sense to think that viruses enter other cells in order to reproduce. Just as it was illuminating when Quine located necessity, possibility, analyticity, meaning, etc., within a family of irreducibly intentional notions,[21] the value analysis, even though it employs a vague value notion, is illuminating if it shows why teleological notions belong in the family of value-centered notions, rather than the family of value-free notions. It is worth remembering that one should expect vague terms to have vague analyses. An analysis of a vague term can be illuminating if it localizes the vagueness so that we can see what causes it and thus see why hard cases are hard. According to the value analysis, the vagueness in teleology is due (at least in part) to vagueness concerning good, and the difficulty of telling what is good for what is one reason why there are borderline cases of teleology.

The value analysis of teleology might remind one of the doctrine that something's good is to be explained by reference to its natural function, end, or *telos*,[22] and this might prompt the worry that the value analysis of teleology is viciously circular. But this worry is not well grounded, for three reasons. First, of course, the value analysis would be susceptible to this circularity only if it relied on the teleological account of value, but there are a number of alternative approaches to value that would not threaten circularity. Second, even the if value theory of teleology *were* combined with a teleological theory of value, the threatened circularity would materialize only if the details of each theory meshed in just the wrong way. But finally, even if the combination should be genuinely circular, that does not entail that the two theories would undermine each other or that their combination would be uninformative. Value and teleology would coexist in an interconnected theoretical system, and so would stand or fall together. But it is no flaw if concepts are theoretically unified in this manner. The combined theoretical structure could still be informative. We can grasp teleological notions such as goal and purpose without recognizing that they involve value, as we can grasp the idea of value without seeing it as involving natural functions. A value-centered theory of teleology and a function-centered theory of value might both be true but not be trivially analytic, and sensible and intelligent people could disagree about them. So, even if teleology and value are part of an inter-defined family of concepts, a value-centered theory of teleology can be a significant and informative theoretical advance.

Someone might think that any value analysis must be wrong because there are cases of value-free teleology. Consider the F-ring of Saturn.[23] Astronomers used to think that Saturn had one large flat ring, but today Saturn is known to have a series of distinct rings, as well as a ringlet called the F-ring. The F-ring is quite narrow and distinct, with sharp edges. Astronomers found this shape puzzling, because over time the dynamics acting on the particles in the ringlet should have spread out the particles into a broad flat ring (like Saturn's other rings). Then two tiny satellites were discovered, one on each side of the F-ring. These satellites explained the F-ring's unusual shape, because the gravitational attraction that they exerted on both sides had the combined effect of forcing the particles in the ringlet into a narrow line, and they have become known as "the shepherd satellites." One might think that the shepherd satellites provide a counterexample to the necessity of the value analysis. It makes little sense to say that preserving this shape is good for the F-ring or for the satellites, and it is simply false that the satellites shepherd the particles *because* this is good for something. At the same time, astronomers some-times apply teleological language to the shepherd satellites and say things like that they attract the particles in order to keep the F-ring narrow and sharp-edged. But such nominal teleology counts for nothing. It is true that the satellites act as *if* they are shepherding the ring particles in order to keep the ring narrow, and one can view the satellite and ring system *as if* preserving the F-ring's distinctive shape were good for the system. But these claims are clearly not literally true. This makes sense from the per-spective of the value approach, since neither the F-ring nor any planetary system containing it is the kind of thing that can be a beneficiary.

4   Variations on Grade Three Teleology

The core idea of the grade three analysis of teleological explanations is explaining things by reference to the goodness of their consequences, or, to be a bit more precise, explanation by reference to *telic structures*, which can be defined as a pair of logically linked propositions, one which cites a means to an end and another which states the goodness of that end. What is it to explain something by reference to a telic structure? Here it helps to treat time more explicitly, and reformulate (G3) slightly, as follows:

(G3*)   At T, A Bs in order to C *iff* at or about time T, the telic structure: [A's Bing is a means to Cing & Cing is good] is causally efficacious with respect to A's Bing.

We can recast (G3*) into the following argument, reminiscent of practical reasoning:[24]

MAJOR PREMISE: Cing is good (at or about T).
MINOR PREMISE: A's Bing is a means to Cing (at or about T).
EXPLANANDUM: A Bs (at T)

The explanandum or conclusion of this argument is an event (state, or condition) for which a teleological explanation is being offered, and the major and minor premises are a telic structure.

It is helpful to express (G3*) in this argument form, because we can immediately see that the argument form is invalid. No reason is given for why the telic structure should entail A's Bing. One could fill this gap by appealing to a background theory about naturalistic *telic mechanisms*, which have the ability to convert the value of an end into an efficient cause of a means to that end.[25] Given an appropriate theory about telic mechanisms, the premises of the argument above would be sufficient to guarantee the truth of the conclusion.

What constraints, if any, must such background theories meet? How could a telic structure be causally efficacious with respect to A's Bing? Flexibility on this matter is necessary if the analysis is to apply to the different kinds of teleology found in agents, artifacts, organisms, and organs. Thus, (G3*) is not a single set of necessary and sufficient conditions, but a schematic theme with different schematic variations. An instance of the schema must not only fill in the schematic variables *A*, *B*, and *C* with specific terms; it must also sketch a background theory that explains how telic structures can have the right sort of causal efficacy. In effect, the analysis contains another schematic variable that ranges over background theories concerning telic mechanisms. A background theory can exploit any mode of explanation, including those involving the mind and those involving evolution by natural selection, as long as the theory in some way gives causal efficacy to telic structures. The full grade three analysis, then, includes the theme and all of its possible variations.

Some limitations must be placed on permissible variations. Not every way in which a telic structure could acquire causal efficacy is sufficient for grade three teleology. Otherwise the door would be open to the familiar problems of deviant causal chains, which plague causal accounts in general (e.g., those involving action, perception, and memory).[26] At present there is no consensus about how to solve the problem of deviant causation in any of the contexts in which it arises, and I have no solution to offer in the context of teleology. Since the difficulty appears to be quite general, it is reasonable to expect that, once a solution is found in one area, it will generalize to others. In the meantime, when I speak of telic structures being causally efficacious, I mean that they are causally efficacious *in the "right" way*, where this phrase is just a place-holder for the eventual solution to the problem.

5   Grade Three Mental Agents

Mental agents can assess the world they are in, detecting and evaluating the outcomes of various courses of action. In addition, their actions can be caused by their assessment that so acting is likely to result in an outcome that they favor. Thus, the mind is a mechanism that can convert the value of a detected end into an efficient cause of a means to that end; agents' actions can be explained by their good consequences.

Minds are not flawless. What is taken to be a good consequence might either not be a consequence or not be good, and genuine good consequences might be undetected. Still, actions of mental agents can be given teleological explanations as long as their actions are caused by the agents' beliefs that the actions are means to good ends. Thus, grade three teleology has one version involving mental agents. Where A is a mental agent, and B and C are features of A:

(G3.1)   A Bs in order to C *iff* A Bs because A believes the following telic structure: [A's Bing is a means to Cing & Cing is good].

So, for example, if I am running in order to get to class on time, this means that I am running because I believe that doing so is a means to getting to class on time and getting to class on time is good. Given common knowledge (our "background theory") about mental agents like A, and given specific knowledge that A believes that Bing is a means to Cing and Cing is good, we can understand why A Bs. That is, we have an adequate explanation of A's Bing.[27]

There is a series of related potential counterexamples to the mental agent version of grade three, which can all be disposed of similarly. Consider someone walking to the store in order to obtain cigarettes. According to (G3.1), this person is walking to the store because he believes that doing so will bring about the good consequence of obtaining cigarettes. However, most smokers know that smoking is bad for them. Some just don't care about their health; others do care but smoke anyway due to the strength of their addiction and the weakness of their will. Such counterexamples are easily answered. Smokers seek cigarettes because they want or desire them, expecting some (weak) benefit from smoking them (such as relief from a certain psychological tension). Weak goods might not be the best thing on balance for someone to pursue, but the smoker still believes that smoking is a means to some weak good (recall the discussion in section 3 above). Similar resoponses can be fashioned for a series of related potential counterexamples, involving sadists, masochists, and suicides. The crux of each response is that mental agents are motivated by seeing some good (at least, a weak good) in their actions.

Some might doubt whether agents always believe the major premise in the telic structure. If A does B in order to C, isn't it possible that A merely desires C without believing that it is good? The analysis can be defended by appeal to the controversial but not implausible thesis that, if A desires C, then A believes that C is a (weak) good, i.e., A sees some good in the prospect of C. This thesis, the Socratic idea that desiring something entails regarding the prospect of having it as good, has been defended and criticized for thousands of years. To provide a new defense is beyond the scope of this paper, so here I will simply assume the thesis.[28]

6 Grade Three Artifacts

There are at least two kinds of teleology concerning artifacts, one concerning used objects and the other concerning designed objects. A rock is sometimes used as a paperweight, and a carburetor is designed to mix air and gasoline. Artifact teleology arises when the structure and behavior of an artifact is controlled by its potential good consequences. That is, a mental agent using or designing an artifact can (within limits) bring it about that the artifact's structure and behavior are controlled by (the agent's assessment of) the value of their effects (consequences). Thus, an artifact can come into existence and be structured in such a way that it brings about good consequences, because of the purposeful activity of the mental agent(s) using or designing it. As noted earlier, mental agents might make mistakes; envisaged consequences or their expected benefits might not materialize, and actual consequential benefits might be overlooked. Still, a feature of an artifact can have a teleological explanation provided that the feature is caused by an agent's belief that good consequences will result.

The explanation behind the teleology in used objects will center around the mental agent using the object. Consider our swimmer, Ralph, who uses swimming goggles. Ralph protects his eyes by using goggles. The goggles keep out the water in order to protect his eyes from chlorine, because Ralph wears them in order to protect his eyes. Or, consider a rock resting on some papers in order to hold them down. The rock is on the papers because someone is using it in order to hold them down, by making the rock rest on the papers. In general, when a mental agent U uses an object A:

(G3.2)   A Bs in order to C *iff* A Bs because U uses A's Bing in order to C.

The explanation of A's Bing appeals to a teleological explanation concerning a mental agent who uses A, which in turn makes reference to the agent's belief in the following telic structure: [A's Bing is a means to Cing & Cing is good]. More complicated versions of (G3.2) are needed for agents using objects in more complicated ways. But each grade three teleological explanation concerning used objects eventually appeals to a teleological explanation involving a mental agent. When a mental agent makes an object have a feature in order to bring about some goal, the object's features acquire a derivative teleological explanation.

Teleology in designed objects is more complicated, because it involves a second mental agent: the designer. The mental states of the designer(s) and the user(s) can be interwoven in different and complicated ways. Consider one relatively simple case. A telephone receiver has a handle in order to enable the user to hold it. The handle came to be there because the agent who designed telephones made the receivers have handles in order to enable telephone users to hold them. Thus, the grade three teleology of the telephone receiver handle can be traced to its designer's grade three teleology with respect to telephone users. In general, if D is an agent who designed object A, and B and C are features of A, then:

(G3.3)   A Bs in order to C *iff* A Bs because D made A B in order to C.

As with used objects, the conditions for designed objects include an embedded mental agent in-order-to statement: D made-A-B in order to C. And, as with used objects, this teleological explanation will involve reference to the designers belief in the following telic structure: [A's Bing is a means to Cing & Cing is good]. Schema (G3.3) is intended to cover only one relatively simple kind of teleology in designed objects. More complicated kinds of designed object teleology might not fit (G3.3) exactly. Some designers might design objects only in order to make money, and not because they believe the objects are suited to bringing about a certain good consequence. But more complicated cases of designed object teleology will fit some variation of (G3.3). In all cases, the designed object becomes subject to grade three teleological explanations as a result of the designer (or some other relevant person) believing (or having some more indirect mental attitude toward) a telic structure; so the teleology in designed objects derives from the teleology in agents. When an agent designs an object in order to bring something about, the object's designed features derivatively acquire a teleological explanation because of the teleological explanation of the designer's actions.

7 Grade Three Selection Processes

Mental and artifact teleology exploit one kind of telic mechanism: the mind, with its (fallible) ability to detect good consequences. A quite different, non-mental telic mechanism can arise in good-promoting selection processes operating in populations of creatures competing for survival. A creature's features might benefit it in various ways, such as promoting its survival. When features typically benefit a type of creature, and those features are inherited when that type of creature reproduces, a good-promoting self-perpetuating process or regularity can become established. That is, there can come to be a process in which the existence of a type of creature is perpetuated (in part) by the good-promoting features that are characteristic of the type. The type of creature with its good-promoting features exists (in part) because those features benefit the creatures. Thus, the features of that type of creature can be explained by reference to the goodness of their consequences.

Good-producing selection processes exist over time. In such a process, the present occurrence of means to good ends can be explained by reference to the fact that in the past they were means to the same ends and those ends were good. Thus, when such a process perpetuates itself, telic structures pertaining to the typical features of types of creatures will be typically true. A good-promoting selection process, then, is a kind of telic mechanism quite different from a mind. It is a diachronic good-producing process in which the telic structure has causal efficacy not because someone believes the propositions in it are true but because they typically *are* true.

Good-producing selection processes share with minds the possibility of going awry. Even if the operation of an organ in a type of creature *typically* produces a consequence that benefits the creature, that consequence might not *always* be beneficial and might not *always* be produced. For example, breathing might typically produce the good consequence of oxygenating the blood. But if the gas being breathed contains no oxygen this will not result, and one can imagine pathological circumstances in which oxygenated blood would provide no benefit at all. For the structures and behaviors typically exhibited by a type of creature to be explained by reference to their good consequences, it is (necessary and) sufficient that the good consequences result normally or typically. As long as a typical structure in a type of creature *normally* or *typically* produces a good consequence, this consequence can explain why that type of creature typically has that structure.

Good-producing selection processes can give rise to grade three teleology for various kinds of entities, including (types of) organisms and their (types of) parts. If A is a type of organism, and B and C are features of A, and A is caused to persist because of a good-producing selection process P, then:

(G3.4)   A Bs in order to C *iff* A Bs because in the process P the following telic structure is typically true: [A's Bing is a means to Cing & Cing is good (for A)].

For the features of parts of types of organisms, almost but not exactly the same conditions must be met. If A is a typical part of an organism of type A', and B and C are features of A, and A' is caused to exist (or persist) because of a good producing selection process P, then:

(G3.5)   A Bs in order to C *iff* A Bs because in the process P the following telic structure is typically true: [A's Bing is a means to Cing & Cing is good (for A')].

These two schemas illustrate only two central instances of the various forms of grade-three teleology that derive from a certain kind of telic mechanism: good-producing selection processes. Analogous but slightly modified schemas would capture telic phenomena involving further types of entities, e.g., species or whole ecosystems. further straightforward modifications would give due attention to the potential relativity of teleology to such factor as the environment.[29] Yet more modifications of the same basic pattern would appropriately delineate certain further special kinds of teleology, such as that of traits that originated for some non-telic reason and only later persisted because of their good consequences.

8   Teleology in Biology

Are there any grade-three good-producing selection processes in biological nature? Natural selection operates in the biological world, and natural selection might look like a good-producing selection process, but natural selection does not meet grade three. This point is easy to overlook, partly because of the scope difference between grade two and grade three.

Natural selection explains why certain features tend to predominate in certain populations in the long run—in particular, those features that promote survival (or reproduction).[30] So, in biological systems natural selection produces survival-promoting selection processes that explain why creatures have survival-promoting features. And survival is surely a good for living creatures, indeed, a paramount good. But natural selection also operates in certain populations of non-living entities and explains certain of their features. For example, natural selection explains certain features found in certain populations of clay crystals and in certain populations of polymers.[31] Now, in contrast with forms of life, survival is neither good nor bad for non-living things. Non-living things are not the kind of thing that can be beneficiaries. (Of course, the survival of non-living things might matter to an interested third party, but that is irrelevant.) So, natural selection over non-living populations is not a good-producing but merely a survival-producing process, and such a situation would exhibit *no* grades of teleology.

Since survival is good (indeed, the paramount good) for forms of life, the scope difference between grade two and grade three comes into play when natural selection operates over living populations. This scope difference concerns whether the goodness of survival plays an essential role *within* the natural selection explanation (grade three), or whether the goodness of survival is merely *conjoined* to the explanation (grade two). To see whether natural selection among living creatures involves grade two or three, notice the parallel with natural selection among non-living entities. We noted above that, in populations of non-living things like crystals, natural selection involves no value. But the kind of explanation that natural selection provides in the non-living case is the *same* as the kind it provides in the biological case. In the biological and non-biological cases alike, natural selection promotes features that contribute merely to a creature's survival. Thus, although it is true that survival is good for living creatures, this is irrelevant to the natural selection. Natural selection is blind to the goodness that supervenes on a biological creature's survival. Since the goodness of survival does not itself play a role in natural selection, biological teleology never surpasses grade two teleology.

There is a relatively clear-cut distinction between traits of an organism that persist in a population *because* these traits contribute to the survival of the organism (or its species, etc.), and traits of an organism that happen to benefit the organism but are *not* explained by these beneficial effects; teleology is often deemed to be appropriate for the former, but not for the latter. This familiar point is not my point here, though. By contrast, I have been arguing that even the former cases are not full-fledged teleology. Even in those cases, it is not a trait's goodness *per se* that explains its presence but merely its contribution to survival (or fitness generally); as far as natural selection goes, any goodness that a fitness-promoting trait might possess is irrelevant.

The value analysis partially supports teleological explanations in biology and partially undercuts them. Many features of organisms perform valuable functions, and so minimally have grade one teleology. Furthermore, the survival-promoting features produced by natural selection happen also to be good-producing features, so grade two teleology in biology is vindicated. On the other hand, grade three teleology in biology does not exist; there are no true full-blooded teleological explanations in biology. Except for the teleology traceable to the mind, the conditions required for grade three explanations are never present in the natural biological world.

9 Why Biological Teleology is Controversial

Teleological explanations in biology have persistently worried biologists and philosophers. Several specifically biological factors fuel the controversy. First, teleology in biology has traditionally been underwritten by deism, and modern biology has no place for a deity. The deism supporting teleology in biology has itself often been underwritten by the argument from design, and no version of this notorious argument can claim general acceptance today. In addition, some think that biological teleology requires that the biological world is a "nice" place, a benevolent Garden of Eden, but this rosy picture of the world reflects only wishful thinking about a biological world that is really "red in tooth and claw." Moreover, some think teleology in biology requires a single global purpose, into which every other local purpose fits, and because of which every other local purpose gets its purpose. The cut-throat competition observable throughout the biological world belies any such global purpose.

Even without a global purpose, some worry that biological teleology still requires preordained goals, fixed for individual species in advance of their evolutionary development, driving evolutionary development toward the realization of those goals. The random, opportunistic, open-ended working of natural selection debars any such pre-ordained species-specific teleology. Finally, teleology in biology has sometimes been underwritten by things like vital forces, and biological science has no place for unscientific, *ad hoc* entities or forces.

There would be good grounds for worrying about any form of biological teleology that involved any of these factors. However, teleological explanation in biology need have none of these defects. For example, teleological explanations underwritten by (hypothetical) good-producing selection processes would not involve deism, the argument from design, a Garden of Eden, a global purpose, pre-ordained purposes, or vital forces. Even so, teleological explanations based on good-producing selection processes would *still* be controversial. Hence, the factors mentioned above do explain why certain kinds of teleological explanations in biology have been criticized, but not why the very idea of any sort of teleological explanation is so objectionable.

The value analysis can attribute much of the controversy over biological teleology simply to the subtle distinction between grade two and grade three teleology. Grade three teleology seems to be present in biology, but only its grade two close cousin really exists. Whenever something is quite different from what it seems to be, confusion and controversy can be expected. But the controversy over biological teleology can be unusually heated, and the value analysis can explain this, too. Grade two biological teleology would entail that (non-explanatory) value considerations apply to the biological world, and the concern to deny any place for value in biological science would prompt suspicion about biological teleology. The deeper commitment to value of grade three teleology provides a further source of controversy. If there were any form of grade three biological teleology, the goodness of biological ends would play an essential role in the explanation of the existence or features of biological organisms. A lingering inclination to reject biological teleology can be traced to the concern to guard against and reject any suggestion that biology gives a specifically *explanatory* role to value.

10   Mentalism Again

Earlier the mental approach was faulted on the grounds that it failed to account for teleological explanations in biology. Now we see that teleological explanations in biology are called into question, for there are no true grade three teleological explanations in biology. This might seem to revive the mental approach; we cannot fault mentalism for failing to cover biological teleology if there is no true biological teleology to cover. However, mentalism (at least, in the form earlier discussed) is not revived, for two reasons.[32]

First, the primary datum that a theory of biological teleology must explain is why so many biological phenomena *seem* teleological. One way to explain this apparent teleology is to claim that it is genuine teleology; another is to claim that, although merely apparent, it is something that is easily mistaken for genuine teleology. (The latter is the sort of explanation provided in section 8.) According to typical formulations of mentalism, all non-mental teleological explanations in biology are false; thus, the first of these explanatory strategies is not available. Moreover, the truth in biology is quite unlike what mentalistic teleological explanations would require. Primitive organisms might have primitive minds, but organs like the heart do not, nor do schools of fish; yet it still seems that the heart pumps blood in order to circulate it and that fish school in order to avoid predators. Therefore, the second explanatory strategy is also unavailable to mentalism. Thus, mentalism cannot explain the striking naturalness of teleological explanations in biology.

Second, even if all *actually true* teleological explanations pertain either to the actions of mental agents or to the artifacts designed and used by them and so are mental, still there are *hypothetical* counterexamples to mentalism. For example, worlds with good-producing selection processes are logically coherent, even if they do not actually exist. And there would be non-mental biological teleology in a hypothetical non-mental world in which, as a matter of some sort of natural law, creatures typically have features because of the goodness they confer. Something like this seems to have been Aristotle's view; he spoke of nature as being a good housekeeper and doing nothing in vain, but intended these regularities to result from something more like natural law than mental agency.[33] Given

Aristotle's "good housekeeper" language, it is no surprise that many commentators feel trapped into (mistakenly) judging Aristotle's teleology to be mentalistic (or, if not mentalistic, then incoherent).[34] The grade three theory, by contrast, shows that it is straightforward to make sense of biological teleology without mental agency. Aristotle's picture of non-mental biological teleology happens not to fit the biological facts, but the picture is nevertheless *possible*. Full grade three teleological explanations *could* be true in non-mental worlds. Thus, the grade three theory does not collapse into mentalism, even if the two are accidentally coextensive.

Mentalism might be thought to be revived merely by the central role given to value considerations in the grade three theory. For one might claim that value notions are themselves mentalistic, being merely Humean projections onto natural objects or events of the sentiments of mental consciousnesses. But this sort of Humean approach to value, even if sound, would not entail the truth of mentalism in teleology—the view that *all* teleology can be reduced to the teleology of mental purposes and goals. If grade three teleology were coupled with a projectivist view of value, the result would be a (partially) projectivist view of teleology. According to such a projectivist view of teleology, all teleology would be traced to the existence of suitable mental sentiments, but those sentiments would not be the mental purposes and goals of mental teleology.

Conclusion

So, where is the good in teleology? This question can be taken in two ways. One interpretation is *why do we seek teleological explanations?* Presumably, we seek teleological explanations in order to obtain a certain kind of knowledge. If the grade three analysis is correct, we seek teleological explanations in order to identify the good consequences that cause their own means. Another interpretation of the question is *what role does value play in teleological explanation?* The simple three grade answer is that there is no simple answer. Value notions play different but thematically related roles in each of three increasingly strong grades of teleology; in addition, value notions play a variety of roles in different instances of each grade. Thus, the three grade analysis is an open-ended set of instances of the three grades, the three core kinds of teleology. A theory of teleology should cover not only actual cases; in fact, it should suggest how to find new hypothetical cases. The three grade analysis does this. To invent a new kind of teleological explanation, simply invent another possible kind of telic mechanism, that is, another way in which telic structures could be causally efficacious.

The value approach to teleology has suffered bad press because the three grades of teleology have not been distinguished. Almost without exception attempts to develop a value approach have followed the good-consequences line, which can reflect only the first grade of evaluative involvement. The limitations of grade one teleology have prompted many to reject the whole value approach. This, I have tried to show, is a mistake. Value plays a role in both grade two and grade three teleological explanations, and in grade three that role is an essential part of the explanation. Failure to appreciate these subtleties makes it impossible to give the different grades of mental and biological teleology their due.[35]

Notes

1. The idea that value plays a central role in teleology is not new. It has ancient roots in Plato and Aristotle, and its modern exponents include Leibniz and Kant. Various versions of a value approach have received contemporary defenses. See F. Ayala, "Teleological Explanations in Evolutionary Biology," *Philosophy of Science* 37 (1970): 1–15; J. Canfield, "Teleological Explanation in Biology," *The British Journal of the Philosophy of Science* 14 (1963–64): 285–295; P. Grice, "Reply to Richards," in *Philosophical Grounds of Rationality: Intentions, Categories, and Ends,* ed. R. Grandy and R. Warner (Oxford: Oxford University Press, 1986), pp. 45–106; C. Hempel, "The Logic of Functional Analysis," in his *Aspects of Scientific Explanation* (New York: Free Press, 1965), pp. 297–330; M. Ruse, *The Philosophy of Biology* (Atlantic Highlands, NJ: Humanities Press, 1973); R. Sorabji, "Function," *The Philosophical Quarterly* 14 (1964): 289–302; T. L. S. Sprigge, "Final Causes," *Aristotelian Society Supplement* 45 (1971): 149–170; Woodfield, *Teleology* (Cambridge: Cambridge University Press, 1976). The view defended in this paper is most like those of Ayala and Woodfield.

Teleology is not the only central philosophical notion that seems to have a value-centered analysis. For example, a diverse range of philosophers hold that, in one way or another, rationality is value-centered rather than value-free. See R. Brandt, *A Theory of The Right and The Good* (New York: Oxford University Press, 1979), D. Gautier, *Morals By Agreement* (New York: Oxford University Press, 1986), B. Gert, *Morality: A New Justification of The Moral Rules* (New York: Oxford University Press, 1988), and J. Rawls, *A Theory of Justice* (Cambridge, MA: Harvard University Press, 1971).

2. C. Boorse, "On the Distinction between Disease and Illness," *Philosophy and Public Affairs* 5 (1975): 49–68, p. 58, n. 13. See also Boorse, "Wright on Functions," *The Philosophical Review* 85 (1976): 70–86, and "Health as a Theoretical Concept," *Philosophy of Science* 44 (1977): 542–573.

3. To minimize tedium, I will henceforth speak elliptically of statements when what I mean are sentences used to express statements.

4. I should emphasize that "in-order-to statement" is a technical term for me, covering just those statements of the form identified above. Not all statements employing the phrase "in order to" are in-order-to statements. For example, during a discussion of Zeno's paradoxes, I might say "In order to reach the goal one must first reach a point midway to the goal." This statement clearly offers no teleological explanation; however, it is not of the form *A Bs in order to C* but (after substituting synonyms and converting into the present tense) of the different form *A must B in order to C,* or, equivalently, *If is to C, then A must B,* instances of which we might call *necessary-condition* statements. Necessary-condition statements are not equivalent to in-order-to statements, as is shown by recasting (1) above into the necessary-condition statement "I must run in order to get to class on time." Even if I must run in order to get to class on time, I might not do so for I might choose to be late. Also, if I am running in order to get to class on time, running might not be the *only* way to do so; driving might be equally effective.

5. See C. J. Ducasse, "Explanation, Mechanism, and Teleology," *The Journal of Philosophy* 22 (1925): 150–154; Woodfield, op. cit. Woodfield's mentalism is not global in scope, but rather applies only to goal-directed behavior.

6. For details, see my "Against Mentalism in Teleology," *American Philosophical Quarterly* 27 (1990): 61–70.

7. *The Mind and Its Place in Nature* (London: Routledge and Kegan Paul, 1925), p. 82. Broad also adds a second condition, that if we investigate the item under the guidance of the hypothesis that it has been designed by a rational mind for certain purposes, this hypothesis is confirmed. This second condition does not materially affect the conclusion of my argument. I discuss Broad's thesis (including the second condition) in greater detail in "Against Mentalism in Teleology," op. cit.

8. This latest version of mentalism appeals to "sufficiently normal" purposes, but not everything makes sense as a possible purpose. Value provides a natural and flexible constraint on possible purposes. If we can understand why (or, at least, we can understand that) someone sees something as *good* for something, then we are inclined to consider it to be a "sufficiently normal" purpose.

9. M. Beckner, *The Biological Way of Thought* (Berkeley: University of California Press, 1968); C. Boorse, "Wright on Functions," op. cit.; R. B. Braithwaite, *Scientific Explanation* (Cambridge: Cambridge University Press, 1953); E. Nagel "Teleology Revisited," *The Journal of Philosophy* 74 (1977): 261–301. See also R. Van Gulick, "Functionalism, Information and Content," *Nature and System* 2 (1980): 139–162.

10. For a more extensive development of these arguments, see my "Goal-Directed Systems and the Good," *The Monist* (1992): 34–49, and my "Naturalism and Teleology," in S. Warner and R. Wagner (eds.), *Naturalism: A Critical Appraisal* (Notre Dame, IN: University of Notre Dame Press, 1992).

11. See J. Bennett, *Linguistic Behavior* (Cambridge: Cambridge University Press, 1976); G. Cohen, *Karl Marx's Theory of History* (Princeton: Princeton University Press, 1978); C. Taylor, *The Explanation of Behavior* (New York: Routledge & Kegan Paul, 1964); L. Wright, "Functions," *The Philosophical Review* 82 (1973): 139–168, and *Teleological Explanations* (Berkeley: University of California Press, 1976). See also R. Millikan, *Language, Thought and Other Biological Categories* (Cambridge, MA: MIT Press, 1984), R. de Sousa, "Teleology and the Great Shift," *The Journal of Philosophy* 81 (1984): 647–653, J. Bigelow and R. Pargetter, "Functions," *The Journal of Philosophy* 84 (1987): 181–196, and M. Matthen, "Biological Functions and Perceptual Content," *The Journal of Philosophy* 85 (1988): 5–27.

12. This example (though not its diagnosis) is due to Robert Van Gulick.

13. There are other differences between these cases, of course; for example, the heart but not the stick is the product of a process of natural selection. For detailed discussion of these issues, see my "Can Biological Teleology be Naturalized?" *The Journal of Philosophy* 88 (1991): 647–655, and my "Naturalism and Teleology," op. cit.

14. For different versions of good-consequence treatments of function statements, see Canfield, op. cit., Hempel, op. cit., Ruse, op. cit., and Sorabji, op. cit.

15. Another example (from Wright, op. cit.) shows that the good-consequence analysis of inorder-to statements for artifacts fails to provide a sufficient condition. Assume for the sake of argument that an accidental consequence of the operation of the sweep second hand of a certain kind of watch is to clean dust from the watch, thereby making the watch more precise. Making the watch more precise is good for those who use the watch, but the sweep second hand does not clean dust in order to make the watch more precise.

16. Some might believe that it would be more accurate to say that in-order-to statements merely "allude to" explanations, since (for example) they do not cite a general law. I will ignore this detail, since my argument does not depend on it.

17. My talk of "three grades" of the involvement of value in teleology is an allusion to W. V. O. Quine's "Three Grades of Modal Involvement," in his *The Ways of Paradox and Other Essays* (New York: Random House, 1966), pp. 156–174.

18. The third grade might more clearly (but less compactly) be expressed as follows: A Bs in order to C iff A's Bing contributes to Cing, Cing is good, and both A's Bing contributing to Cing and the goodness of Cing explain A's Bing.

19. The three grades have not before been distinguished, to my knowledge. Most contemporary value approaches (for example, Canfield, op. cit., Hempel, op. cit., Ruse, op. cit., and Sorabji, op. cit.) have used the first grade as an analysis

of having a function. Ayala, op. cit., and Woodfield, op. cit., defend analyses of teleological explanation that give value an explanatory role, but it is unclear whether their analyses are intended to be grade two or grade three.

20. See Canfield, op. cit., and Ruse, op. cit.

21. "Two Dogmas of Empiricism," reprinted in W. V. O. Quine, *From a Logical Point of View* (New York: Harper and Row, 1953), pp. 20–46.

22. Various versions of this thesis have commanded wide agreement throughout the history in philosophy, perhaps starting with Aristotle, *Nicomachean Ethics*, bk. I. Rawls, to pick a contemporary example, seems to hold a version of this view; see op. cit., §61. For a wealth of further references, see Rawls, p. 400, n. 2.

23. See J. Elliot and R. Kerr, *Rings* (Cambridge, MA: MIT Press, 1984).

24. Again, perhaps this is best described as the form of an explanation sketch, while the form of an actual explanation would require another premise—a general law.

25. Other accounts of teleology appeal to distinctively telic laws, for the same sorts of reason. See, for example, Bennett, op. cit., Cohen, op. cit., and R. de Sousa, op. cit.

26. Discussion of the problem originated with R. Chisholm's "The Descriptive Element in the Concept of Action," *The Journal of Philosophy* 61 (1964): 613–624; subsequent discussion includes D. Davidson, "Freedom to Act," in his *Essays on Actions and Events* (New York: Oxford University Press, 1980), C. Peacock, "Deviant Causal Chains," *Midwest Studies in Philosophy* 4 (1979): 123–156, J. Searle, *Intentionality* (Cambridge: Cambridge University Press, 1983), and Woodfield, op. cit.

27. One might object to the mental instance of grade three teleology on the grounds that causal connections between (a) things that are concrete and physical, and (b) things that are abstract and mental, are impossible, incoherent, or objectionably mysterious. For a response to this kind of objection, see my "Cartesian Interaction," *Midwest Studies in Philosophy* 10 (1986): 483–502. I will ignore a number of details of practical reasoning that are peripheral to my main argument, such as which courses of action are open to an agent and how an agent ranks competing preferences. For a treatment of these details, see Bennett, op. cit.

28. Recent defenders of the Socratic thesis include A. Goldman, *A Theory of Human Action* (Princeton: Princeton University Press, 1970), and A. Woodfield, op. cit. Some traditional support can be found in Aristotle, *Nicomachean Ethics*, Bk, 1, ch. 1; Aquinas, *Summa Theologica*, Bk. I, Quest. 5, Art. 4, Quest. 20, Art. 1, Quest 82, Art. 2, Reply 1; and J. S. Mill, *Utilitarianism*, ch. 4. Traditional critics include Bishop Butler, Sermons 1 and 11, and D. Hume, *Enquires Concerning the Human Understanding and Concerning the Principles of Morals*, Appendix II ("Of Self-Love"). The case against the thesis comes from examples such as a drug addict who wants to quit, believes that his drug taking does not provide even a weak good, but still desires the drugs and acts on the basis of

this desire. My defense of the thesis emphasizes that the addict surely sees taking the drugs in some kind of positive light, if the drug taking activity is purposeful.

29. For a careful delineation of many such factors, see W. C. Wimsatt, "Teleology and the Logical Structure of Function Statements," *Studies in the History and Philosophy of Science* 3 (1972): 1–80.

30. My discussion here of natural selection is of necessity quick and superficial. For much more detailed discussions, see my "Can Biological Teleology be Naturalized?" and "Naturalism and Teleology," op. cit., as well as my "Measurement of Evolutionary Activity, Teleology, and Life" (with Norman Packard), in C. Langton, C. Taylor, D. Farmer, and S. Rasmussen, eds., *Artificial Life II* (Redwood City, CA: Addison Wesley, 1991), pp. 431–461. For an recent account of selection theories in general, see L. Darden and J. Cain, "Selection Type Theories," *Philosophy of Science* 56 (1989): 106–129.

31. For an explanation of how evolution takes place in populations of crystals, see A. G. Cairns-Smith, "The First Organisms," *Scientific American* 252, no. 6 (June 1985): 90–100, and *Seven Clues to the Origin of Life* (Cambridge: Cambridge University Press, 1985). For a more detailed discussion of some of the philosophical implications of non-biological evolution, see my "Can Biological Teleology be Naturalized?" and "Naturalism and Teleology," op. cit.

32. For additional reasons, see my "Against Mentalism in Teleology," op. cit.

33. For example, *Parts of Animals, passim*. For a recent discussion on the role of value in Aristotle's theory of teleology, see J. Cooper, "Aristotle on Natural Teleology," in *Language and Logos: Studies in ancient Greek philosophy presented to G. E. L. Owen*, ed. M. Schefield and M. Nussbaum (Cambridge: Cambridge University Press, 1982); A. Gotthelf, "The Place of Good in Aristotle's Natural Teleology," in *Proceedings of the Boston Area Colloquium in Ancient Philosophy*, vol. 4, ed. J. Cleary and D. Shartin (Lanham, MD: University Press of America, 1989); C. Kahn, "The Place of the Prime Mover in Aristotle's Teleology," in *Aristotle on Nature and Living Things*, ed. A. Gotthelf (Pittsburgh: Mathesis Publications, 1985); and A. Woodfield, op. cit.

34. For example, W. K. C. Guthrie, *Aristotle: An Encounter* (Oxford: Oxford University Press, 1981), p. 107, and Sir David Ross, *Aristotle* (London: Metheun, 1964), p. 186.

35. Earlier versions of this paper were presented at the first George Myro Memorial Conference at the University of California at Berkeley, Reed College, the University of Illinois at Urbana-Champaign, Vassar College, the University of Edinburgh, the University of St. Andrews, the University of Hull, and the Pacific Division of the American Philosophical Association. I thank the audiences on those occasions for their comments. Special thanks to George Bealer, Hugo Bedau, Jonathan Bennett, Janet Broughton, Alan Code, Stanley Eveling, Bob Fogelin, Bernie Gert, Paul Grice, Mark Hinchliff, Marvin Levich, John Llewelyn, Ruth Millikan, Jim Moor, David Reeve, Alex Rosenberg, Walter Sinnott-Armstrong, George Smith, Neil Thomason, Gerry Wakefield, Alan White, Carol Voeller, Steve Yablo, and especially George Myro.

III

Critical Developments

10

In Defense of Proper Functions

Ruth Garrett Millikan

Several years ago I laid down a notion that I dubbed "proper function" (Millikan 1984a) which I have since relied on in writing on diverse subjects. I have never paused to compare this notion with other descriptions of "function" in the literature, or to defend it against alternatives. That may seem a large oversight, amounting even to irresponsibility, and I wish to take this opportunity to remedy it.

It does not seem to be so much the details of the definition of "proper function" that need defense as its basic form or general plan, which looks to the *history* of an item to determine its function rather than to the item's present properties or dispositions. At any rate, it is this historical turn in the definition that I propose to defend. To understand this defense, you will not need to know the details of the definition I have given. Let me just say this much about it.

The definition of "proper function" is recursive. Putting things very roughly, for an item A to have a function F as a "proper function," it is necessary (and close to sufficient) that one of these two conditions should hold. (1) A originated as a "reproduction" (to give one example, as a copy, or a copy of a copy) of some prior item or items that, *due* in part to possession of the properties reproduced, have actually performed F in the past, and A exists because (causally historically because) of this or these performances. (2) A originated as the product of some prior device that, given its circumstances, had performance of F as a proper function and that, under those circumstances, normally causes F to be performed by *means* of producing an item like A. Items that fall under condition (2) have "derived proper functions," functions derived from the functions of the devices that produce them. Because the producing devices sometimes labor under conditions not normal for proper performance of their functions, devices with derived proper functions do not always have normal structure, hence are not always capable of performing their proper functions—a fact, I claim, that is of considerable importance.

This disjunctive description is extremely rough and ready. To make it work, "reproduction" must be carefully defined, the kind of causal-historical "because" that is meant carefully described, "normal conditions" defined, and various other niceties attended to. (The full description of proper functions consumes two chapters of Millikan 1984a.) But this rough description should make clear what I mean by saying that the definition of "proper function" looks to history rather than merely to present properties or dispositions to determine function. Easy cases of items having proper functions are body organs and instinctive behaviors. A proper function of such an organ or behavior is, roughly, a function that its ancestors have performed that has helped account for proliferation of the genes responsible for it, hence helped account for its own existence. But the definition of "proper function" covers, univocally, the functions of many other items as well, including the functions of learned behaviors, reasoned behaviors, customs, language devices such as words and syntactic forms, and artifacts. Moreover, if my arguments in (Millikan 1984a) are correct, explicit or conscious purposes and intentions turn out to have proper functions that coincide with their explicit or conscious contents. I have built a naturalist description of intentionality on the notion "proper function" (Millikan 1984a, 1986a, 1989, 1990).

I do have an excuse for having delayed my defense of the notion "proper function." The fact is that it is not crucial for the uses to which I have put the notion, whether or not its definition is merely stipulative or, if it is not merely stipulative, in what sense it is not. The point of the notion "proper function" was/is mainly to gather together certain phenomena under a heading or category that can be used productively in the construction of various explanatory theories. The ultimate defense of such a definition can only be a series of illustrations of its usefulness, and I *have* devoted considerable attention to such illustrations (Millikan 1984a, 1984b, 1986a, 1989, 1990, 1993). However, although it makes no material difference for the uses to which I have put the definition whether it is or is not merely stipulative, I believe that it is *not* merely stipulative, and that it is clarifying to understand the sense in which it is not. Besides, even if "proper function" were to be taken as a merely stipulative notion, the best way to understand it, though not to evaluate it, would surely be to see how it compares with other notions of function that have been described in the literature.

Some writers on function, teleology, and related matters have been explicit that they were attempting to provide conceptual analyses of certain idioms in current usage. For example, Andrew Woodfield states at the outset of his book *Teleology* (1975) that his project is to provide necessary and sufficient conditions for application of various kinds of sentences containing "in order to" and equivalent phrases. Many other writers simply take for granted that the project is conceptual analysis. Giving a germane example, Larry Wright (1976, p. 97); Christopher Boorse (1976, p. 74); Ernest Nagel (1977, p. 284); and Bigelow and Pargetter (1987, p. 188) each argue against an account of biological function that presupposes evolution by natural selection on the grounds that Harvey didn't know about natural selection when he proclaimed the discovery of the heart's function, or that evolutionary theory would have to be conceptually true to play any such role in the definition of function.[1] Such criticisms are valid only if the project is an analysis of the *concept* of function.

Now I firmly believe that "conceptual analysis," taken as a search for necessary and sufficient conditions for the application of terms, or as a search for criteria for application by reference to which a term has the *meaning* it has, is a confused program, a philosophical chimera, a squaring of the circle, the misconceived child of a mistaken view of the nature of language and thought. (Not to appear opinionated. But this prejudice is painstakingly defended in Millikan [1984a, chaps. 6, 8, and especially 9] so I have, I think, a right to it.) Still, I think that Woodfield and Wright, especially, have done good jobs of putting large portions of the area of this particular circle into a square, whether or not they have used only compass and rule in doing so. That is, theories of meaning to one side, each has done a fine job of collecting and systematizing various things that, without doubt, often *are* in the backs of people's minds when applying notions like "in order to" and "function," a fine job of spelling out analogies that commonly lubricate our transitions with these terms from one sort of context to another. Luckily there is no need to compete with Woodfield and Wright. My purpose, my program, is an entirely different one from that of conceptual analysis. An indication of this is that I do need to assume the truth of evolutionary theory in order to show that quite mundane functional items such as screwdrivers and kidneys are indeed items with proper functions. It is true, of course, that common persons make no such assumption when attributing functions to these items, nor does my thesis imply that they do.

It is traditional to contrast three kinds of definition: stipulative, descriptive, and theoretical. Descriptive definitions are thought to describe marks that people actually attend to when applying terms. Conceptual analysts take themselves to be attempting descriptive definitions. Theoretical definitions do something else, exactly *what* is controversial, but the phenomenon itself—the existence of this kind of definition—is evident enough. A theoretical definition is the sort the scientist gives you in saying that water is HOH, that gold is the element with atomic number 79 or that consumption was, in reality, several varieties of respiratory disease, the chief being tuberculosis, which is an infection caused by the bacterium *bacillus tuberculosis*. Now I do have a theory about what theoretical definitions are, a theory about how the theoretical definition of "theoretical definition" should go. Unfortunately this theory rests upon a theory of meaning that rests in turn on the notion "proper function," the very notion under scrutiny. But assuming that you at least countenance the *phenomenon* of theoretical definition, let me say that my definition of "proper function" may be read, roughly, as a theoretical definition of function.[2] It may be read as a theoretical definition of function in the context "The/a function of _____ is _____" (the function of the heart is to pump blood), though *not* in the context "_____ functions *as* a _____" (the rock functions as a paperweight). The definition of proper function may also be read as a theoretical definition of "purpose."

Now jade turned out to be either of two compounds, nephrite or jadite, these being chemically quite distinct, and there are two quite different kinds of acidity and several alternative ways to count genes, if current chemical and genetic theory are correct. Similarly, there would be no reason to suppose *in advance* that "has a function" must correspond to a unitary non-disjunctive kind. But my claim is that "has a function" *does* as a matter of fact correspond, in a surprising diversity of cases, to having a *proper* function. Further, the various properties, the various analogies, which are influential in leading us to speak of quite diverse categories of items as having "functions" are properties or analogies that are, characteristically, *accounted for* by the fact that these items have coincident *proper* functions. It does not follow, nor is it my claim, that there are no *logically* possible cases in which analogy might lead us to speak of an item's function when the item in fact had no proper function. Nor does it follow that every *logically* possible case in which an item has a proper function is a case we would recognize offhand as a case of having a function. The technique of testing a definition by a search through possible worlds, by ingenious construction of fictional counter-examples, is not appropriate for theoretical definitions. There are also logically (or, at least, conceptually) possible worlds in which water does not turn out to be HOH.

A particularly glaring example of the failure of my definition to cover logically possible cases that strike many as legitimate cases of having purpose or function is the example of accidental doubles. According to my definition, whether a thing has a proper function depends on whether it has the right sort of history. Take any object, then, that has a proper function or functions, a purpose or purposes, and consider a double of it, molecule for molecule exactly the same. Now suppose that this double has just come into being through a cosmic accident resulting in the sudden spontaneous convergence of molecules which, until a moment ago, had been scattered about in random motion. Such a double has no proper functions because its history is not right. It is not a reproduction of anything, nor has it been produced by anything having proper functions. Suppose, for example, that this double is your double. Suddenly it is sitting right there beside you. The thing that appears to be its heart does not, in fact, have circulating blood as a proper function, nor do its apparent eyes have helping it to find its way about as a proper function, and when it scratches where it itches, the scratching has no proper function.

Contrast this historical notion of proper function with any description of function that makes reference only to *current* properties, relations, dispositions or capacities of a thing. Contrast it, for example, with any of the various contemporary descriptions that have been offered of purposive behavior as goal-directed, or with descriptions of purposiveness as involving negative feedback mechanisms, or of the purposive as that which tends toward the "good" of some creature, or contrast it with Jonathan Bennett's (1976) description of purposive action in terms of dispositions to act given dispositions to "register" situations in which "instrumental predicates" apply, or with Robert Cummins' (1980, 1984) description of the function of an item within a system admitting of a functional analysis. According to each of these conceptions of function, if anything has a function, of course its double must have a function too— the same function. To many this seems an obvious truth, one that any respectable theory of function should entail. So in the case of your double's heart, eyes and scratchings, all of these contemporary definitions seem to be *headed* in the right direction, my definition in the wrong direction. What am I to say about that?

What I'm going to say, rather brazenly, is that such cases are like the case of fool's gold, or better, since the case is fictional, like the case of Twin Earth water. Perhaps lots of people have taken fool's gold for gold, people in perfectly good command of their language. And if suddenly transplanted to Twin Earth, you would take XYZ to be water. Similarly, even though many people would be prone to say it did have a purpose, that apparent heart in your double's body really would not have a purpose. It would merely display enough *marks* of purposiveness to fool even very sophisticated people. Without any question, there has never in fact existed anything within several orders of magnitude of the complexity of your fictional double, anything that was as neatly engineered to further its own survival and reproduction, that did not also have a history of the right sort to bestow upon its various parts the relevant proper functions. Nor do there *in fact* exist complicated goal-directed items, or items displaying complicated negative feedback mechanisms, or items that do anything like "registering" situations, or items with interesting Cummins functions, that do not *in fact* have corresponding proper functions. Having the *right sort* of current properties and dispositions is in point of *fact*, in *our* world, an infallible index of having proper functions. If you like, it is criterial—as criterial, say, as the red of the litmus paper is of acidity. But it is not turning litmus paper red that *constitutes* acidity, nor is it having the right sort of current properties and dispositions that *constitutes* a thing's having a purpose. To the degree that each of these contemporary descriptions in terms of current properties or dispositions is successful, each describes only a *mark* of purposiveness, not the underlying structure.

The definition of "proper function" is intended as a theoretical definition of function or purpose. It is an attempt to describe a unitary phenomenon that lies behind all the various sorts of cases in which we ascribe purposes or functions to things, which phenomenon normally *accounts for* the existence of the various analogies upon which applications of the notion "purpose" or "function" customarily rest. My claim is that actual body organs and systems, actual actions and purposive behaviors, artifacts, words and grammatical forms, and many customs, etc., all have proper functions, and that these proper functions correspond to their functions or purposes ordinarily so called. Further, it is *because* each of these has a proper function or set of proper functions that it has whatever marks we tend to go by in *claiming* that it has functions, a purpose, or purposes.[3]

I have said that the definition of "proper function" is intended to explain what it is for an item to *have* a function or purpose, but not what it is for an item to function *as* something. Robert Cummins (1980, 1984) has given us a definition of function that is probably best construed as a theoretical definition, but a theoretical definition of "function as," in some contexts of use, rather than of "function" meaning purpose. Cummins' project is to explicate what "contemporary natural scientists" are describing when they offer a certain kind of explanation of the performance of an item within the context of a system. Very roughly, what these scientists do, according to Cummins, is to explain why the system as a whole has the capacity to do some complex or sophisticated task by appealing either to its capacity or to the capacities of its parts to do a series of simpler tasks which add up, flow chart style, to the original complex capacity. Cummins calls this kind of explanation "functional explanation," and claims that the various elementary capacities appealed to in such explanation correspond to "functions," within the system, of the elements having these capacities.

This notion of function is a highly illuminating one. But it does not correspond, nor does Cummins take it to correspond, to that basic sense of "function" that hooks function to purpose. For example, according to Cummins' definition it is, arguably, the function of clouds to make rain with which to fill the streams and rivers, this in the context of the water-cycle system, the end result to be explained being, say, how moisture is maintained in the soil so that vegetation can grow. Now it is quite true that, in the context of the water cycle, clouds function to produce rain, function *as* rain producers; that *is* their function in that cycle. But in *another* sense of "function," the clouds have no function at all—because they have no purpose.

Cummins explicitly waves aside all reference to "purposes" and all "appeals to the intentions of designers and users" in describing a thing's function, at the same time acknowledging that such appeals are of course made in many contexts in which we apply the term "function" (Cummins 1980, p. 185). By Cummins' definition, in order to *have* a function an item must actually *function* in a certain way, function *as* something or other, or at least must have a disposition or capacity so to function: "... if something *functions as* a pump in a system ... then it must be capable of pumping ..." (Cummins 1980, p. 185, emphasis mine). But it is of the essence of purposes and intentions that they are not always fulfilled. The fact that we appeal to purposes and intentions when applying the term "function" results directly in ascriptions of functions to things that are not in fact capable of performing those functions; they neither function as nor have dispositions to function as anything in particular. For example, the function of a certain defective item may be to open cans; that is why it is called a can opener. Yet it may not function *as* a can opener; it may be that it won't open a can no matter how you force it. Similarly, a diseased heart may not be capable of pumping, of functioning *as* a pump, although it is clearly its function, its biological purpose, *to* pump, and a mating display may fail to attract a mate although it is called a "mating display" because its biological purpose is to attract a mate.

There is another way of viewing the definition I have given of "proper function," which throws the spotlight not on purpose but on *function categories*. Every language contains nearly innumerable common nouns and noun phrases under which things are collected together in accordance *not* with current properties, activities or dispositions, but in accordance with function. For example, consider the categories thermometer, can opener, heart, kidney, greeting ritual, mating display, fleeing behavior, stalking behavior, adverb, noun, indicative mood sentences, and word for green. Anything falling in one of these categories is what it is, falls in the category it does, by reference to function. One way to focus on the problem that the definition of "proper function" is designed to solve is to ask how items that fall under function categories are grouped into types.

Now an obvious fact about function categories is that their members can always be defective—diseased, malformed, injured, broken, disfunctional, etc.,—hence unable to perform the very functions by which they get their names.[4] Nor will it do (as a surprising number of people have done) merely to point out that the typical or normal items falling in a function category actually do or can perform the function that defines the category. The problem is, how did the atypical members of the category that cannot perform its defining function *get* into the same function category as the things that actually can perform the function? Besides, it is not always true that typical items falling in a function category perform that function. It is quite possible, for example, that the typical token of a mating display fails to attract a mate, and that the typical distraction display fails to distract the predator.

Nor is mere similarity to other items that perform a certain function either necessary or sufficient, by itself, to bring an item under a function category. No matter how similar a piece of driftwood is to an oar, this similarity does not, by itself, make it into an oar. No matter how similar the mating display of one fish may be to the aggression display of another, this does not make the mating display into an aggression display. And no matter how similar the scratches that the glacier left on the rock are to token of the English word "green," they do not thereby compose a token of a word for green. Also consider: exactly what rules would articulate the *kind* of similarity to functioning members of a category the non-functioning members should have? For example, exactly *what* sorts of (current) properties must an item have in common with some functioning token or other of a can opener in order to count as a "can opener that doesn't work"? The question is absurd on its face.

Indeed, a thing that bears no resemblance to any can opener previously on earth—suppose it has been designed in accordance with a totally new principle—may still *be* a can opener, and may be one despite the fact that it doesn't work. Remember the train brake that Christopher Robin made that "worked with a string sort of thing"? "It's a very good brake, but it hasn't worked yet," said Christopher. What's amusing about that is not that Christopher claims it's a brake, but that he claims it's a very *good* brake.

There is a tendency, I think, to believe that the phenomenon of defective members of function categories is a superficial phenomenon, that it is only by some sort of extension or loosening up of basic criteria in accordance with which things are placed in function categories that defective members are admitted as members at all. But note how different the notion "defective" is from, say, the notion "borderline case" or the notion "case only by courtesy." Monographs may be only borderline cases of books, or may be books only by courtesy, but surely monographs are not *defective* books. The notion "defective" is a normative notion. The problem is, what makes the defective item fall under a *norm*? Surely, not just that it *reminds* one of things that do serve a certain function, so that it makes one wistful?

That members of function categories can be defective is coordinate with purposes as, essentially, things that may not get fulfilled. Function categories are *essentially* categories of things that need not fulfill their functions in order to have them. Just as the characteristic mark of intentionality is that intentional items can be false, unsatisfied, or seemingly "about" what does not exist, so the characteristic mark of the purposive, of that which has a function, is that it may *not* in fact fulfill that purpose or serve that function. For example, your randomly created double exhibits no purposive behaviors and has no purposive parts because there is no way that any of his/her states or parts could be *defective* or might *fail*. That creature of accident, wonderful as he or she may be, falls under no norms.

The intimate connection between function category and purpose and the essential connection of these with norms, hence with possible failure, is easily obscured, however, when we turn to the analysis of purposive *behaviors*. This is because the vast majority of our categorizations of behaviors, the vast majority of our simple descriptions of behaviors, employ success verbs rather than verbs of trying. We tend to categorize behaviors according to the purposes they actually *achieve*. Indeed, we often ignore purpose altogether, classifying behaviors in accordance *merely* with effects of these behaviors, whether purposive effects or not. For example, I can bump you with my elbow either purposefully or by accident. On the other hand, merely trying to bump you and failing doesn't count as bumping at all. Of course verbs of trying do exist in the language. Consider "fleeing," "stalking," "fishing," "hunting," "looking," "bidding for attention," and, of course, any ordinary action verb prefaced by "trying to." *These* action categories are function categories, defined by reference to purposes rather than achieved effects. What makes a behavior fall into one of *these* categories?

The question is seldom tackled head on. Rather, investigators typically begin with the question, "What makes purposive behavior purposive?" (or, say, "goal directed?") and then take as their *paradigms* of purposive behavior not trying behaviors but behaviors described, as it is most natural to describe behaviors, by success verbs. Only later do they attempt to loosen up or stretch the model they have already built for successful purposive behaviors to cover the unsuccessful cases (if they ever recognize these cases at all). The result is that unsuccessful trying behavior is described as though it were a loose or borderline case of purposeful behavior, purposive only by courtesy. Or purposiveness is described as though it admits of degrees, the distinction between the purposive and the nonpurposive appearing to need drawing at an arbitrary place. Let me give an example.[5]

Early in his discussion of goal-directed behavior, Wright tells us that

... successful teleological behavior, directed behavior that actually achieves its goal, provides us with the best paradigm; it gives us the sort of case from which all others can be seen as natural derivatives. (Wright 1976, p. 37)

Soon after, Wright gives us the following formula:

S does B for the sake of G iff:
i. *B* tends to bring about *G*
ii. *B* occurs because ... it tends to bring about *G*. (p. 39)

Context makes it clear that this formula is a modification of a simpler formula covering only successful purposive behavior and containing "does bring about" in place of the looser "tends to bring about." Wright comments that the phrase "tends to bring about" in his formula "represents the entire 'family'" of which other members are "is the type of thing that brings about" (compare: is similar to items that *do* function to bring about—the wistfulness again), "is required to bring about," "is in some way appropriate for bringing about" (1976, p. 39), etc. (Later Wright seems to imply that "might easily be mistaken for something that would bring about" may have to be included in this "family" too (1976, p. 49). A loosening up indeed!) The unpacking that Wright then produces of "*B* occurs because it tends to bring about *G*," in this context, is dispositional. What he describes under this heading is a strong *correlation* between behaviors to which *S* is disposed and behaviors that "tend to bring about *G*." His claim is that the existence of such a correlation, *taken alone*, licenses us to infer, or to say, that the behavior occurs *because* of the tendency. This "because," it is important to notice, is *not* a causal-historical "because," but a special teleological "because."

Now suppose that we set aside problems that may arise with other members of the "tends to bring about" family. And suppose that we set aside questions about the reference class within which the statistics for "tends to" are to be gathered. (More about that in a moment.) We must still answer these questions. First, how *strong* a correlation must there be between what *S* is disposed to do and behaviors that "tend to bring about *G*" for us to attribute purposiveness to *S*? Second, how *strong* does a tendency have to be to count as a tendency, to count as helping to strengthen the correlation? And how would one decide either of those questions but arbitrarily? This kind of *intrinsically* fuzzy and arbitrary distinction between the purposeful and the purposeless—quite different, notice, from admission of borderline cases between clear paradigms—is just wrong. It not only is bad theory, it is not good conceptual analysis, not an accurate reflection of how most people *think* of purposiveness. But where writers acknowledge at all that purposive behavior may fail, this kind of fuzzy result is typical.[6]

Because investigators have assumed that the paradigm cases of purposive behavior are cases of successful behavior, they have been led to give descriptions of purposiveness that are variations on dispositional themes; to give one example, led to locate purposiveness in mechanisms, such as feedback mechanisms, which will produce dispositions to reach some goal or state. True, Wright takes the dispositions that evidence a behavior as being purposive to license a peculiar sort of inference to a peculiar sort of *explanation*, namely, the behavior occurs *because* of its tendency to lead to the goal. But Wright gives this kind of "because" or "on account of" no explication besides saying that it *is* the kind of "because" we make inferences to when we discover dispositions of the sort he describes. Possibly something like this is correct on the level of conceptual analysis. But on the level of *theory*, it leaves us with no useful distinction between an animal's having the right dispositions, and its having them in accordance with the right explanation, no useful distinction between the dispositions that count as *evidence* for teleological structure, and that which they evidence—the teleological structure itself.[7] My position, by contrast, is that although discovery of the sorts of mechanisms and/or dispositional structures that Wright and other theorists describe usually does license inference (inductive yet empirically certain inference) to a peculiar sort of explanation, this explanation is a straightforward *historical* explanation. Things just don't turn up with inner mechanisms or with dispositions like that unless they have corresponding proper functions, that is, unless they have been preceded by a certain kind of history. Moreover, being preceded by the right kind of history is *sufficient* to set the norms that determine purposiveness; the dispositions themselves are not necessary to purposiveness.

My claim has been that accounts of purpose or function in terms of present disposition or structure run afoul exactly when they confront the most central issue of all, namely, the problem of what failure of purpose and defectiveness are. But what leads me to conclude that *historical* analysis is what is needed instead? The fact that a historical analysis *works*, of course. The historical analysis I have given does cover all of the actual cases in which we ascribe functions to things. But prior to that, there is a strong clue that suggests that a good look at the historical dimension is needed for this kind of analysis.

Notice that talk of functional mechanisms and of dispositions that characterize purposive items is always talk accompanied, implicitly, by ceteris paribus clauses. My desk lamp has a disposition to give off light when its switch is depressed, but not of course when unplugged, when under water or when at 1000 degrees Fahrenheit. The mouse may have a disposition to take measures that will remove it from the vicinity of the cat, but not if under water, at 1000 Fahrenheit, in the absence of oxygen, while being sprayed with mace, or just after ingesting cyanide. Indeed, there are thousands of stressful conditions under which the mouse might be placed, which would extinguish its escaping behavior; you merely have to be sadistic enough to think of them. Nor can we plead that the mechanism must be under conditions such that it operates "properly," for "operating properly" is merely operating as it is "supposed" to operate, that is, in accordance with its design or purpose.

Increasingly I find in the literature on purpose (and on various subjects connected in one way or another with the related phenomenon of intentionality) handwaving, when things get rough, toward the relevant ceteris paribus clauses under the heading "normal conditions." On pain of circularity, "normal conditions" cannot mean "conditions under which the thing operates properly." What then *are* "normal conditions"? Are they average conditions? Where? Not throughout the universe, surely, for the average conditions there are being in nearly empty space at nearly absolute zero. Average conditions on earth? Conditions on earth have varied enormously throughout its history and they vary from place to place.

More central, note that what count as normal conditions for mouse behavior, shark behavior, robin behavior, earthworm behavior, and tapeworm behavior are quite different. Being underwater *is* a normal condition for shark behavior. The dispositions that express goal-directedness toward, say, obtaining food, are dispositions defined against quite different background conditions in the cases of the various species.

Are normal conditions for a mouse, perhaps, just conditions that mice, on the average, are in? Then if we tossed all mice but Amos into outer space, our listing of Amos' "dispositions under normal conditions" would have to change, the main one left to him being, I suppose, to explode.

Perhaps normal conditions under which Amos' dispositions are to be described are those under which his design is optimal for survival and proliferation? But *those* conditions include living in a world without cats. Also, of course, the wonder drug that prevents cell aging must be in Amos' water just as oxygen just be in his air; if Amos happens to be diabetic, someone who is disposed to administrator insulin must be available; if Amos is neurotic, some useful reward must be found for his neurotic behavior.

To explain what "normal conditions" are is surely going to take us on an excursion into history. At the very least, we must make a reference to something like conditions in which mice have *historically* found themselves, or better, found themselves when their dispositions actually aided survival. And if a listing of Amos' relevant dispositions depends on an implied reference to his species, then the question of what makes him fall into the category "mouse" needs to be raised. But the question of Amos' species is itself a question that diverts us through history. Mice must be born of mice. Consider: if a seeming mouse were born of a fish, what would set the "normal conditions" for manifestation of its relevant disposition?

But a more telling question, perhaps, than, Why look at history when trying to describe what functions and purposes are? is the question, Why *not* look at history? Why is there so much resistance to looking at history? There is an univocal answer to this question, I believe, and bringing it into the light of day proves very instructive.

We take consciously intentional action to be a paradigm of purposiveness. And that is correct; it *is* one paradigm (among others). But the idea that a consciously intentional action could have the purpose it does *not* by reference to anything merely present, let alone anything present in *consciousness*, runs strongly against the grain. Indeed, it runs against one of the most entrenched beliefs of both philosophers and laymen. This is the belief that the consciousness of one's own intentions is an *epistemic* consciousness, that is, in the case of one's own explicit intentions at least, what one intends is *given*, simply and *wholly* given, to consciousness. But historical facts, certainly facts about one's evolutionary history, clearly are not simply *given* to consciousness. Hence, what one's explicit conscious intentions are could not possibly depend on facts about one's history. Q.E.D.

The belief that the intentional contents of one's explicit intentions are "given" to consciousness is just one strand of a tangle of entrenched beliefs which I have called "meaning rationalism" (Millikan 1984a). Meaning rationalism, in its various forms, has gone unquestioned in the philosophical tradition to such a degree that, to my knowledge, no arguments have ever been adduced to support it. However, a large portion of Wittgenstein's *Philosophical Investigations* is devoted to an attempt to *dispel* the notion that what one intends, especially when one intends to follow a rule, is given in what appears before consciousness. And a considerable portion of the Wilfrid Sellars corpus is built on the motif that *nothing* is, epistemically, given to consciousness. More recently, Hilary Putnam (1975) and then Tyler Burge (1979) have argued that what one means by a word is not, certainly not always, determined by the contents of one's head, but by relations *between* one's head and the world. But it is hard to see how *any* relation between one's head and the rest of the world could be a relation that is simply and wholly given to consciousness. If these philosophers are right, and meaning something or intending something or purposing something depends on relations *not* packed inside an epistemic consciousness, then why are historical relations not as good candidates for this position as any other relations?

Acknowledgments

I am grateful to John Troyer, Peter Brown, and Jonathan Bennett, and to the members of the philosophy departments at Dartmouth and at Johns Hopkins, for helpful comments on earlier drafts of this essay.

Notes

1. Karen Neander's "Teleology in Biology," from which I originally got the page references in the above passage to Wright, Boorse and Nagel, contains a brilliant defense of the "etiological" account of function while remaining within the tradition of conceptual analysis.

2. More accurately, the definition is intended to express the "sense" of this notion rather than describing its "intensions," where "sense" and "intensions" are interpreted as described in Millikan (1984a). That is, according to my theoretical definition of "theoretical definition," what a theoretical definition analyzes is (Millikanian) sense.

3. Suppose that there really is a planet on which something as complex and apparently functional as your double is created by accident. I don't mean to rest my case wholly on the overwhelming unlikelihood of such an event. Similarly, there *may* be some very queer circumstances under which litmus paper turns red in a ph-neutral environment. Then the "criteria" commonly used to determine purposiveness or acidity are fallible indices, but the natural phenomena that correspond to these notions remain the same. For a theory of the relation of intension to extension that supports this kind of claim, see Millikan (1984a, 1986b, 1989, 1990).

4. Defective sentences and words? The easy examples are mispronunciations. For example, the child that pronounces "sin" like "thin" *mispronounces* the word "sin." She does not correctly pronounce the word "thin."

5. I have a special reason for using Wright's views as my foil in the following paragraphs. Various published remarks to the contrary, there is no overlap at all between Wright's analysis of function and mine. A reason for the confusion, I believe, is Wright's peculiar usage of the term "etiological" which does *not*, in his vocabulary, make reference to causes or origins—see below and, for example, note 7.

6. According to my own definition of "proper function," borderline cases do exist. These are always cases either of *derived* proper functions, or of functions of members of "higher order reproductively established families" (Millikan 1984a, chap. 1), and it is not the failure of the functional device itself but a partial failure of its producer which results in the vagueness.

7. Wright's discussion of goal-directed behavior differs, in this respect, from his analysis of the functions of body organs. In the case of body organs, he reads the "because," in "X is there because it does (results in) Z" (1976, p. 81) more like a causal "because," but still not as a causal-historical "because." Wright says that the formulation "because X does Z" does *not* reduce to "because things like X have done Z in the past" (pp. 89–90). Rather, we are asked to accept that X might be there *now* because it is true that *now* X does or Xs do result in Z. How the truth of proposition about the present can "cause" something else to be the case *at present* is not explained.

References

Bennett, J. (1976), *Linguistic Behavior*. Cambridge: Cambridge University Press.

Bigelow, J., and Pargetter, R. (1987), "Functions," *The Journal of Philosophy* 84: 181–196.

Boorse, C. (1976), "Wright on Functions," *The Philosophical Review* 85: 70–86.

Burge, T. (1979), "Individualism and the Mental," *Midwest Studies in Philosophy*, vol. 4, pp. 73–121.

Cummins, R. (1975), "Functional Analysis," *Journal of Philosophy* 72: 741–765.

————. (1980), "Functional Analysis," reprinted in part in N. Block (ed.) *Readings in Philosophy of Psychology*. Cambridge, Mass.: Harvard University Press, pp. 185–190.

————. (1984), *Psychological Explanation*. Cambridge, Mass.: Bradford Books/ MIT Press.

Millikan, R. G. (1984a), *Language, Thought, and Other Biological Categories*. Cambridge, Mass.: Bradford Books/MIT Press.

————. (1984b), "Naturalist Reflections on Knowledge," *Pacific Philosophical Quarterly* 65: 315–334.

————. (1986a), "Thoughts without Laws; Cognitive Science with Content," *Philosophical Review* 95: 47–80.

————. (1986b), "The Price of Correspondence Truth," *Noûs* 20: 453–468.

————. (1989), "Biosemantics," *Journal of Philosophy* 86 (No. 6).

————. (1990), "Truth Rules, Hoverflies, and the Kripke-Wittgenstein Paradox," *Philosophical Review* 99: 325–353.

————. (1993), "What Is Behavior? A Philosophical Essay on Ethology and Individualism in Psychology," chap. 7 of *White Queen Psychology and Other Essays for Alice*, Cambridge, Mass.: MIT Press.

Nagel, E. (1977), "Teleology Revisited," *The Journal of Philosophy* 84: 261–301.

Neander, K. (manuscript), "Teleology in Biology" (Wollongong University, Australia).

Putnam, H. (1975), "The Meaning of 'Meaning,'" *Philosophical Papers*, vol. 2. Cambridge: Cambridge University Press.

Woodfield, A. (1975), *Teleology*. Cambridge: Cambridge University Press.

Wright, L. (1976), *Teleological Explanation*. Berkeley: University of California Press.

11

Functions as Selected Effects: The Conceptual Analyst's Defense

Karen Neander

1 Introduction

This essay defends the etiological theory of proper functions, according to which, roughly speaking, biological proper functions are effects for which traits were selected by natural selection.[1] According to this theory, for instance, hearts have the proper function of pumping blood, because pumping blood is what hearts did that caused them to be favored by natural selection. This theory is now familiar enough (indeed it is fast becoming the consensus) but while it is mostly accepted that something of the sort happens to capture the actual reference class, the theory has been rejected as conceptual analysis. It is only because conceptual analysis itself has plummeted in the popularity stakes that the popularity of the etiological theory has, inversely, been allowed to rise. It is this to which I object.

The supposedly problematic feature of the etiological theory as conceptual analysis is that (true to its name) it makes a trait's function depend on its history, more specifically (and supposedly worse) on its evolutionary history. There are three standard objections to this which I will mention now and discuss later. (i) It seems blatantly inaccurate historically. When Harvey announced the function of the heart in 1616 he knew nothing of Darwin's theory, and so he clearly didn't mean, or have in mind, that the circulation of blood was the effect for which hearts were selected by natural selection (Wright 1976, 97; Boorse 1976, 74; Nagel 1977, 284). (ii) Defining the notion of a "proper function" in terms of natural selection begs empirical and theological questions, and is "analytically arrogant" because it would seem "... to suggest that it is impossible by the very nature of the concepts—logically impossible—that organismic structures and processes get their functions by the conscious intervention (design) of a Divine Creator" (Wright 1976, 96–97. See also Boorse 1976, 74; and Bigelow and Pargetter 1987, 188). (iii) It is strongly counterintuitive that a creature which lacked a history would thereby lack functions. Suppose we discovered that the whole lion species freakishly coalesced into existence one day, without evolution or design of any kind, "by an unparalleled saltation" (Boorse 1976, 74). Or consider "... the possible world identical to this one in all matters of laws and particular matters of fact, except that it came into existence by chance (or without cause) five minutes ago" (Bigelow and Pargetter 1987, 188). Surely, we can correctly ascribe biological functions to any such complex, intricately integrated organisms, despite their lack of history and their accidental genesis. Or so the argument goes.

Ruth Millikan has defended an etiological theory of proper functions from these objections (1989). However, she does so by turning her back on conceptual analysis. In her view, to use her own rhetoric, conceptual analysis is "... a confused program, a philosophical chimera, a squaring of the circle, the misconceived child of a mistaken view of the nature of language and thought" (1989, 290).[2] Millikan argues that the standard objections to the etiological theory are thus undermined by the fact that they presuppose that our task is conceptual analysis.

While Millikan's defense throws into stark relief our need to be clear about our analytic aims, I believe it is wrong to base our defense of the etiological theory on a rejection of conceptual analysis. In my view, the sins of conceptual analysis do not measure large enough for it to be discounted as a witness in this case. Moreover, if we are clear about our analytic aims, we can confront head-on these standard objections and successfully defend the etiological theory as conceptual analysis. Toward this, I have found it useful to follow Millikan in making a distinction between conceptual analysis and something she calls "theoretical definition." The first section of this paper begins with a discussion of this meta-analytic distinction, and explains why I believe conceptual analysis cannot be disregarded. Following this, I elaborate on the etiological theory being defended and then consider each of the standard objections in turn.

2   Conceptual Analysis or Theoretical Definition, or Both?

The difference between conceptual analysis and theoretical definition is hard to state in uncontroversial terms, but the basic gist of the difference, as I will be using this meta-analytic distinction in this paper, is as follows.

Conceptual analysis is an attempt to describe certain features of the relationship between utterances of the term under analysis, and the beliefs, ideas, and perceptions of those who do the uttering. It involves trying to describe the criteria of application that the members of the linguistic community generally have (implicitly or explicitly) in mind when they use the term. In contrast, a theoretical definition is an attempt to explain some aspect of the thing referred to, or some aspect of the relationship between utterances of the term and the actual world. Which particular aspect depends on one's semantic theory and on what one thinks is most central for semantic theory. Millikan argues (this is a crude rendition) that a theoretical definition should describe what a term refers to that explains the use of the term and why the term has survived and continued to be used (1984). She also suggests that theoretical definitions should describe the underlying phenomenon that explains the surface analogies by which we have recognized that things are things of a kind (1989, 293). Thus "water is a liquid with the molecular structure HOH" and "gold is the element with the atomic number 79" are theoretical definitions. (Liquid HOH is the underlying phenomenon to which we usefully refer, and it also explains water's opacity, its thirst quenching powers, and so on.)

Some would add two further features, which I have purposely omitted, to my characterization of conceptual analysis. They would add that conceptual analysis is thought to constitute a search for meaning, and that it is a search for necessary and sufficient conditions for the application of the term under analysis. Add these, and I too will probably abandon ship. Since Millikan includes these features in her characterization of conceptual analysis (1989, 290) some of our disagreement may be merely verbal. However, her defense of the etiological theory requires us to reject conceptual analysis as I have more weakly characterized it, and my weaker characterization describes an analytic exercise that is more readily defended.

Dispute over whether conceptual analysis or theoretical definition best captures meaning (properly so called) may be largely responsible for the rejection of conceptual analysis. So let me concede a point or two. It may be that meanings are not inside people's heads, not wholly, and perhaps not even partly. And since conceptual analysis is an attempt to understand what goes on inside speakers' heads—in that it is a search for the criteria of application that people generally have in mind—it may be that, if we are searching for meaning, conceptual analysis is a confused approach. But we might be interested in something other than, or in addition to, meaning. So let us define "conceptual analysis" as a search for the criteria of application that people generally have in mind when they use the term under analysis, and discuss the merits of such a search, leaving the issue of meaning aside. Perhaps attempting to understand how biologists implicitly understand the notion of a "proper function" can be fruitful, whether or not this gives us the meaning of the term.

Conceptual analysis has also often been associated with a search for necessary and sufficient conditions, and the idea that such conditions are required has long been disreputable. But such a requirement is no more inherent to the aims of conceptual analysis than it is to the aims of theoretical definition. The criteria of application that the relevant linguistic community generally has in mind might be better expressed in terms of a family resemblance, similarity to prototypes, or Minskian frames.[3] Indeed, necessary and sufficient conditions are probably more likely to be found for theoretical definitions (consider those given for "water" and "gold").

Admittedly, the criteria of application that people actually use are often vague, shifting, highly context-sensitive, highly variable between individuals, and often involve perceptual data of a kind that is inaccessible, at least to philosophical methods. This is certainly daunting. But it does not follow that we cannot discover anything useful about these criteria of application. Conceptual analysts can describe those criteria of application that people standardly apply, most of the time, in the most standard contexts. Also, when the relevant linguistic community consists of specialists, and the term under analysis is one of their specialist terms, and is also abstract (nonperceptual) and embedded in well-articulated theory, the severity of each of these factors will be greatly reduced. Since this is the case for contemporary biologists and their notion of a "proper function," we have more reason to expect success in this case than in most.[4] The problems with conceptual analysis are many, but they amount to a need to concede that it will be rare, if ever, that a definitive, exhaustive and universal analysis can be given. This is not to say that some insight into relevant criteria of application cannot be achieved in many cases. Since philosophers seem to have been providing insights along these lines, stronger argument is needed to show that they cannot have been doing so.

Of course it would be futile trying to do conceptual analysis if we didn't have any use for it; but we do. We need it to clarify thought and communication, which are the time-honored reasons for engaging in the activity. While this point epitomizes the unoriginal, we may need to remind ourselves that theoretical definition cannot entirely replace conceptual analysis in this respect. This point can be made most vividly with respect to nonreferring terms, but it also applies more generally.

Given that conceptual analysis is an attempt to describe what people think they are referring to, we can provide a conceptual analysis for "witch," "entelechy" and "phlogiston" because we can explain what people thought witches, entelechies and phlogiston were. A theoretical definition, in contrast, is (roughly) an attempt to describe the things referred to, so where reference fails there is no theoretical definition. If we wanted to persuade a Medieval theologian that witches don't exist, no theoretical definition could refine our understanding of his or her understanding of "witch." Good communication and debate with this benighted individual demands that we know more than a theoretical definition can tell us (which is just that witches don't exist). To successfully engage in debate we would need to know what precisely a witch was supposed to be. Admittedly, we are not likely to meet many theologians of this kind, but similar sorts of debates occur often enough in philosophy. For instance, to persuade us that there is no pain, Daniel Dennett once argued that our criteria of application are inconsistent (Dennett 1978). He engaged (as he must) in conceptual analysis to do this. Similar styles of argument can also be found regarding libertarian free-will, to give one other example.

Nor is theoretical definition independent of conceptual analysis, since the latter is required to delimit the scope of the former. Furthermore, when the relevant linguistic community is very well informed about what they speak of, we should expect conceptual analysis and theoretical definition to closely correspond. For example, while it was once false as conceptual analysis that water meant "liquid with the molecular structure HOH," the criteria of application have changed and kept abreast with our knowledge. Now a conceptual analysis of water would have to include that water is HOH. I could be deceived into thinking that some other clear and thirst-quenching liquid was water. But if I learned that it was not HOH, I would then deny that it was water. So being HOH is now my criterion of water. Closer to the task at hand, if we were to offer a conceptual analysis of "water" *as used by modern Western chemists* we would certainly have to mention that the molecular structure HOH was a defining characteristic *for them.* Given that modern biologists are in possession of a fairly full and correct set of beliefs about biological proper functions (more so than we are anyway) it is highly likely that conceptual analysis and theoretical definition will closely correspond in this case too.

The moral for Millikan's defense of the etiological theory is this: if the etiological theory is implausible as conceptual analysis, then it is thereby implausible as theoretical definition. We can gain no immunity from the standard objections to the etiological theory by insisting that we are only interested in theoretical definition. When the relevant linguistic community thoroughly understands the nature of what it speaks of, there is a strong prima facie case in favor of its having integrated this understanding into the basis of its linguistic competence. Our understanding of the world, and the conceptual framework we employ in talking about it, are virtually the same thing.

I see conceptual analysis and theoretical definition as complementary and interdependent, describing different but related aspects of language. In keeping with the views expressed in this section, the claims made in the next, where I explain the etiological theory in a little more detail, are intended to hold true from the standpoint of both theoretical definition and conceptual analysis. In later sections, the differences between these two modes of analysis are again the focus of attention, for different criticisms and justifications are appropriate depending on our analytic aims.

3   Proper Functions as Selected Effects

In biology, as well as in other contexts, the most explicit use of the notion of a "proper function" is in sentences that express function attributions, such as "It is normal for penguins to be myopic on land since the proper function of their optic lenses is sharp underwater vision," or "A function of the hypothalamus is the monitoring of blood oxygen and sugar levels." Such function attributions are also made regarding artifacts (the function of an armchair is to provide comfortable seating, the function of the carburetor is to mix fuel with air, and so on). Especially in biology, the notion of a proper function has other less obvious but essential roles that derive from these basic function attributions (more on this in the last section). However, these explicit function attributions are a natural place to start when we seek to understand the notion. The problem is to discover what exactly is attributed when we attribute a proper function (or more loosely a function) in sentences of the form "The/a proper function of X is Z."

My claim is that the proper function of a trait is to do whatever it was selected for. In so far as our organismic structures and processes are the result of selection they are the result of natural selection,[5] so I claim that the biological proper function of such an item is to do whatever items of that type did that caused them to be favored by natural selection. More carefully expressed, and assuming that the unit of selection is a genotype, the idea is as follows (where the relevant items are evolved biological parts and processes):

It is the/a proper function of an item (X) of an organism (O) to do that which items of X's type did to contribute to the inclusive fitness of O's ancestors, and which caused the genotype, of which X is the phenotypic expression, to be selected by natural selection.

For example, it is the function of your opposable thumb to assist in grasping objects, because it is this which opposable thumbs contributed to the inclusive fitness of your ancestors, and which caused the underlying genotype, of which opposable thumbs are the phenotypic expression, to increase proportionally in the gene pool. In brief, grasping objects was what the trait was selected for, and that is why it is the function of your thumb to help you to grasp objects.

There are certain constraints on the proper functions of evolved traits that do not apply to most artifacts, and which I have built into the offered definition. Differences between biological and artifact functions arise because of the different nature of the selection processes involved in the two cases. Biological proper functions belong primarily to types and only secondarily tokens because natural selection does not operate on individuals or their biological parts and processes. A particular piece of genetic material, or a particular instance of a trait (your thumb, Reagan's nose) cannot be selected by natural selection which operates over whole populations. Also, biological functions of evolved traits must be selected on the basis of actual past performances (by, or in, ancestral organisms) of the effect that becomes the function. In contrast, because artifacts are generally the result of a very different sort of selection process—the intentional design of an agent with forethought—they can have idiosyncratic functions, like the details of James Bond's cars. And intentional selection can be made without any performance of the function-effect ever being realized (either in the past, present or future) because the mere hint of a hope of a desired effect is enough to inspire intentional selection by an agent with forethought.

My main concern in this paper is to address the three standard objections to the etiological theory which I mentioned in the introduction. Hopefully, this brief elaboration of the theory will suffice for this limited purpose, along with just one additional point. It is debatable whether an etiological theory of biological functions must explicitly mention natural selection. Since this feature of the theory is found particularly objectionable by some people, this is worth noting. It might be thought that the theory could be expressed more neutrally, in terms of selection *simpliciter*, for instance. The idea might be that a proper function of a trait was, simply, whatever effect that trait was selected for, leaving open what kind of selection process was causally operative. This might have the double advantage of providing a univocal account for biological and artifactual functions, as well as being neutral with respect to evolution and Creationism.[6] While I haven't the space to explore this possibility at length, my view is that this is indeed the vaguer, unifying, every-day notion of a "proper function" from which the biological notion is derived. Nonetheless, the peculiarities of natural selection impose certain constraints upon a more detailed and precise analysis of the biological notion (as mentioned above), constraints which do not apply to the everyday notion employed for artifacts. Moreover, the standard objections to overtly preferring natural selection in a conceptual analysis of biological functions are misguided, or so I will now attempt to show (in sections 4 and 5).

4   The Historical Accuracy of an Evolving Notion

The first of the objections to be considered is the weakest of the three. It charges the etiological theory with historical inaccuracy, since the theory claims that the notion of a "proper function" in biology should be analyzed in terms of natural selection. Presumably, logically derivative notions must be chronologically antecedent notions. So, the etiological theory is charged with the implication that natural selection was dis-covered before the notion of a (biological) "proper function" was used, and this is plainly false. Harvey, for one, proclaimed the proper function of the heart two hundred years in advance of Darwin's *On the Origin of the Species*.

First note that Harvey's ignorance of natural selection is not directly relevant to a theoretical definition of "proper function" (Millikan 1989, 290). Water would be HOH whether or not anyone knew it, and sim-ilarly, proper functions might happen to be effects for which traits were selected by natural selection, whether or not Harvey or anyone else was aware of the fact. But now what is the reply of the conceptual analyst who is, after all, concerned with the criteria of application that people have in mind?

Although conceptual analysis is often equated with something called "ordinary language philosophy" it need not be an analysis of ordinary as opposed to specialist language. We can attempt to understand how members of a scientific community, like contemporary biologists, use a term by searching for their criteria of application. This would still be conceptual analysis, not theoretical definition, as I have defined them, and this seems to be the task before us here. It is a relevant philosophical commonplace, therefore, that scientific terms are shaped by background theories. So how biologists understand the notion of a "proper function" might shift significantly with dramatic changes in these background theories. In other words, it is unproblematic if Harvey's notion of a "proper function," before the Darwinian Revolution, was different from the closely related notion used by biologists today, after the Darwinian Revolution. Scientific notions are not static. Harvey obviously did not have natural selection in mind when he proclaimed the function of the heart, but that does not show that modern biologists do not have it in mind. Just so, Newton did not have the theory of General Relativity in mind when he talked of time, but that does not show that modern physicists do not have it in mind when they speak of time, or space-time.

Of course I do not claim that modern biologists have natural selection consciously or explicitly in mind when they use the notion of a "proper function." If that were required, conceptual analysis could be done by deed poll. That a grammatical rule might come as a surprise to us is not proof that we do not employ it, and by the same token, no explicit knowledge is required here either. What matters is only that biologists implicitly understand "proper function" to refer to the effects for which traits were selected by natural selection.

Larry Wright was the first (to my knowledge) to propose and be persuaded by this objection. Yet as he himself suggested, Harvey will have supposed that biological parts and processes were the result of *some* sort of selection process (such as design by God). Add that scientific concepts are not set in stone—that radical shifts in background theory can modify a scientific community's understanding of its terms—and this first objection in more than adequately answered.

5   Analytic Aims and Pretensions

The second objection is more subtle. Analyzing the biological notion of a "proper function" in terms of natural selection is said to be "analytically arrogant" because it begs empirical and theological questions. More precisely, the etiological theory seems to entail the following disjunct: natural selection theory is true or there are no proper functions. Given that there obviously are proper functions, this entails that the theory of natural selection is true as a matter of logic rather than contingent fact.

Again, begin by noting that this objection has no direct bite against the etiological theory offered as theoretical definition. Briefly, theoretical definitions are legitimately in the business of answering empirical questions (Millikan 1989, 290). The scientist begs no empirical questions in telling us that water is HOH, or that gold is the element with atomic number 79. Nor does the philosopher who suggests that pain is, as a matter of fact, C-fibers firing. These may be false empirical claims, but they are not empirical questions begged. Conceptual analysis, however, is in the business of conceptual, not empirical truths; so it seems to be in the firing line of this objection, at least until we take a more careful look.

The etiological theory, offered as conceptual analysis, entails a slightly, yet significantly, different disjunct, which is as follows: natural selection theory is true or there are no proper functions *in the modern biologists' sense of "proper functions."* By comparison, if I were to offer a conceptual analysis of "time" as understood by Newtonian physicists, I suppose my analysis would refer to some axioms of Newtonian physics, such as that time is absolute. Now we can ask whether this would entail that either time is absolute or there is no time. And it is obvious that it does not. Rather it entails that either time is absolute or there is no time in the Newtonian physicist's sense of "time." Since conceptual analysis is aimed at discovering the criteria of application that people have in mind, and since these can and often do change over time, conceptual truths will always be relative to a linguistic community at a given time.

The persuasive force of this second objection stems from our belief that there would be proper functions even in the unlikely event that the theory of natural selection was falsified. However, this belief statement is ambiguous between opaque and transparent (de dicto and de re) readings of the proposition involved. The proposition involved is compatible with my claims if what we believe is that our reference to proper functions will not radically fail, as did our reference to phlogiston, entelechies and witches. But this is just to say that, even if the theory of natural selection is falsified, and the etiological theory *as theoretical definition* is thereby shown to be untrue, some other theoretical definition of proper functions will be true. It would not follow that the etiological theory as conceptual analysis would also be falsified, simply because the truth of that does not depend upon the truth of evolutionary theory. The relevant truth for conceptual analysis is that biologists today *believe* that a theory of natural selection is true, and they do. By comparison again, Newtonian physics was indeed falsified, and so the Newtonian theoretical definition of "time" was falsified. But some other theoretical definition of "time," or at least of "space-time" is presumably true. (Time still exists.) Further, the falsification of Newtonian physics leaves untouched the claim that Newtonian physicists understood "time" in the light of Newtonian theory. All that matters for the conceptual analysis is that the physicists of the day believed that time was absolute, and they did.

6   The Status of Hypothetical Examples

Both objections just discussed are directed against the explicit mention of natural selection in the analysis. The third and last objection has a broader aim. It is aimed at showing that "function" is an ahistorical notion. The essence of the objection is the claim that we could ascribe proper functions to creatures that were just like actual creatures, except for the fact that they have no history and are not the result of any kind of selection process, either natural or supernatural. Suppose, for example, that we discovered that all lions came into being as the result of "one unparalleled saltation."

Let us again begin by considering how the etiological theory fares when offered as a theoretical definition. The theoretical definer might reply that theoretical definitions are not obliged to cater for the merely hypothetical, since they are concerned with matters of fact (Millikan (1989, 292–293). This may be too quick, however. If a theoretical definition is supposed only to describe the actual reference class, then hypothetical cases are beside the point. But if it is supposed to describe the reference class in a way that reveals how the notion has been useful and has survived, then something more is required. A theoretical definition should then elucidate the theoretical role that the notion has enjoyed, and hypothetical cases can be instrumental in this respect. It is arguable that the mere possibility of Martian pain, for example, was a legitimate problem for the early mind-brain identity theorists—if we take them to have been offering a theoretical definition of this kind—when they asserted that pain is (as a contingent matter of fact) C-fibers firing (or some other type of neurophysical state). Whether or not Martians (or any other aliens) actually exist, the mere possibility of Martian pain highlighted the fact that moral theory requires a notion of pain that transcends the biological peculiarities of creatures on this planet. It is arguable that a theoretical definition of pain expressed in abstract functional terms, which explicitly allows for the possibility of radical variation in physical instantiation, better elucidates the theoretical role of "pain." (This was not the case with "water," since the theoretical role of "water" requires chemical specificity. Were Twin Earth water to exist, it would not really be water if it was not HOH.)[7]

Both theoretical definer and conceptual analyst are in the same boat for this objection. Conceptual analysis notoriously involves a parading of examples past our intuitions. If we are trying to discover the criteria of application upon which we implicitly rely, this is not a bad place to start. However, since these criteria are not—by hypothesis—already explicitly known to us, we might make mistakes about how we would consistently "go on in the same way," especially when the examples become wildly divorced from actual facts and natural laws. Intuitions should not therefore be considered the final arbiters for the conceptual analysts either. They too must be constrained by the theoretical role of the notion being analyzed. After all, if we successfully use a notion to perform a given theoretical role, we must implicitly understand the notion in a way that allows it to perform that role. As I will argue in the next section, wayward intuitions about instant lions might have to be revised in the light of the fact that biology has and needs a notion of a "proper function" that is normative.

But first let's tamper with the intuition pump to see how robust these intuitions really are.[8] Suppose there are no lions. Then suppose that half a dozen lions pop into existence, we know not how. Having stared at them in stupefied amazement for some time, we eventually begin to wonder about their wing-like protuberances on each flank. We ask ourselves whether these limbs have the proper function of flight. Do they? When we discover that the lions cannot actually fly because their "wings" are not strong enough, we are tempted to suppose that this settles the mater, until we remember that organismic structures are often incapable of performing their proper function because they are deformed, diseased, atrophied from lack of use, or because the creature is displaced from its natural habitat (the lions could perhaps fly in a lower gravitational field). On the other hand, often enough there are complex structures that have no functions, for instance, the vestigial wings of emus and the human appendix. The puzzle is where among these various categories are we to place the lions' "wings." I contend that we could not reliably place them in any category until we knew or could infer the lions' history. And if we were to somehow discover that the lions had no history, and were the result of an accidental and freak collision of atoms, they would definitely not belong in any of our familiar functional categories. They are not then dysfunctional either because of disease, deformity, lack of use, or because they are exiled from their natural environment. All of these require a past. Nor did they once have a function that they have now lost. Without history the usual biological/functional norms do not apply.[9]

This point requires more argument, and luckily we need not allow the debate to further "decay into the dull thud of conflicting intuitions" (to use Bigelow and Pargetter's nice expression; 1987, 196). In the next section, I argue that ahistorical theories of functions do not provide a notion of "proper function" which is appropriately normative, and so they cannot accommodate the theoretical role required of the notion in biology.

7   Biological Norms and Their Role in Biology

The notion of a "proper function" is the notion of what a part is *supposed* to do. This fact is crucial to one of the most important theoretical roles of the notion in biology, which is that most biological categories are only definable in functional terms (Beckner 1959, 112–118). For instance, "heart" cannot be defined except by reference to the function of hearts because no description purely in terms of morphological criteria could demarcate hearts from non-hearts. Biologists need a category that ranges over different species, and hearts are morphologically diverse: fish have a single pump with only one auricle, but amphibians and most other reptiles have the single heart with two auricles, and while many reptiles have the ventricle partly partitioned, only crocodiles, birds and mammals have the two separate ventricles. Highly significant, moreover, is that for the purposes of classifying hearts, what matters is not whether the organ in question manages to pump blood, but whether that is what it is supposed to do. The heart that cannot perform its proper function (because it is atrophied, clogged, congenitally malformed, or sliced in two) is still a heart.

Some theories which imply that instant lions (and piggyback traits) would have proper functions do not capture the distinction between what an item does and what it is supposed to do, and so they do not describe a notion of a "proper function" that is capable of generating these biological categories which embrace both interspecies and pathological diversity. A brief discussion of the three most interesting of these alternative theories (mentioning just one problem with each regarding biological norms) will further illustrate my point that the biological norms are determined by the history of traits.

(i)   According to one important theory of functions, they are causal contributions to complexly achieved overall activities of the containing organism or system (Cummins 1975). According to this theory, functions are relative to our interests, and both the boundaries of the containing system, and which of its overall activities we focus upon, will vary with our current concerns. Now instant lions could be analyzed as complex causal systems, in effect organized toward the achievement of certain activities—like hunting, consuming deer, birthing cubs, and so on. So instant lions would have proper functions if this theory accurately described what proper functions were. But note that this kind of causal analysis can be done on any complex causal system (plate movements that culminate in earthquakes, intragalactic motion, for examples). So, this theory bestows proper functions on instant lions only at the expense of bestowing them upon a vast range of systems to which we do not normally attribute them.

We could usefully refer to Cummins's notion of a "function" as a "causal role function." This notion has important uses. Causal role functions have played an important part in functionalist theories of mind, for instance (although some of us have argued that we should "put the function back into functionalism" by employing the more biological notion of a "proper function").[10] However, causal role functions are not equivalent to proper functions. While this account offered by Cummins allows us to choose among causal roles according to our interests, causal role functions are nonetheless purely descriptive (not normative) in the sense that the causal role function of X is some subset of X's actual causal roles. Yet in biology the proper function of a trait is not necessarily anything it actually does. By way of illustration of the difficulties with this account, consider that one thing that many organisms do, unfortunately, is die of cancer. Dying of cancer is complexly achieved; it involves chromosome replication and cell reproduction in the growth of tumors, and as the tumors spread, they replace healthy tissue and functional integrity progressively deteriorates. If a tumor presses on an artery to the brain, this is an actual causal role the tumor has in this pathological process, and one in which we are very interested. However, this causal role is not the tumors' proper function; tumors simply don't have proper functions. Moreover, in this process of deterioration, proper functioning will be disrupted, and so the proper functions of many diseased organs will cease to be among their actual effects. The brain may fail to coordinate bodily movements, for instance, in which case this will be a proper function of the brain that is not an actual causal contribution by the brain to any complexly achieved overall activity of the system.

(ii)   Most writers, not so long ago, wanted to specify certain goals to which proper biological functions causally contribute. William Wimsatt (1972, 62—65) suggested that the goal is provided by background theories and Christopher Boorse (1975) suggested that they are given by the context of enquiry in combination with these background theories. The "biologically given goal" has been variously identified as the survival and reproduction of the individual organism within physiology (ibid., 57) and as the fitness, or long range survival of the evolutionary unit, which is "basically a temporally defined concept of a breeding population" (Wimsatt 1972, 6, f.n. 11). Perhaps the goal should be specified as the inclusive fitness of the individual organism. In any case, the general idea was that a proper function was a causal contribution to the specified goal.

So far described, such theories suffer the same problem as that encountered in (i). The problem is that if a biological part dysfunctions badly enough, it is no longer capable of making any causal contribution to fitness (or whatever), so it lacks a proper function if having a proper function means having a causal contribution to fitness. This is unacceptable because items that are dysfunctional are dysfunctional precisely because of their incapacity to perform their proper function.

An appeal to statistical normality is therefore introduced so that proper functions are defined as contributions to fitness that are "species-typical," or "standard," or "statistically normal" within a class of organisms (see Wimsatt 1972, 50; and Boorse 1976, 557).[11] Although a particular kidney might fail to filter blood, kidneys typically manage to perform their proper function, and so they standardly contribute to the fitness of organisms of the type. On this statistical account, therefore, a dysfunctional kidney does have a proper function that it fails to perform because it fails to contribute what kidneys standardly contribute toward fitness. However, biological norms cannot be reduced to statistical norms, as we can see by noting that dysfunction can become widespread within a population through epidemics or major environmental disasters. A statistical definition of biological norms implies that when a trait standardly fails to perform its function, its function ceases to be its function; so that if enough of us are stricken with disease (roughly, are dysfunctional) we cease to be diseased, which is nonsense.[12]

(iii) Much the same problem arises for the closely related "propensity theory" that Bigelow and Pargetter (1978) have proposed, in which a biological character has a function if it has a disposition that is apt for selection, or that enhances survival systematically, in a creature's "natural habitat." While this theory distinguishes proper functions from actual causal roles (the creatures may not actually be in their natural habitat, for instance), it links them instead to actual dispositions and categorical bases for these dispositions. But the problem is that dysfunctional traits do not actually have the disposition to perform their proper function. The impaired kidney does not have a disposition that is apt for selection, or a disposition to enhance survival systematically, so according to the propensity theory it does not have a proper function. But, once again, it is precisely because it has a function that it is incapable of performing that it counts as dysfunctional.[13]

To conclude this section and the essay, for a trait to have a proper function is not for it presently to have any actual causal role, statistically typical contribution to fitness, or disposition. Instead, a trait has a proper function if there is something that it is supposed to do. According to my etiological theory, a trait is supposed to do whatever it was selected for by natural selection. So understood, we can easily accommodate the above deviations from biological norms. It is not the function of tumors to press on arteries to the brain because tumors have not been selected for that effect. It is the function of kidneys (both normal and abnormal) to filter wastes from blood because that is what kidneys did in ancestral organisms that caused them to be favored by natural selection (and this fact remains true even if renal failure becomes universal). Functional norms seem to be determined by a history of selection, and this would explain why the biological categories that are based on functional norms are stable, relative to current vagaries in demographic, environmental, or pathological features. Given that biologists use the notion of a proper function to generate these stable biological categories, biologists must implicitly understand the notion of a "proper function" as an historical notion.

Acknowledgments

I am grateful to many people for their helpful comments on earlier drafts of this essay, especially Paul Griffiths, William Lycan, Ruth Millikan, Robert Pargetter, Huw Price, Elliott Sober, and Kim Sterelny.

Notes

1. I defended this theory in my unpublished but widely circulated paper, "Teleology in Biology," first read at the A.A.P. Conference, Christchurch, New Zealand, 1980, and later in my Ph.D. dissertation, La Trobe, 1983. Some of the material in this paper appears in these works. See also Millikan (1984, 1989) who independently developed a very similar theory. The most notable forerunner of the etiological theory was provided by Wright (1973, 1976). His work greatly influenced me.

2. Millikan defends these strong words to some extent in Millikan (1984, see especially chap. 9). I address her criticisms of conceptual analysis in section 2 of this paper.

3. Minskian frames and prototypes are popular but controversial models among Artificial Intelligence researchers and cognitive scientists, respectively. For seminal papers on these, see Minsky (1974) and Rosch (1973, 1975).

4. Another reason why it might be thought that conceptual analysis is impossible is that our concepts are endlessly interwoven. But this point, interesting as it is, is out of place here. A given conceptual analysis does not have to stand on its own. It connects with this whole web of ideas by analyzing one notion in terms of other, implicitly understood, but unanalyzed (or analyzed elsewhere) notions.

5. Elliott Sober (1984, 147–155) argues that natural selection does not explain the traits of individuals, as opposed to the distribution of traits in a population. His argument is based upon an illuminating distinction between what he calls "selectional explanations" and "developmental explanations." According to Sober, developmental explanations explain how an individual comes to possess a trait, whereas selectional explanations can only explain why a population has a certain distribution of individuals with a given trait. I have elsewhere argued (Neander 1988) that it is a mistake to suppose that the two forms of explanation are mutually exclusive. I argue that a full developmental explanation of an individual's trait will often contain a selectional explanation. Very roughly, the idea is that you and I have opposable thumbs because we come from a long line of ancestors *all of whom had opposable thumbs*, and no countervailing mutation intervened to prevent you and me from inheriting the trait. By Sober's own lights, a selectional explanation is needed to explain why all of our recent ancestors had opposable thumbs, and this selectional explanation is thus part of the full developmental explanation of why we have opposable thumbs.

6. Wright attempted to provide an analysis that was both univocal and neutral in these respects (1976, 73–98) but this led him into difficulties (Boorse 1976, 72).

7. Moreover, the objection need not be based on purely hypothetical cases. A variation of the same objection appeals to piggyback traits. For example, it is believed that some llamas, due to a certain property of their hemoglobin molecules, possessed the capacity to survive in higher altitudes than their conspecifics, before there was any need to move to higher ground and therefore before this capacity enhanced fitness. Some of us may be tempted to say that these hemoglobin molecules had the proper function of enabling the llamas to survive in high altitudes before the migration up the mountains actually took place. (I am grateful to an anonymous referee of this journal for this point and this example.)

8. Our intuitions regarding piggyback traits are even less robust. Indeed, some piggyback traits are taken in the literature to be paradigm cases of nonfunctions: the heart's thumping acting as a diagnostic aid, and the bridge of the nose keeping up our spectacles, for instance. (Nor should it be said that these cases can be ignored because they involve artifacts and are therefore "artificial." Consider the English tit that evolved a beak specially adapted for sipping from milk bottles.)

9. I make this and several of the following points in my dissertation (Neander 1983, 73–140). See also Millikan (1989, 294ff.).

10. See especially Sober (1985).

11. Thus these theories are also forced to acknowledge that biological proper functions apply primarily to types and only secondarily to tokens, but the fact does not fall neatly out of the logic of the theory; it is introduced to ward off a plethora of counterexamples.

12. Christopher Boorse and William Lycan have both suggested, in correspondence, that biological norms might be statistically relative to a larger time-slice of a population (such as the past millennium or two). I have argued elsewhere that the suggestion will not work (Neander 1983, 85), but I need not pursue the idea here since I am here concerned with theories that might support the claim that "proper function" is an ahistorical notion.

13. In another paper (Neander 1991), I further argue that the propensity theory paradoxically fails to capture the forward-looking nature of proper functions, and so fails to endow proper functions with explanatory power, contrary to the positive claims which Bigelow and Pargetter make on behalf of their theory (1978, 181–182, 189–191).

References

Beckner, M. (1959), *The Biological Way of Thought*. New York: Columbia University Press.

Bigelow, J. and Pargetter, R. (1987), "Functions," *The Journal of Philosophy* 84: 181–196.

Boorse, C. (1976), "Wright on Functions," *The Philosophical Review* 85: 70–86.

Cummins, R. (1975), "Functional Analysis," *The Journal of Philosophy* 72: 741–765. [Excerpt reprinted in N. Block (ed.), (1980), *Readings in Philosophy of Psychology*. Cambridge, MA: Harvard University Press, pp. 185–190.]

Dennett, D. (1986), *Brainstorms: Philosophical Essays on Mind and Psychology*. Fourth Printing. Cambridge, MA: MIT Press.

Millikan, R. G. (1984), *Language, Thought and Other Biological Categories: New Foundations for Realism*. Cambridge, MA: MIT Press.

―――. (1989), "In Defense of Proper Functions," *Philosophy of Science* 56: 288–302.

Minsky, M. (1974), "A Framework for Representing Knowledge." Cambridge, MA: MIT AI Lab Memo 306. [Excerpts reprinted in P. Winston (ed.), (1975), *The Psychology of Computer Vision*. New York: McGraw-Hill, pp. 211–277; other excepts reprinted in J. Haugeland (ed.), (1981), *Mind Design*. Cambridge, MA: MIT Press, pp. 95–128.]

Nagel, E. (1977), "Teleology Revisited," *The Journal of Philosophy* 84: 261–301.

Neander, K. (1983), "Abnormal Psychobiology." Ph.D. Dissertation, La Trobe University.

————. (1988), "Discussion: What Does Natural Selection Explain? Correction to Sober," *Philosophy of Science* 55: 422–426.

————. (1991), "The Teleological Notion of 'Function,'" *Australasian Journal of Philosophy* 69: 454–468.

Rosch, E. (1973), "Natural Categories," *Cognitive Psychology* 4: 328–350.

————. (1975), "Cognitive Representations of Semantic Categories," *Journal of Experimental Psychology* 104: 192–233.

Sober, E. (1984), *The Nature of Selection: Evolutionary Theory in Philosophical Focus.* Cambridge, MA: MIT Press.

————. (1985), "Panglossian Functionalism and the Philosophy of Mind," *Synthese* 64: 165–193.

Wimsatt, W. (1972), "Teleology and the Logical Structure of Function Statements," *Studies in the History and Philosophy of Science* 3: 1–80.

Wright, L. (1973), "Functions," *The Philosophical Review* 82: 139–168.

————. (1976), *Teleological Explanations: An Etiological Analysis of Goals and Functions.* Berkeley and Los Angeles: University of California Press.

12

Function without Purpose: The Uses of Causal Role Function in Evolutionary Biology

Ron Amundson and George V. Lauder

1 Introduction

Philosophical analyses of the concept of biological function come in three kinds. One kind defines the function of a given trait of an organism in terms of the history of natural selection which ancestors of the organism have undergone. In this account the function of a trait can be seen as its evolutionary purpose, with purpose being imbued by selective history. A second approach is non-historical, and identifies the function of a trait as certain of its current causal properties. The relevant properties are seen either as those which contribute to organism's current needs, purposes, and goals (Boorse 1976) or those which have evolutionary significance to the organism's survival and reproduction (Ruse 1971; Bigelow and Pargetter 1987). A third approach has been articulated and defended by Robert Cummins (1975, 1983), mostly in application to psychological theory. Cummins's view is unique in that neither evolutionary nor contemporary purposes or goals play a role in the analysis of function. It has received little support in the philosophy of biology, even from Cummins himself. Nevertheless, we will show that the concept is central to certain ongoing research programs in biology, and that it is not threatened by the philosophical criticisms usually raised against it. Philosophers' special interests in purposive concepts can lead to the neglect of many crucial but non-purposive concepts in the science of biology.

Karen Neander recently and correctly reported that the selective view of function is "fast becoming the consensus" (Neander 1991, p. 168). Larry Wright showed in his canonical (1973) paper on selective function that an intuitively pleasing feature of the view is that citing a trait's function would play a role in explaining how the trait came to exist. Concepts of function similar to Wright's were hinted at by the biologists Francisco Ayala (1970) and G. C. Williams (1966), and later endorsed by the philosophers Robert Brandon (1981; 1990), Elliott Sober (1984), Ruth Millikan (1989), and Karen Neander (1991), among others. (For a good review of the history of philosophical discussions of function see Kenneth Schaffner, 1993, chapter 8.)

The evolutionary, selective account of function is commonly termed the "etiological concept" since functions are individuated by a trait's causal history. In the present context the term "etiological" may lead to confusion, so we will refer rather to the *selected effect* (SE) account of function. The Cummins style of account will be designated the *causal role* (CR) account (following Neander 1991, p. 181).

Given the consensus in favor of SE function among philosophers of biology, it is surprisingly difficult to find an unequivocal rejection of Cummins's alternative. This may stem from a recognition that some areas of science (medicine, physiology, and perhaps psychology) require other kinds of function concepts. It does seem generally accepted, however, that SE function is the concept uniquely appropriate *to evolutionary biology*. It is this position which we will attempt to refute.

Ruth Millikan (1989) and Karen Neander (1991) have recently presented arguments in favor of selected effect concepts of biological function. We will pay special attention to these papers for two reasons. First, they express positions on the nature of philosophical analysis which we find valuable, and which we will use in defending CR function. Second, they examine Cummins's account of function in detail. Some of the ideas they develop are shared with other advocates of SE function, and many are novel; all are worthy of analysis. (Unless otherwise stated, all references to Millikan and Neander will be to those papers.)

Millikan examined the source of the criticisms which philosophers had made against the SE theory, and found them to be based in the philosophical practice of conceptual analysis. She declared this practice "a confused program, a philosophical chimera, a squaring of the circle ... " among other crackling critiques (p. 290). The search for necessary and sufficient conditions for the common sense application of terms was not what Millikan was about. Neander similarly rejected conceptual analysis of ordinary language as the goal of the philosophical analysis of function. Indeed, in rereading the debates on function of the 1970s, one is struck by the concern shown by philosophers for consistency with ordinary language. Millikan and Neander replace the old style of ordinary language analysis with somewhat different alternatives. Millikan was interested in a *theoretical definition* of the concept of function, a concept which she labels "proper function." Neander instead focussed on a conceptual analysis, but not the traditional kind based on ordinary language. Rather, Neander intended to analyze *specialists'* language—in this case the usage of the term *function* in the language of evolutionary biology. "What matters is only that biologists implicitly understand 'proper function' to refer to the effects for which traits were selected by natural selection" (p. 176).[1] While each writer intended the analysis of function to be relativized to a theory (rather than to ordinary language), Neander intended the relevant theory to be evolutionary biology, while Millikan located her analysis in the context of her own research project involving the relations among language, thought, and biology (Millikan 1984, 1993).

While the intended status of their resulting analyses differed, Millikan's and Neander's approaches had similar benefits for the SE analysis of function (and also, as we will presently argue, for the CR analysis of function). Both approaches tied function analyses to actual theories, in this way eliminating many ordinary-language based counterexamples to SE function. Theoretical definitions, such as "Gold is the element with atomic number 79," need not match ordinary usage, but instead reflect current scientific knowledge about the true nature of the subject matter. The use of bizarre counterfactuals such as Twin Earth cases and miraculous instantaneous creations of living beings (e.g. lions) were a mainstay of earlier criticisms of SE function. These kinds of cases are irrelevant to evolutionary theory and to the vocabulary of real world evolutionary scientists. Appeals to pre-Darwinian uses of the term "function" (e.g. William Harvey said that the function of the heart was to pump blood) are equally irrelevant. After all, Harvey didn't know the atomic number of gold any more than he knew the historical origin of organic design. Nonetheless gold is (and was) the element with atomic number 79, and (by the SE definition) the heart's blood-pumping function is constituted by its natural selective history for that effect.

We fully approve of these moves. Taking the contents of science more seriously than is philosophically customary is exactly what philosophers of science ought to be doing. We will not question the philosophical adequacy of Millikan's or Neander's approaches, nor their defenses of SE theory against its philosophical critics. We will, however, call into question the common SE functionalist's belief that evolutionary biology is univocally committed to SE function. We will show that the rejection of ordinary language conceptual analysis immunizes Cummins-style CR function against some very appealing philosophical critiques—critiques expressed by Millikan and Neander themselves. We will show that a well articulated causal role concept of function is in current use in biology. It is as immune from Millikan's and Neander's critiques of CR function as their own SE accounts are from ordinary language opposition.

The field of biology called functional anatomy or functional morphology explicitly rejects the exclusive use of the SE concept of function. To be sure, there are other biological fields in which the SE concept is the common one—ethology is an example. The most moderate conclusion of this semantic observation is only a plea for conceptual pluralism, for the usefulness of different concepts in different areas of research. But further conclusions will be stronger than mere pluralism. We will defend CR function from philosophical refutation. We will show that a detailed knowledge of the selective history (and so the SE function) of specific anatomical traits is much more difficult to achieve than one would expect from the intuitive ease of its application. Finally, we will demonstrate the ineliminability of CR function from certain key research programs in evolutionary biology.

2 Adaptation and Selected Effect Functions

First, a specification of the selected-effect concept of function:

The function of X is F *means*

a. X is there because it does F,

b. F is a consequence (or result) of X's being there. (Wright 1973, p. 161. Variables renamed for consistency.)

Wright intended his analysis to apply equally to intentional and natural selection. When the context is restricted to evolution, and natural selection accounts for the "because" in (a), something like Neander's definition results.

It is the/a proper function of an item (X) of an organism (O) to do that which items of X's type did to contribute to the inclusive fitness of O's ancestors, and which caused the genotype, of which X is the phenotypic expression, to be selected by natural selection. (Neander, p. 174. Cf. Millikan, p. 228.)

Not surprisingly, there are very closely related concepts within evolutionary biology, particularly the concept of *adaptation*. During the 20th century there has been some semantic slippage surrounding the term. Describing an organic trait as adaptation has meant either (1) that it benefits the organism in its present environment (whatever the trait's causal origin), or (2) that it arose via natural selection to perform the action which now benefits the organism. That is, the term adaptation has sometimes but sometimes not been given an SE, historical meaning. G. C. Williams gave a trenchant examination to the concept of adaptation, referring to it as "a special and onerous concept that should be used only where it is really necessary" (1966, p. 4). In particular, Williams thought it important to distinguish between an adaptation and a fortuitous benefit. These ideas inspired the "historical concept" of adaptation, according to which the term was restricted to traits which carried selective benefits and which resulted from natural selection for those benefits. Terms such as "adaptedness" or "aptness' came to be used to designate current utility, covering both selected adaptations and fortuitous benefits (Gould and Vrba 1982). The onerous term adaptation was reserved for traits which had evolved by natural selection. Robert Brandon recently declared the historical definition of adaptation "the received view" (1990, p. 186). Elliott Sober described the concept as follows:

X is an adaptation for task F in population P if and only if X became prevalent in P because there was selection for X, where the selective advantage of X was due to the fact that X helped perform task F. (Sober 1984, p. 208. Variables renamed for consistency.)

Sober's task F is precisely what SE theorists would call the function of trait X. Moreover, Williams, Sober, and Brandon, like Millikan and Neander, all refer to a benefit produced by X as the function of X just when that benefit was the cause of selection for X. In other words, for a trait to be an adaptation (historically defined) is *precisely* for that trait to have a function (selected-effect defined). A trait *is* an adaptation when and only when it *has* a function. The two terms are interchangeable. If a law were passed against the SE concept of function, its use in biology could be fully served by the historical concept of adaptation.

3   Functional Anatomy and Causal Role Functions

The major philosophical competitors to the SE concept of function refer to contemporary causal powers of a trait rather than the causal origins of that trait. Most of these non-selective analyses also advert to the (contemporary) purposes or goals of a system. The goals are presumed knowable prior to addressing the question of function, so that identifying a trait's function amounts to identifying the causal role played by the trait in the organism's ability to achieve a contemporary goal. Robert Cummins (1975) introduced a novel concept of function in which the specification of a real, objective goal simply dropped out. Since neither current benefits and goals nor evolutionary purposes were relevant, evolutionary history was also irrelevant to the specification of function. Cummins focused on functional analysis, which he took to be a distinctive scientific explanatory strategy. In functional analysis, a scientist intends to explain a *capacity* of a system by appealing to the capacities of the system's component parts. A novel feature of Cummins's analysis is that capacities are not presented as (necessarily) goals or purposes of the system. Scientists choose capacities which they feel are worthy of functional analysis, and then try to devise accounts of how those capacities arise from interactions among (capacities of) the component parts. The functions assigned to each trait (component) are thus relativized both to the overall capacity chosen for analysis and the functional explanation offered by the scientist. Given some functional system s:

X functions as an F in s (or: the function of X in s is to F) relative to an analytical account A of s's capacity to G just in case X is capable of F-ing in s and A appropriately and adequately accounts for s's capacity to G by, in part, appealing to the capacity of X to F in s. (Cummins 1975, p. 762. Variables renamed for consistency.)

Cummins's assessments of function do not depend on prior discoveries of the purposes or goals served by the analyzed capacities, as do other non-SE theories of function.[2] This creates a problem for Cummins. Prior, extrinsic information about system goals would narrow the list of possible functions to those which *can* contribute to the already-known goal. With no extrinsic criteria to delimit the list of relevant causal properties, Cummins needs some other method of constraining the list of causal powers which are to be identified as functions. Indeed, the problem of constraint gives rise to the most frequent challenge to Cummins's approach; examples will be discussed below. Critics find it easy to devise whimsical "functional analyses" which trade on the lack of external constraint, and which appear to show Cummins's definition of function to be too weak to distinguish between functions and mere effects. To make up for the loss of the external constraint of goal-specificity, Cummins offers internal criteria for assessing the scientific significance of a proffered functional analysis. A valuable (as opposed to a trivial) functional analysis is one which adds a great deal to our understanding of the analyzed trait. In particular, the scientific significance or value of a given functional analysis is judged to be high when the analyzing capacities cited are *simpler* and *different in type* from the analyzed capacities. An analysis is also of high value when it reveals a high degree of *complexity of organization* in the system. Functional analyses of very simple systems are judged to be trivial on these criteria. "As the role of organization becomes less and less significant, the [functional] analytical strategy becomes less and less appropriate, and talk of functions makes less and less sense. This may be philosophically disappointing, but there is no help for it" (ibid., p. 764). Philosophical disappointment in this messy outcome could be alleviated by requiring an independent specification of goals and purposes prior to any functional analysis. But, as we shall see, such philosophical serenity would carry a high cost for scientific practice.

Cummins's account is of special interest because of its close match to the concepts of function used within functional anatomy. His emphasis on causal capacities of components and the absence of essential reference to overall systemic goals is shared by the anatomists. This is somewhat surprising, since Cummins's chief interest was in functional analysis in psychology. He did assert (without documentation) that biology fit the model, but has written nothing else on biological function (ibid. p. 760). Other philosophers have recognized non-SE uses of function in biology. Boorse cited physiology and medicine as supporting his goal oriented causal role analysis (Boorse 1976, p. 85). Brandon acknowledged the non-historical use in physiology, but disapproved. "I believe that ahistorical functional ascriptions only invite confusion, and that biologists ought to restrict the concept [to] its evolutionary meaning, but I will not offer further arguments for that here" (Brandon 1990, p. 187 n. 24). The wisdom of this counsel will be assessed below.

The classic account of the vocabulary of functional anatomy was given by Walter Bock and Gerd von Wahlert (1965). These authors referred to "the form-function complex" as an alternative to the customary contrast between the two—form *versus* function. This was not merely an attempt at conciliation between advocates of the primacy of form over function and advocates of the converse. Rather, it was a reconceptualization of the task of anatomists, especially evolutionary anatomists. Bock and von Wahlert stated that the form and the function of anatomical traits were *both* at the methodological base, the lowest level, of the functional anatomist's enterprise. The rejection of the contrast between form and function (its replacement with the form-function complex) amounted to a rejection of the SE concept of function itself. In the functional anatomist's vocabulary, form and function were both observable, experimentally measurable attributes of anatomical items (e.g. bones, muscles, ligaments). Neither form nor function was inferred via hypotheses of evolutionary history. The form of an item was its physical shape and constitution. The function of the same item was "all physical and chemical properties arising from its form ... providing that [predicates describing the function] do not mention any reference to the environment of the organism" (ibid. p. 274). This denial of reference to environment eliminates not only the SE concept of evolutionary function, but also the non-historical notion of function as a contribution to contemporary adaptedness or other goal-achieving properties. These implications were intended. Concepts involving biological importance, selective value, and (especially) selective *history*, (and therefore Darwinian adaptation), are all at higher and more inferential levels of analysis than that of anatomical function. The intention was not to ignore these higher levels, but to provide an adequate functional-anatomic evidentiary base from which the higher levels can be addressed.

The level of organization above the form-function complex is the character complex. A character complex is a group of features (typically anatomical items themselves seen as form-function complexes) which interact functionally to carry out a common *biological role*. When we reach the biological role, we find ourselves in more familiar Darwinian territory. The biological role of a character complex (or of a single trait) is designated by "that class of predicates which includes all actions or uses of the faculties (the form-function complex) of the feature by the organism in the course of its life history, provided that these predicates include reference to the environment of the organism" (ibid. p. 278). At last we find reference to that organism/environment relation which constitutes adaptedness or fitness. The further inference to the SE advocate's concept of evolutionary function involves an additional assertion that the trait's present existence is not fortuitous, but the result of a history of natural selection controlled by the same benefits which the trait now confers in its biological role.

So a chain of inference from anatomical function to evolutionary function involves several steps and additional (i.e. non-anatomical) kinds of data. An evolutionist may not feel the need to start from the anatomical base, of course. Given a simple trait with a known biological role, the evolutionist might feel justified in ignoring anatomical details. But in highly integrated character complexes with long evolutionary histories (e.g. the vertebrate jaw or limb) it is arguably perilous to ignore anatomical function (Wake and Roth 1989).

In one way, Bock and von Wahlert's concept of function is even more radical than Cummins's. Cummins assigns functions only to those capacities of components which are actually invoked in a functional explanation, those which are believed to contribute to the higher level capacity being analyzed. Bock and von Wahlert include *all possible* capacities (causal powers) of the feature, given its current form. Some of these capacities are utilized and some are not. Both utilized and unutilized capacities are properly called functions. The determination of unutilized functions may require experiments which are ecologically unrealistic, but this is still a part of the functional anatomist's job. Bock and von Wahlert suggest that a functional anatomist might want to experimentally study the functional properties of a muscle at 40 percent of its rest length, even when it is known that the muscle never contracts more than 10 percent during the life history of the organism (1965, p. 274). The relevance of unutilized functions depends on the sort of question being asked. Other anatomists attend primarily to utilized functions. "The study of function is the study of how structures are used, and functional data are those in which the use of structural features has been directly measured. Functions are the actions of phenotypic components" (Lauder 1990, p. 318). Bock's special interest in unutilized functions comes from his interest in the phenomenon of preadaptation (or exaptation) (Bock 1958; Gould and Vrba 1982). It is often the unutilized functional properties of traits which allow them to be "coopted" and put to new uses when the evolutionary opportunity arises.

Apart from the issue of unutilized functions, Cummins's concept of function matches the anatomists'. Functional anatomists typically choose to analyze integrated character complexes which have significant biological roles. An anatomist might choose to analyze the crushing capacity of the jaw of a particular species. Cummins's s is the jaw, and G the capacity to crush things. In the analysis the anatomist might cite the capacity of a particular muscle (component X) to contract, thereby bringing two bones (other components of s) closer together. If the citation of that capacity of X fits together with other citations of component capacities into an "appropriate and adequate" account of the capacity of the jaw to crush things, then it is proper on Cummins's analysis to say that the function (or a function) of that muscle is to bring those two bones closer together.

We can also apply Cummins's evaluative suggestions to such an analysis. In a valuable functional analysis, the analyzing capacities will be simpler and/or different in type from the analyzed, and the system's discovered organization will be complex. Suppose the capacity to crush of the hypothetical jaw derives from the extremely simple fact that objects between the two bones are subjected to the brute force of muscle X forcing the bones together. Here the "organization" of the system is almost degeneratively simple, and the force of the muscle hardly simpler or different in kind from the crushing capacity of the jaw. A functional analysis of very low value. On the other hand, suppose that the jaw is a complex of many elements, muscle X is much weaker than the observed crushing capacity, the crushing action itself is a complex rolling and grinding, the action of muscle X moves one of its attached bones into a position from which the bone can support one of the several directions of motion, and that this action must be coordinated with other muscle actions so that it will occur at a particular time in the crushing cycle. Here X's function is much simpler than the analyzed capacity, is different in kind (moving in one dimension in contrast to the three dimensional motion of the jaw) and the organization of components which explains jaw action is complex indeed. A functional analysis of high value.

As in Cummins's account, functional anatomical analyses make no essential reference to the benefits which the analyzed capacity might have, nor to the capacity's evolutionary goal or purpose. While the decision to analyze the jaw may have been motivated by a knowledge of its biological role (the fish eats snails), that knowledge plays no part in the analysis itself. The biological role of the jaw system does not influence the function which the component muscle is analyzed to have. The discovery of a new biological role (perhaps the jaws are also used in producing mating sounds) may suggest new situations under which to examine the function of muscle X, but even such a discovery would not alter the estimated function(s). Even more remote from functional analysis are hypotheses regarding selective pressures, or any other explanations of why the jaw has its present capacities. Neither Cummins nor a functional anatomist intends to explain the origin of a muscle when stating its function in the jaw.

4   Criticisms of Causal Role Function

The purpose of this section is to evaluate some critical commentary on Cummins's concept of causal role function, and to assess the extent to which it might call into question the use of CR function in functional anatomy. In Millikan's case at least, it would be inaccurate to read the comments as a general critique of CR function. She has herself made use of Cummins-like concepts in other contexts (1993, p. 191). In (1989) her intent was to discuss purpose and dysfunction, concepts to which CR function doesn't apply. Nevertheless, in discussing purposive function both Millikan and Neander make claims for its importance which would appear to subordinate CR function to SE function. It is these implications which we must examine.

Millikan and Neander each amply demonstrated that the most common philosophical objections to SE function lost their force when the theory of SE function was understood not as ordinary language conceptual analysis, but as an explication of current scientific theory. Their own criticisms of CR function, however, seem to assume that the opposition theory is exactly what they deny their own theories to be—a good old-fashioned ordinary language conceptual analysis. If CR function theory is treated as an explication of the practices of science, those criticisms fail in exactly the way Millikan and Neander show their own philosophical opponents to fail. In other words, their criticisms of CR function rely on giving SE and CR functions unequal treatment—one as theoretical definition and the other as ordinary language analysis.

First, a minor example of the unequal treatment of the SE and CR theories. Millikan, Neander, and Sober each point out that Cummins's CR theory counterintuitively allows reference to "functions" in nonbiological (or biologically uninteresting) systems. These are examples of the whimsical Cummins functions mentioned above, made possible by Cummins's abandonment of goal specification. Millikan offers the "function" of clouds as making rain in the water cycle (p. 294), Neander the "function" of geological plate movements in tectonic systems (p. 181), Sober the "function" of the heart (via its mass) to allow an organism to have a certain weight (1993, p. 86). These are indeed counterintuitive results. But the criticism simply does not apply to the real world of scientific practice. By Cummins's own evaluative criteria (and given the facts of the real world) functional analyses of these systems would have no interest. Analyzing capacities would not be significantly simpler or different in type from analyzed capacities (are plate movements simpler than earthquakes?) nor would the system's organization be notably complex. (The geological structures which result in earthquakes might be complex, but the "organization" of these structures *vis a vis* their explanation of the capacity of the earth to quake is not.) Real world scientists do not perform Cummins-like functional analyses outside the organic and artifactual domains (or on non-organized properties like body weight). Millikan herself elegantly explains why this should be so. In defense of SE function she observed that the only items *in our world* with interesting Cummins functions are items with proper (SE) functions (p. 293). In our world, all of the interesting causal role functions have a history of natural selection. Instant lions would have no such history, but they do not exist in our world. Earthquakes and rainfalls are in our world, but have no such history, and so no complex functional organization. Such imaginative counterexamples might be telling against conceptual analyses of ordinary language function concepts. But they count neither for nor against CR or SE function theories, so long as those theories are *each* seen as science-based rather than conceptual analyses of ordinary language.

A second and more complex criticism involves the so-called normative role of function ascriptions and the problem of pathological malformations of functional items. Neander considers it the responsibility of a theory of biological function to categorize organic parts such that the categories are able to "embrace both interspecies and pathological diversity" (Neander p. 181). Millikan endorses at least the latter, and other SE theorists have been concerned with variation and dysfunction as far back as Wright (1973, pp. 146, 151). According to these theorists, only SE function can categorize parts into their proper categories irrespective of variation and malformation. It does so by defining "function categories." CR function (like other non-historical theories) cannot define appropriate function categories, and so is unable both to identify diseased or malformed hearts as hearts, and to identify the same organ under different forms in different species.

On pathology, Millikan points out that diseased, malformed, and otherwise dysfunctional organs are denominated by the function they would serve if normal. "The problem is, how did the atypical members of the category that cannot perform its defining function *get* into the same function category as the things that actually can perform the function?" (Millikan p. 295. *Cf.* Neander p. 180–181.) A CR analysis of a deformed heart which cannot pump blood obviously cannot designate its *function* as pumping blood, since it doesn't have that causal capacity. On the other hand, even the organism with the malformed heart has a selective history of ancestors which survived because *their* hearts pumped blood. So the category "heart" which ranges over both healthy and malformed organs must be defined by SE, not CR, function. On interspecies diversity of form:

The notion of a "proper function" is the notion of what a part is *supposed* to do. This fact is crucial to one of the most important theoretical roles of the notion in biology, which is that most biological categories are only definable in functional terms. For instance, "heart" cannot be defined except by reference to the function of hearts because no description purely in terms of morphological criteria could demarcate hearts from non-hearts. (Neander p. 180)

The claim that biological categories must be defined by SE functional analyses is a significant challenge to CR functional analysis. If SE function is truly the basis of biological classification, then CR functional analyses must either (1) deal with undefined biological categories, or (2) depend on prior SE functional analyses for a classification of biological traits. We will now argue that SE functionalists are simply mistaken in this claim. SE functions are not the foundation for the classification of basic biological traits. To be sure, CR function does not define basic categories either. The classifications come from a third, non-functional source.

Consider Neander's claim that "most biological categories are only definable in functional terms." Hardly a controversial statement, especially in the philosophical literature. Nevertheless it is utterly false. Perhaps most *philosophically interesting* biological categories are functional (depending on the interests of philosophers). But a glance in any comparative anatomy textbook rapidly convinces the reader (and appalls the student) with the ocean of individually classified bones, ligaments, tendons, nerves, etc., etc. We do not mean simply to quibble over a census count of functional versus anatomical terms in biology. Rather, we wish to argue for the importance, often unrecognized by philosophers, of anatomical, morphological, and other non-purposive but theoretically crucial concepts in biology. In this case the relevant conceptual apparatus belongs to the field of comparative anatomy.

Many body parts can be referred to either by anatomical or functional characterizations. The human kneecap is a bone referred to as the patella. "Kneecap" is a (roughly) functional characterization; a kneecap covers what would otherwise be an exposed joint surface between the femur and the tibia. "Patella" is an anatomical, not a functional, characterization. The patella in other vertebrates need not "cap" the "knee" (for example, in species in which it is greatly reduced) and some species might conceivably have their knees capped by bones not homologous to the patella. The category *patella* is not a function category but an anatomical category. *Kneecap* is a function category. To call a feature a wing is to characterize it (primarily) functionally. To call it a vertebrate forelimb is to characterize it anatomically. The wings of butterflies and birds have common functions but no common anatomy.

The concept of *homology* is central to the practice of evolutionary biology. It is arguably as important as the concept of *adaptation*. Anatomical features which are known (at their naming) to be homologically corresponding features in related species are given common names. A traditional Darwinian definition of homology refers to the common derivation of body parts: "A feature in two or more taxa is homologous when it is derived from the same (or a corresponding) feature of their common ancestor" (Mayr 1982, p. 45). This definition has recently come under scrutiny, and a more openly phylogenetic definition (most clearly explicated by Patterson 1982) is often preferred. (See Hall 1994 for discussions of homology.) On this concept, homologous traits are those which characterize natural (monophyletic) clades of species. Thus, the wing of a sparrow is homologous to the wing of an owl because the character "wing" (recognized by a particular structural configuration of bones, muscles, and feathers) characterizes a natural evolutionary clade (birds) to which sparrows and owls belong. Wings of sparrows are not homologous to wings of insects because there is no evidence that a clade consisting of birds + insects constitutes a natural evolutionary unit. This remains true even if "wing" is characterized functionally, as "flattened body appendage used in flight." Whatever the favored definition of homology, one feature of the concept is crucial: *the relation of homology does not derive from the common function of homologous organs.* Organs which are similar in form not by virtue of phylogeny but because of common biological role (or SE function) are said to be *analogous* rather than homologous. The wings of insects and birds are analogous— they have similar SE functions, and so evolved to have similar gross structure. The forelimbs of humans, dogs, bats, moles, and whales, and each of their component parts—humerus, carpals, phalanges—are homologous. Morphologically they are the same feature under different forms. Functionally they are quite distinct.[3]

Comparative anatomy, morphology, and the concept of homology predate evolutionary biology. They provided Darwin with some of the most potent evidence for the fact of descent with modification. (This alone demonstrates the importance of other-than-adaptational factors in evolutionary biology.) So the evolutionary definition of homology mentioned above is a theoretical definition. As with other theoretical definitions, it is subject to sniping from practitioners of conceptual analysis. A philosopher could argue (pointlessly) that "homology" cannot *mean* "traits which characterize monophyletic clades," since many 1840s biologists knew that birds' wings were homologous to human arms but disbelieved in evolution (and so disbelieved that humans and birds shared a clade). SE advocates' usual reply to the William Harvey objection is applicable here. Just as Harvey could see the marks of biological purpose without knowing the origin or true nature of biological purpose, pre-Darwinian anatomists could see the marks of homology without knowing the cause and true nature of homology itself.

But if anatomical items are not anatomically categorized by function, how are they identified? There are several classical (pre-Darwinian) ways of postulating homologies. Similarity in structure may suggest homology. Second, the "principle of connectedness" states that items are identical which have identical connections or positions within an overall structural pattern. Third, structurally diverse characters may be recognized as homological by their common developmental origin in the embryo. Mammalian inner ear bones and reptile jaw bones can be seen (if you look *very* carefully) to arise out of common embryological elements. If you look closer yet, the reptilian jaw bones can be seen to be homologous to portions of the gill arches of fish. The important point is that if anatomical parts had to be identified by their common biological role or SE function, all interesting homologies would be invisible. Darwin would have lost crucial evidence for descent with modification.

The fact that anatomical or morphological terms typically designate homologies shows that they are not functional categories. There is some casual use of anatomical terms by biologists, especially when formal analogies are striking. Arthropods and vertebrates each have "tibias" and "thoraxes" but the usage is selfconsciously metaphorical between the groups; dictionaries of biology have two separate entries. The anatomical unit is, e.g., the *vertebrate tibia*.

There is indeed a set of important biological categories which group organic traits by their common biological roles or SE functions. The most general of these apply to items which have biological roles of broadly significant in the animal world that they are served by analogous structures in widely divergent taxa. Among such concepts are gut (and mouth and anus), gill, gonad, eye, wing, and head (but not skull, an anatomical feature only of vertebrates). Also in the group is that all-time favorite of philosophical commentators on function—the heart. These are presumably what Neander had in mind as typical "biological categories," and they are reasonably regarded as "function categories" in Millikan's sense. They are analogical (as opposed to homological) in implication. Narrower function categories occur also (e.g. kneecap and ring finger) but are of limited scientific interest.

The importance of the above function categories comes from the fact that they all apply to features which result from evolutionary convergence—the selective shaping of non-homologous parts to common biological roles. It might be argued that homologous organs or body parts can be categorized by function as well. For example, *kidney* is not listed among the above function terms. Kidneys do all perform similar functions, but properly-so-called (i.e. by scientific biological usage) they exist only as homologs in vertebrates. Analogous organs exist in mollusks, but are only informally called kidneys. "The excretory organs are a pair of tubular metanephridia, commonly called kidneys in living species" (Barnes 1991, p. 345). But isn't "kidney" a function category? Well, kidneys do all perform common functions (in vertebrates). But they are also homologous. This means that we could identify all members of the category "kidney" by morphological criteria alone (morphological connectedness and developmental origin). So, at least in that sense, "kidney" is not a function category, or at least not *essentially and necessarily* a function category. Unlike hearts, kidneys can be picked out by anatomical criteria alone. Identifying the function of kidneys amounts to discovering a (universal) functional fact about an anatomically defined category.

Even full-fledged, cross-taxon functional categories like "heart" can often be given anatomical readings within a taxon. That is, *the vertebrate heart* can be treated as an anatomical category like the kidney. Vertebrate hearts, like kidneys, do have common functions. But they are identifiable within the taxon by their anatomical features alone. For example, mammalian heart muscle (as well as that of many other vertebrates) has a unique structure with individual cardiac muscle cells connected electrically in specialized junctional discs. The histological structure of mammalian cardiac muscle could not be mistaken for any other tissue. Thus, it is incorrect to suggest that hearts that characterize natural evolutionary clades cannot be characterized by anatomical criteria. This situation will obtain just when all of the members of the functional category are homologous within the taxon. Since all *vertebrate* hearts are homologous, they can be identified by anatomical criteria, notwithstanding the name they share with their molluscan analogs. Similarly, tetrapod hearts can be defined by unique anatomical features as can amniote hearts, and mammal hearts. The nested phylogenetic pattern (vertebrates : tetrapods : amniotes : mammals) is thus mirrored in the nested set of anatomical definitions available for vertebrate hearts. This is not surprising as it is nested sets of similarities that provides evidence of phylogeny. On the other hand, *insect wing* cannot be treated as an anatomical category, for the simple reason that the wings of all insect taxa are probably not homologous.

Again, the point is not to quibble over the word-counts of biological concepts which are function categories and those which are not. The question is this: Do the observations of Millikan, Neander, and other SE advocates on function categories imply that CR functional anatomists will be dependent on SE functionalists in order to characterize their subject matter? Does the existence of biological function categories mean that a reliance on causal role function will leave functional anatomists unable to identify dysfunctional hearts as hearts, a malformed tibia as a tibia? Is it true, as Neander reports, that "no description purely in terms of morphological criteria could demarcate hearts from non-hearts"?

These claims, taken as critiques of CR functional anatomy, are almost completely groundless.[4] Morphologists are able to identify anatomical items by anatomical criteria, ignoring SE function, and do so frequently. Are hearts impossible to define by "morphological criteria alone"? It is hard to know what Neander means by this. Criteria actually *used by morphologists*, e.g. connection, microstructure, and developmental origin, certainly *are* capable of discriminating between hearts and non-hearts within vertebrates. Perhaps by "morphological criteria" Neander has in mind the gross physical shapes of organs. To be sure, hearts have quite different shapes and different numbers of chambers in different vertebrate species. But no practicing morphologist uses gross shape as the "morphological criterion" for an organ's identity. Even a severely malformed vertebrate heart, completely incapable of pumping blood (or serving any biological role at all), could be identified as a heart by histological examination.

Complaining about the absence of necessary and sufficient gross physical characteristics for a morphological identification of *vertebrate heart* is surely an unwarranted philosophical intrusion on science. Such an argument should only be offered by someone practicing the "confused program, philosophical chimera" of ordinary language conceptual analysis. Morphologists can get along quite well without providing necessary and sufficient conditions for hearthood which would satisfy conceptual analysts. There is no doubt that the philosophers among us could play the conceptual analyst's game, and dream up a bizarre case in which a miraculously-deformed vertebrate's heart happened to have bizarre embryonic origins and histology, and histology, and was located under the poor creature's kneecap. The organism, if real, would baffle the anatomists just as the instant lion would baffle Darwin. But post-ordinary language philosophers do not indulge in that style of philosophy. Anatomy *as it is practiced* requires no input from SE functionalists or from biological students of adaptation in order to adequately classify and identify the structures and traits with which it deals.

SE functionalists are not the only philosophers whose emphasis on purposive function is associated with an underappreciation of anatomical concepts. Daniel Dennett shows the same tendency. Dennett argued for the indeterminacy of (purposive) functional characterizations. He brought up Stephen Jay Gould's famous example of the panda's thumb. Gould (1980) had observed that the body part used as a thumb by the panda was not anatomically a digit at all, but an enlarged radial sesamoid, a bone from the panda's wrist. Dennett's comment: "The panda's thumb was no more *really* a wrist bone than it is a thumb" (Dennett 1987, p. 320). The problem with this claim is that while "thumb" is a functional category, "radial sesamoid" (or "wrist bone") is an anatomical one. Even if Dennett is correct about functional indeterminacy, anatomical indeterminacy would require a separate argument, nowhere offered. Dennett's arguments for functional indeterminacy involved the optimality assumptions he claimed were present in all functional ascriptions. Such arguments carry no weight in anatomical contexts. Such an unsupported application of a point about function to an anatomical category reflects the widespread philosophical presumption that biology is almost entirely the study of purposive function. (See Amundson 1988, 1990 on Dennett's defenses of adaptationism.)

To be fair, we must acknowledge that Millikan and Neander, like other SE functionalists, were primarily interested in *purposive* concepts of function, not in *all possible* function concepts. And it is true that SE function provides an analysis of purpose which is lacking in CR function. But their interests in purpose can lead SE functionalists to overestimate the value of purposive concepts. It is simply false that anatomists require purposive concepts in order to properly categorize body parts. Anatomical categorizations of biological items already embrace interspecies and pathological diversity without any appeal to purposive function. Anatomical distinctions are not normally based on CR function *either*, to be sure. Functional anatomists *per se* do not categorize body parts. Rather they study the capacities of anatomical complexes which have already been categorized by comparative anatomists. Causal role functional anatomy proceeds unencumbered by demands to account either for the categorization or the causal origins of the systems under analysis.

5   The Eliminability of Causal Role Functions

In this and the following two sections we will consider whether CR functions, as studied in functional anatomy, can be eliminated from evolutionary biology in favor of SE functions. We will find them ineliminable.

First, let us consider the simplest case. Is it possible that there is a one-to-one correspondence between SE functions and CR functions? Perhaps CR functions just *are* SE functions seen through jaundiced non-historical and non-purposive lenses. To examine this possibility let us suppose that we could easily identify which character complexes serve their present biological roles in virtue of having been selected to do so. (Not at all a trivial assumption, as will soon be seen.) What would be the relation between the biological role(s) played by a character complex (e.g. a jaw) and the CR functions which characterize the actions of its component parts? Bock and von Wahlert offer an answer. "Usually ... the biological roles of the individual features are the same as those of the character complex" (Bock and von Wahlert 1965, p. 272). Taking the jaw as a character complex which has as one of its biological roles the mastication of food, each component muscle, bone, etc. of the jaw shares in the food mastication biological role.

But if the biological roles, and hence the SE functions, of the components of a character complex are the same as those of the overall complex itself, the CR functions of the components cannot be the same as their SE functions. All components of a complex have the *same* biological role/SE function, but each plays a *different* causal role within the character complex. So on this account SE functions cannot replace CR functions. Perhaps this result is to be expected. Bock and von Wahlert are, after all, functional anatomists. But if advocates of SE function hope to oppose this result, and refute the special significance of CR function, they presumably must argue that the activities of each component of a character complex is individually subject to the SE definition of function.

One consideration which might tempt an SE advocate in this direction is Millikan's observation, mentioned above, that all items in this world with functional complexity have undergone histories of natural selection. (Or, in the case of artifacts, were created by organisms which have such a history.) Notice, however, that the generalization *Functionally complex items have selective histories* does not by itself imply that a positive selective influence was responsible for every causal property of every component of the functional complex. Bock and von Wahlert could accept the generalization but still distinguish biological role from CR function.

Indeed, there are many reasons to reject the identification of CR functions as merely non-historically-viewed SE functions. For example, some functional anatomists wish to examine *unutilized* CR functions; clearly an unutilized function is not one which can be selected for. Further, the identification of CR with SE functions would define preadaptations (or exaptations) out of existence. But the question of the existence of currently utilized but unselected-for preadaptations (exaptations) or other selectively unshaped causal properties must be decided on the basis of evidence, not by definitional fiat.

We will not further belabor this implausible position; perhaps no SE advocate would take it anyhow. The point of this and the previous section is only that CR functions cannot be definitionally or philosophically eliminated. More interesting questions remain. Why do anatomists *need* to deal with causal role functions? Why can't they get along with purposes and selected effects?

6   Applicability of Selected Effect Function to Research in Functional Anatomy

A major concern of practicing functional anatomists is the utility of concepts such as function and biological role. In day-to-day research, how are functions to be identified and compared across species, and how, in practice, are we to identify the biological role of a structure? By specifying that function is that effect for which a trait was selected, SE functionalists have placed anatomists in a difficult position. In order to be able to label a structure with a corresponding function, a functional morphologist must be able to demonstrate first, that selection acted on that structure in the population in which it arose historically, and second, that selection acted specifically to increase fitness in the ancestral population by enhancing the one specific effect that we are now to label a function of the structure. There are at least three areas in which practical difficulties arise in meeting these conditions.

First, as biologists have long recognized (e.g., Darwin 1859, chapter 6), structures may have more than one function, and these functions may change in evolution. If such change occurs, are we to identify the function of a structure as the effect for which it was first selected? If selection changes to alter the SE function of a structure through time, how are functional morphologists to identify which SE function should be applied to a structure? A recent example that points out some of the difficulties of an SE concept of function in this regard is the analysis of the origin of insect wings performed by Kingsolver and Koehl (1985). Although efforts to estimate the past action of selection (as discussed below) are fraught with difficulty, Kingsolver and Koehl used aerodynamic modeling experiments in an effort to understand the possible function of early insect wings. Do short-winged insect models obtain any aerodynamic benefit from the short wings? In other words, is it likely that selection acted on very small wings to improve aerodynamic efficiency and enhance the utility of the small wings for flight, eventually producing larger-winged insects? If so, then it would be possible to argue that the SE function of insect wings is flight. However, Kingsolver and Koehl (1985, p. 488) found that short insect wings provided no aerodynamic advantage, and argued that "there could be no effective selection for increasing wing length in wingless or short-winged insects...." These authors did find, however, that short wings provided a significant advantage for thermoregulation; short wings specifically aided in increasing body temperature, which is important for increasing muscle contraction kinetics and allowing for rapid movements. Based on these data then, one might hypothesize that insect wings originated as a result of selection for improved thermoregulatory ability, and that only subsequently (when wings had reached a certain threshold size) did selection act to improve flight performance.

If we identify the function of insect wings as that effect for which they were *first* selected, then we would say that the function of insect wings is thermoregulation. It might be argued that in fact, the earliest wing-like structures actually are not proper wings, and that modern insect wings really do have the SE function of flight because at some point there was selection for improved flight performance. But this fails to recognize the size continuum of morphological structures that we call insect wings, the fact that large wings even today are used in thermoregulation, the structural homology of large and small wings, and the virtual impossibility of identifying the selection threshold in past evolutionary time. If we cannot identify the threshold, we will not know when to change the SE function of wings from thermoregulation to flight. Examples such as this illustrate the difficulty of assuming that the present day roles or uses of structures are an accurate guide to inferring past selection and hence SE function.

The SE theory of function does not rule out the existence of changing patterns of selection on a given structure nor the existence, in principle, of several SE functions for one structure. However, the complexities of this common biological situation for the association of an SE function with a specific structure have not been adequately addressed or appreciated.

Second, there are enormous practical difficulties in determining just what the selected effect of a structure was in the first place. Many structures are ancient, having arisen hundreds of millions of years ago. During this time, environments and selection pressures have changed enormously. How are we to reconstruct the ancient selected effect? The example of insect wings given above represents a best case scenario in which we are able to make biophysical models and use well-established mathematical theories of fluid flow to estimate the likely action of selection. But many structures (particularly in fossils) are not amenable to such an analysis. Even with modern populations, studies designed to show selection on a given trait are difficult and are subject to numerous alternative interpretations and confounding effects (Endler 1986; Arnold 1986). Functional morphologists do not have the luxury of simply asserting that the SE function of structure X is F (as philosophers so regularly do with the heart): there must be direct evidence that selection acted on structure X for effect F.

Third, there is considerable difficulty in determining that selection is acting (or acted) on *just* the structure of interest, even in extant taxa. Such difficulties are, for all practical purposes, insurmountable when dealing with fossil taxa or ancient structures. For the SE function of a structure to be identified, it is critical to be able to show that selection acted on that particular structure. However, as has been widely documented (e.g., Falconer 1989; Rose 1982), selection on one trait will cause manifold changes in many other traits through pleiotropic effects of the gene(s) under selection. Thus, selection for increased running endurance in a population of lizards may have the concomitant effect of increasing heart mass, muscle enzyme concentrations, body size, and the number of eggs laid, despite the fact that selection was directed only at endurance.

In fact, many phenotypic features are linked via common developmental and genetic controls, and this pattern of phenotypic interconnection makes isolation of any single trait and its selected effect very difficult (Lauder et al. 1993). If biologists had a ready means of locating the specific trait that is (or was) being acted on by selection, then the SE definition of function would be easy to apply. In actuality, due to pleiotropy, one typically sees a response in many traits to any particular selective influence. In laboratory selection experiments, the selected effect is known, and it is relatively easy to separate the selected trait from correlated responses. But in wild populations, one observes changing mean values of numerous traits in response to selection, and it is extremely difficult to separate the individual trait that is responding to selection from those that are exhibiting a correlated response.

It is also important to recognize that in extant species, the selected effect may be easier to identify than the trait acted upon by selection. This might seem counterintuitive at first, since so many studies of adaptation proceed by first identifying a trait, and only then searching for its selective advantage(s). The difficulty of identifying the trait arises because of the correlation of the many biological traits that influence selected effects or organismal performance, and the hierarchical nature of physiological causation. Consider one powerful method for the study of selection in nature: the analysis of cohorts of individuals in a population and their demographic statistics by following individuals through time (Endler 1986). For example, if one marks individual insects in a population and measures their fitness (e.g., mating success) and their performance on an ecologically relevant variable (say, maximum flight duration) one might well find that the mean flight duration increases in the population through time due to selection against individuals that cannot remain aloft long enough to successfully mate. (Such selection might be demonstrated using the statistical methods proposed by Arnold [1983; Arnold and Wade 1984; Lande and Arnold 1983].) Here we have strong evidence that selection is operating, and an identified selected effect (increased flight duration). But what is the trait X on which selection is acting? Suppose, as we mark the individual insects, we also take a number of measurements of morphology (such as body size, eye diameter, wing length and area). We can now examine these morphological variables to see if we observe changes in these population means that are correlated with changes in flight duration. If we find that only one variable, wing area, showed an increase in mean value that was correlated with the increase in flight performance through time, then we might we willing to conclude that wing area was trait X, the trait for which the SE function is "increasing flight duration."

Unfortunately, an example of this type would be truly exceptional. The common result is that *many* variables are usually correlated with changes in performance and fitness. It is almost certain, in fact, that many aspects of muscle physiology, nervous system activity, flight muscle enzyme concentrations and kinetics, and numerous other physiological features would show correlated change in mean values with the increase in flight duration. In addition, body length and mass are likely to show positive correlations, as are wing length, area, and traits that have no obvious functional relevance to flight performance (such as leg length). If we cannot identify the causal relationships among these correlated variables to single out the one that was selected for, we will be unable to assign a trait X to the SE function already identified. We have a SE function, but we do not know which trait to hang it on. The fact that pleiotropic effects are so pervasive in biological systems causes severe problems in applying the definition of SE function.

Two issues relate to the analysis of traits that might be selected for in an example such as the one discussed above. First, we might choose only to measure traits on individuals which *a priori* physiological and mechanical considerations suggest should bear a functional relationship to the demonstrated performance change. Thus, we might decide not to measure variables such as leg length since it is difficult to identify a physiological model in which increasing leg length would cause increased flight duration. Choosing variables based on an *a priori* model will certainly help narrow the universe of possible traits, but the remaining number of physiologically and mechanically relevant traits will still be very large. A second complexity in picking the trait that has been selected for arises from the hierarchical nature of physiological processes. A change in a performance characteristic (such as flight duration) may result from changes at many levels of biological design (Lauder 1991): muscle mass and insertions could change, muscle contraction kinetics could change by changing the proportion of different fiber types, enzyme concentrations within fiber types could be altered, and many features of the nervous system could be transformed. These different types of physiological traits have a hierarchical relationship to each other (in addition to a possible pleiotropic relationship) that represents a causal chain: changes at any one or more of these levels of design could account for a performance change at the organismal level. Yet, each of these features must be a distinct trait X in the SE definition, and we are unlikely in most cases to be able to identify the particular trait, or particular combination of traits, that was selected for. Of course, flight duration itself might well be considered as a trait, subject to selection and the same hierarchical patterns of underlying physiological variation as any other trait. In this case, the very same difficulties would obtain: we would need to be able to document selection *on that trait* (flight duration) in order to apply the SE concept of function.

These considerations show why anatomists are rarely able to identify which of the causal role functions of a given trait are its SE functions— that is, which (if any) are the effects for which the trait was selectively favored. But, as the next section will show, anatomists can not afford to abandon CR functions simply because SE function assignments are unavailable. Important research programs are at stake.

7 Research Programs in which Causal Role Function Is Central

Several aspects of current research in functional and evolutionary morphology make crucial and ineliminable use of the concept of CR function. Anatomists often write on "the evolution of function" in certain organs or mechanical systems, and may do so with no reference to selection or to the effects of selection (e.g., Goslow et al. 1989; Lauder 1991; Liem 1989; Nishikawa et al. 1992). Rather, in these papers functional morphologists mean to consider how CR functions have changed through time, in the same manner that morphologists have traditionally examined structures in a comparative and phylogenetic context to reconstruct their evolutionary history. Indeed, a significant contribution of functional anatomy (which has blossomed in the last twenty years by adopting physiological techniques to measure CR functions in different species) has been to treat functions as conceptually similar to structures. For example, Lauder (1982) and others (e.g., Wake 1991; Lauder and Wainwright 1992) have argued that CR functions may be treated just like any other phenotypic trait, and analyzed in an historical and phylogenetic context to reveal the evolutionary relationship between structure and function.

So, like SE functionalists, CR functional anatomists and morphologists are interested in history. But unlike SE functionalists, anatomists do not *define* a trait's function by its history. CR function is non-historically defined. The historical interests of evolutionary morphologists are not directed towards the evolutionary mechanism of selection or the analysis of adaptation. The relation between the approaches to history taken by SE functionalists and anatomical functionalists parallels the two major explanatory modes used in the analysis of organismal structure and function. These have been termed the *equilibrium* and the *transformational* approaches (Lauder 1981; Lewontin 1969). Studies of organismal design conducted under the equilibrium view study structure in relationship to environmental and ecological variables. Such analyses are appropriate for investigating current patterns of selection and for interpreting biological design in terms of extant environmental influences. The goal of equilibrium studies is to understand extrinsic influences on form (such as temperature, wind velocity, or competition for resources), and these studies are designed to clarify current patterns of selection and hence adaptation (Bock 1980; Gans 1974). Equilibrium studies tell us little about the history of characters, however (Lewontin 1969), as the very nature of the methodology presumes (at least a momentary) equilibrium between organismal design and environmental stresses.

Many studies in functional morphology, especially in the last ten years, have adopted the transformational approach (Lauder 1981) in which historical (phylogenetic) patterns to change in form are explicitly analyzed for the effects of intrinsic design properties. Here, the focus is not on adaptation, selection, or the influence of the environment, but rather on the effect that specific structural configurations might have on directions of evolutionary transformation. For example, a functional morphologist might ask: does the possession of a segmented body plan in a clade have any consequences for subsequent evolutionary transformation in design? Under a transformational research program one might examine a number of lineages, each of which has independently acquired a segmented body plan, to determine if subsequent phylogenetic diversification within each lineage shows any common features attributable to the presence of segmentation (regardless of the different environmental or biophysical influences on each of the species). In fact, segmentation, or more generally, the duplication or repetition of parts, appears to be a significant vehicle for the generation of evolutionary diversity in form and function by allowing independent specialization of structural and functional components (Lauder and Liem 1989). An exemplary transformational study is Emerson's (1988) analysis of frog pectoral girdles in which she showed that the initial starting configuration of the pectoral girdle in several clades was predictive of subsequent changes in shape. This transformational regularity occurred despite the different environments inhabited by the frog species studied. Transformational analyses by functional morphologists are historical in character: they focus on pathways of phylogenetic transformation in design which result from the arrangement of structures and the causal roles of those structures.

Functional morphologists also view organismal design as a complex interacting system of structures and functions (Liem and Wake 1985; Wake and Roth 1989). Indeed, the notion of "functional integration," which describes the interconnectedness of structures and their CR functions, is central to discussions of organismal design and its evolution. The extent to which individual components of morphology can be altered independently of other elements without changing the (CR) functioning of the whole is one aspect of this current research (Lauder 1991). Given a structural configuration involving many muscles, bones, nerves, and ligaments, for example, all of which interact to move the jaws in a species, one might ask what effect changing the mass of just one muscle will have on the action (CR function) of the jaws as a whole. Some arrangements of structural components will have limited evolutionary flexibility due to the necessity of performing a given function such as mouth opening: even minor alterations in design may have a deleterious effect on the performance of such a critical function. This implicates CR functions as an agent of evolutionary constraint. We could also inquire about possible components in a functionally integrated system that might theoretically be changed while maintaining the function of the whole system: do predicted permitted changes correspond to patterns of evolutionary transformation actually seen? The comparison of predicted and actual pathways of transformation is but one part of a larger effort to map a theoretical "morphospace" of *possible* biological designs. By defining basic design parameters for a given complex morphological system, a multidimensional morphospace may be constructed (e.g., Bookstein et al. 1985; Raup and Stanley 1971). Comparing this theoretical construct with the extent to which actual biological forms have filled the theoretically possible space allows the identification of fundamental constraints on the evolution of biological design. A frequent finding is that large areas of the theoretically possible morphospace are unoccupied, and explaining this unoccupied space is a key task of functional and evolutionary morphology.

For these reasons, it is difficult to envision how the concept of a CR function, so integral to both transformational analysis and functional integration, could be eliminated from the conceptual armamentarium of functional morphologists without also eliminating many key research questions.

8   Conclusion

Our rejection of some of Millikan's and Neander's conclusions should not disguise our strong agreement with their stance on the relation between the practices of science and philosophy. We heartily agree that conceptual analyses of ordinary language are inappropriately used to critique the concepts of a science. Indeed, most of our defenses of CR function against ordinary language conceptual analysis are versions of the ones used first by Millikan or Neander as they defended SE function against the same opponent. We differ from them not on the proper uses of philosophy, but on the needs and practices of biology.

We are more pluralistic than most philosophical commentators on function. We do not consider the SE concept of function, or its near-synonym the historical concept of adaptation, to be biologically or philosophically illegitimate. Our reservations about the application of purposive concepts in biology are primarily epistemological. As Williams said of adaptation, SE function in biology is "a special and onerous concept that should be used only where it is really necessary." Causal role function in anatomy, if less philosophically fertile than selected-effect function, is on much firmer epistemic footing. It also happens to be ineliminably involved in ongoing research programs. This alone ought to establish its credentials.

Given comparative anatomy to categorize its subject matter, and ecological or ethological studies of biological role to suggest which character complexes to analyze, functional anatomy is subject to none of the conceptual analyst's critiques of CR function. It is just as immune from philosophical refutation as Millikan's and Neander's science-based theory of SE function. The adequacy of each account is to be assessed not by its ability to fend off the facile imaginations of conceptual analysts, but to deal with real world scientific issues.

Finally, a recent recommendation from Elliott Sober.

If function is understood to mean adaptation, then it is clear enough what the concept means. If a scientist or philosopher uses the concept of function in some other way, we should demand that the concept be clarified. (Sober 1993, p. 86)

We submit that Sober's challenge has now been met.

Acknowledgements

We received valuable comments on an earlier version of this essay from Elliott Sober, Ruth Millikan, Robert Brandon, and an anonymous referee. Kenneth Schaffner generously supplied a prepublication copy of the chapter we cited, and helpful observations on various function concepts. The work was supported by NSF grants SBE–9122646 (to RA) and IBN91–19502 (to GVL).

Notes

1. We take it that Neander intends her analysis to reflect biologists' use of the term "function," not necessarily their use of the concept defined by Millikan as "proper function." Both Millikan (p. 290, note #1) and Neander (p. 168, note #1) refer to Neander's widely circulated but unpublished "Teleology in

Biology." In that paper Neander referred only to the biological concept of "function" (i.e., not to "proper function") except when she needed to distinguish between "a part's proper function and things which it just happens to do fortuitously." (Neander manuscript, p. 11)

2. A minority of commentators interpret Cummins as surreptitiously introducing goals and purposes by choosing for analysis only traits which are already known to be purposive (Rosenberg 1985, p. 68; Schaffner 1993, p. 399 ff.). We interpret Cummins as fully agnostic with regard to purpose, which is why the criticisms being considered are worthy of discussion. Rosenberg appears to be the only philosopher who supports Cummins's account of function for evolutionary biology; he does so partly because of this purposive reading. Whatever Cummins's original intentions, we intend CR function to be both non-historical and non-purposive in its applications.

3. Note that even extremely similar traits may arise by convergent evolution, and that the final test of homology is not similarity but rather congruent phylogenetic distribution of the putative homology with other characters providing evidence of monophyly. Thus, the eye of a squid and the eye of vertebrates are very similar in many (but not all) features. The non-homology of squid and vertebrate eyes does not rest on the differences noted between the eyes (virtually all homologous characters have some differences), but rather on the fact that very few other traits support the hypothesis that squids + vertebrates constitute a natural evolutionary lineage. The phylogenetic relationships among species thus provides the basis on which we make decisions about the homology of individual characters. For similar reasons, our statements to the effect that (homologous) traits *characterize* taxa should not be taken to mean that those traits are logically necessary or sufficient conditions for a species's membership in a taxon. Snakes are tetrapods notwithstanding their leglessness. The phylogenetic distribution of other traits than legs makes it clear that snakes are members of the same monophyletic group as more typically-legged tetrapods. See Sober (1993, p. 178) for a caution against appearances of essentialism in discussion of phylogenetic classification.

4. There is one felicitous application of Neander's claim about the inadequacies of morphological criteria to designate hearts. Since the category "heart" is used across major taxonomic differences, a vertebrate taxonomist unfamiliar with mollusks might well not be able to use *vertebrate* morphological criteria to identify a *molluscan* heart. And, to get only slightly bizarre, it is possible to imagine discovering a new taxon of animals which has organs functionally identifiable as hearts, but which fit the morphological criteria for hearts of no known taxon. We agree with the SE functionalist's point in this rather limited set of cases.

References

Amundson, R.: 1988, "Logical Adaptationism," *Behavioral and Brain Sciences* 11, 505–506.

Amundson, R.: 1990, "Doctor Dennett and Doctor Pangloss: Perfection and Selection in Psychology and Biology," *Behavioral and Brain Sciences* 13, 577–584.

Arnold, S. J.: 1983, "Morphology, Performance, and Fitness," *American Zoologist* 23, 347–361.

Arnold, S. J.: 1986, "Laboratory and Field Approaches to the Study of Adaptation," in M. E. Feder and G. V. Lauder (eds.), *Predator-Prey Relationships: Perspectives and Approaches from the Study of Lower Vertebrates*, University of Chicago Press, Chicago.

Arnold, S. J., and M. J. Wade: 1984, "On the Measurement of Natural and Sexual Selection: Theory," *Evolution* 38: 709–719.

Ayala, Francisco J.: 1970, "Teleological Explanations in Evolutionary Biology," *Philosophy of Science* 37: 1–15.

Barnes, R. D.: 1991, *Invertebrate Zoology*, Harcourt Brace Jovanovich, Fort Worth TX.

Bigelow, J., and R. Pargetter: 1987, "Functions," *The Journal of Philosophy* 84: 181–196.

Bock, W. J.: 1958, "Preadaptation and multiple evolutionary pathways," *Evolution* 13: 194–211.

Bock, W. J.: 1980, "The Definition and Recognition of Biological Adaptation," *American Zoologist* 20: 217–227.

Bock, W. J., and G. von Wahlert: 1965, "Adaptation and the form-function complex," *Evolution* 19: 269–299.

Bookstein, F., B. Chernoff, R. Elder, J. Humphries, G. Smith, and R. Strauss: 1985, *Morphometrics in Evolutionary Biology*, Academy of Natural Sciences, Philadelphia.

Boorse, C.: 1976, "Wright on Functions," *Philosophical Review* 85: 70–86.

Brandon, R. N.: 1981, "Biological Teleology: Questions and Explanations," *Studies in History and Philosophy of Science* 12: 91–105.

Brandon, R. N.: 1990, *Adaptation and Environment*, Princeton University Press, Princeton NJ.

Cummins, R.: 1975, "Functional Analysis," *Journal of Philosophy* 72: 741–765. [Excerpts reprinted in Ned Block (ed.), Readings in Philosophy of Psychology, MIT Press, Cambridge, MA.]

Cummins, R.: 1983, *The Nature of Psychological Explanation*, MIT Press, Cambridge, MA.

Darwin, C.: 1859, *On the Origin of Species*, John Murray, London.

Dennett, D. C.: 1987, *The Intentional Stance*, MIT Press, Cambridge, MA.

Emerson, S.: 1988, "Testing for Historical Patterns of Change: a Case Study with Frog Pectoral Girdles," *Paleobiology* 14, 174–186.

Endler, J.: 1986, *Natural Selection in the Wild*, Princeton University Press, Princeton NJ.

Falconer, D. S.: 1989, *Introduction to Quantitative Genetics*, 3rd ed., Longman, London.

Gans, C.: 1974, *Biomechanics: an Approach to Vertebrate Biology*, J. B. Lippincott, Philadelphia.

Goslow, G. E., K. P. Dial, and F. A. Jenkins: 1989, "The Avian Shoulder: an Experimental Approach," *American Zoologist* 29: 287–301.

Gould, S. J.: 1980, *The Panda's Thumb*, W. W. Norton and Company, New York.

Gould, S. J., and E. S. Vrba: 1982, "Exaptation—A Missing Term in the Science of Form," *Paleobiology* 8: 4–15.

Hall, B. K. (ed.): 1994, *Homology*, Academic Press, San Diego.

Kingsolver, J. G., and M. A. R. Koehl: 1985, "Aerodynamics, Thermoregulation, and the Evolution of Insect Wings: Differential Scaling and Evolutionary Change," *Evolution* 39: 488–504.

Lande, R., and S. J. Arnold: 1983, "The Measurement of Selection on Correlated Characters," *Evolution* 37: 1210–1226.

Lauder, G. V.: 1981, "Form and Function: Structural Analysis in Evolutionary Morphology," *Paleobiology* 7: 430–442.

Lauder, G. V.: 1982, "Historical Biology and the Problem of Design," *Journal of Theoretical Biology* 97: 57–67.

Lauder, G. V.: 1990, "Functional Morphology and Systematics: Studying Functional Patterns in an Historical Context," *Annual Review of Ecological Systems* 21: 317–40.

Lauder, G. V.: 1991, "Biomechanics and Evolution: Integrating Physical and Historical Biology in the Study of Complex Systems," in J. M. V. Rayner and R. J. Wootton (eds.), *Biomechanics in Evolution*, Cambridge University Press, Cambridge.

Lauder, G. V., A. Leroi, and M. Rose: 1993, "Adaptations and History," *Trends in Ecology and Evolution* 8, 294–297.

Lauder, G. V., and K. F. Liem: 1989, "The role of historical factors in the evolution of complex organismal functions," in D. B. Wake and G. Roth (eds.), *Complex Organismal Functions: Integration and Evolution in Vertebrates*, John Wiley and Sons, Chichester.

Lauder, G. V., and P. C. Wainwright: 1992, "Function and History: The Pharyngeal Jaw Apparatus in Primitive Ray-finned Fishes," in R. Mayden (ed.), *Systematics, Historical Ecology, and North American Freshwater Fishes*, Stanford University Press, Stanford.

Lewontin, R. C. 1969. "The Bases of Conflict in Biological Explanation," *Journal of the History of Biology* 2 35–45.

Liem, K. F.: 1989, "Respiratory Gas Bladders in Teleosts: Functional Conservatism and Morphological Diversity," *American Zoologist* 29 333–352.

Liem, K. F., and D. B. Wake: 1985, "Morphology: current approaches and concepts," in M. Hildebrand, D. M. Bramble, K. F. Liem, and D. B. Wake (eds.), *Functional Vertebrate Morphology*, Harvard Univ. Press, Cambridge, MA.

Mayr, E.: 1982, *The Growth of Biological Thought*, Harvard University Press, Cambridge, MA.

Millikan, R. G.: 1984, *Language, Thought, and Other Biological Categories*, Cambridge, MA: MIT Press.

Millikan, R. G.: 1989, "In Defense of Proper Functions," *Philosophy of Science* 56 288–302.

Millikan, R. G.: 1993, *White Queen Psychology and Other Essays for Alice*, Cambridge, MA: MIT Press.

Neander, K.: 1991, "Functions as Selected Effects: The Conceptual Analyst's Defense," *Philosophy of Science* 58 168–184.

Neander, K.: manuscript, "Teleology in Biology" (Woollongong University, Australia).

Nishikawa, K., C. W. Anderson, S. M. Deban, and J. O'Reilly: 1992, "The Evolution of Neural Circuits Controlling Feeding Behavior in Frogs," *Brain, Behavior, and Evolution* 40 125–140.

Patterson, C.: 1982, "Morphological Characters and Homology," in K. A. Joysey and A. E. Friday (eds.), *Problems of Phylogenetic Reconstruction*, Academic Press, London.

Raup, D. M., and S. M. Stanley: 1971, *Principles of Paleontology*, W. H. Freeman and Co., San Francisco.

Rose, M. R.: 1982, "Antagonistic Pleiotropy, Dominance, and Genetic Variation," *Heredity* 48 63–78.

Rosenberg, A.: 1985, *The Structure of Biological Science*, Cambridge: Cambridge University Press.

Ruse, M.: 1971, "Function Statements in Biology," *Philosophy of Science* 38: 87–95.

Schaffner, K.: 1993, *Discovery and Explanation in Biology and Medicine*, Chicago: University of Chicago Press.

Sober, E.: 1984, *The Nature of Selection*, MIT Press, Cambridge MA.

Sober, E.: 1993, *Philosophy of Biology*, Westview Press, Boulder CO.

Wake, D. B., and R. Roth: 1989, *Complex Organismal Functions: Integration and Evolution in Vertebrates*, John Wiley and Sons, Chichester.

Wake, M. H.: 1991, "Morphology, the Study of Form and Function, in Modern Evolutionary Biology," *Oxford Surveys in Evolutionary Biology* 8 289–346.

Williams, G. C.: 1966, *Adaptation and Natural Selection*, Princeton University Press, Princeton.

Wright, L.: 1973, "Functions," *Philosophical Review* 82 139–168.

13

Functions and Goal Directedness

Berent Enç and Fred Adams

I

Recent work on the functions of organs (and of artifacts) and on related issues in the philosophy of mind has brought into focus a contrast between two ways of approaching these issues. The first way consists in construing functions etiologically. According to this construal, something has a function only if that thing has the appropriate type of causal antecedents. The second way, on the other hand, takes functions to be forward-looking dispositions, so that a system is said to have a function only if the system has the propensity for producing certain types of consequences.[1]

Our chapter has three stages. In the first stage, we rehearse the main features of the etiological theory and show how the puzzles that surround functions are a special case of a more general issue in philosophy. In the second stage, we present the forward-looking theory; we bring out some of the difficulties involved in developing an acceptable version; and we discuss the consequences of choosing the forward-looking propensity account as opposed to the etiological account of functions. In the final stage, we draw analogies between these two accounts of functions and corresponding accounts of goal-directed behavior or of teleology in general. The object of this third stage is to hint at some of the ramifications of these analogies to current issues in the philosophy of psychology.

II

When we inquire about the function of an object, we usually ask something roughly like, "What is this object for?" Traditionally, the canonical form of the answer is, "The function of this type of object C in this type of system S is to do such and such," (where C can be an organ, a character, part of an artifact or a gadget, and S is a system that includes C as a component). The answer seems to attribute to the object in question the property of *having the function of doing such and such*. Attributions of this kind appear to be genuinely explanatory, and the appearance persists in spite of the lack of consensus as to *what* exactly they explain.

The first inclination here might be to focus on the fact that attributing a function to a character is (in part) attributing a disposition to it, where the disposition is expressed as a subjunctive conditional: Were such and such conditions to be met, the character would bring about such and such results. And there is nothing mysterious about the explanatory use of dispositions. For a character to be such that were certain conditions to obtain it would do such and such is for that character to possess structural features that satisfy the truth conditions of the subjunctive conditional.

However, it would be a mistake to maintain that this is all that there is to functions, for attributing a function to a character is not *just* attributing a disposition to it: Sugar is such that were certain conditions to hold it would dissolve, but sugar does not have the function of dissolving (under those conditions). *The challenge* to philosophers interested in the analysis of functions has been that of finding what, *in addition* to dispositions, is being attributed to a character when we attribute to it a function. As we intend to show below, some of the most plausible ways of meeting this challenge have indeed made it hard to understand how the proposed additional element contributes to scientific explanation. We suggest that the explanatory role of this "additional element" is the source of the problems involved in function attributions.

Various approaches to meeting what we have called "the challenge" may be classified into three main categories: (1) *Eliminative views*: These views identify functions merely with the activities (or dispositions) of a character that happen to interest the investigator, and they effectively reject the intuition that a real difference exists between dispositions and functions. (2) *Etiological views*: These identify functions with dispositions that have a particular type of *causal history*. On some versions, the history involves the essential causal role of an (intentional) representation of the goal state. On other versions, the dispositions in question are required to have been "selected" by the advantageous consequences they have had for their possessors. (3) *Forward-looking views*: These analyze functions in terms of the *propensity* they have for producing in the future a certain type of consequence (e.g., a survival enhancing consequence).

We tend to think that eliminative theories fly in the face of common sense. Without arguing, we will simply ignore, for the purposes of this paper, any position that involves skepticism about functions, or attempts to reduce functions to the pragmatics of a context. We have in the past individually contributed to the defense of different versions of the etiological view (see, for example, Adams 1979 and Enç 1979). Our objective in this paper is not to criticize or defend any specific version of such a view, but rather to examine the general approach contained in that view to help us see the "problem" that surrounds functions more clearly and to identify the source of teleology that needs to be explained in any (non-eliminative) account of function attributions.

The issue that faces us is a special case of a more general question in philosophy. On any etiological account of functions, when a function is attributed to a character, a disposition is associated with the character, and a story is implied about *how* the character acquired this disposition. On an etiological account, the property of having the function of doing something is constituted by two components: one, a dispositional component (a disposition that is realized in some structural property of the character), and second, a historical component (an account that specifies the etiology of the disposition). Both components are essential to the identity of the property. For example, on one etiological account, the claim that the function of the heart is to circulate the blood can be understood in the following way: (i) the heart has the disposition to produce certain activities, which result in the circulation of the blood, and (ii) hearts in general have a structure that realizes this disposition *owing to the fact* that the exercise of this disposition in past organisms has resulted in advantageous consequences to them (see Enç 1979; and Millikan 1984, 1989). Here (ii) is independent of (i) in that *qualitatively identical* structures may have come about either fortuitously or due to causes other than that of "selection by consequences." On an etiological account, if a qualitatively identical structure *had* come about in a way that violated (ii), that structure would *not* have the function of circulating the blood. This is why the question of the explanatory usefulness of function attributions to a character seems so mystifying. Usually, if attributing a property to a character explains something, it explains it owing to the fact that the possession of that property confers a causal power to the character. Now, it seems that a complete story of the causal role of the character that is attributed to the functional property can be told by adverting to the disposition component of that property—to the structure that realizes that disposition. But the story of the emergence of that structure, interesting as it may be on its own merits, seems to contribute *nothing* to the causal power of the character.[2] All this may be summarized as a dilemma: *Either* the eliminative account is true and functions are just dispositions, in which case functions explain in just the way dispositions explain (highly counterintuitive), *or* there is a historical component to functions, in which case that component is explanatorily impotent and hence functions do not explain at all (equally counterintuitive).

In different terminology, we can say that the function of a character that contains a historical component as a part of its identity cannot supervene fully on the structural properties of the character: Its dispositional component may thus supervene, but the property of having acquired that disposition in the required way will not. As a result, when one looks for the "causal powers" of the function, it would be natural to look at the structural properties that realize the disposition; and there will appear to be a mystery as to how, regarding two identical structures, one of which came about a certain way, the other of which did not, the first can possess a causal power that the second lacks. So the question about the scientific usefulness of function attributions basically boils down to the following question: Given that the function of a character has two components (an etiological and a dispositional one), how does the etiological component, which does *not* supervene on the structural properties of the character, causally explain anything that is not also explainable by the dispositional component, which *does* supervene on its structural properties?[3]

Perhaps we can arrive at a tentative answer to this question by asking a preliminary question: What precisely does the etiologically conceived property of having the function of doing such and such explain? For example, we say, uncontroversially, "the function of the heart is to circulate the blood." If this is an explanans, what is its proper explanandum? Here are some wrong answers:

(i)   Why does this individual heart circulate blood?

(ii)   Why does this individual heart dilate and contract?

(iii)   Why is this individual heart here?

These are wrong answers roughly for the reasons summarized in the previous paragraph: The consequences of the activities of past hearts, which functions are said to invoke on the etiological approach, are irrelevant to what causes a particular heart to be where it is or to do what it does. We can give a full explanation of why a particular heart dilates and contracts, why it circulates the blood, by citing purely physiological facts. We can give an equally complete explanation of why the particular heart is where it is by citing facts about the genetic coding and about the ontogenesis of the individual organism. These are explanations that employ only facts that are supervenient on the structural features of the organism, and whatever the past history of these facts might be, they cannot, it seems, help distinguish the causal role of structures that possess such histories from that of identical structures that do not.

A different type of explanandum may seem to do better:

(iv)   Why are heartlike organs widely prevalent in the population?

(v)   Why have hearts survived in the history of the population?

(vi)   Why do hearts continue to circulate the blood in organisms of this type?

A tendency exists in certain circles in the philosophy of science to regard such selectionist explanations with suspicion—see, for example, Bigelow and Pargetter (1987), where they claim that etiological accounts render functions impotent toward explananda like (iv)–(vi). This is not right. To say, for example, that hearts have survived (have persisted, or perhaps more accurately, are prevalent) in the population *because* the effect they have had of circulating the blood has contributed to their survival (persistence, or prevalence) is not to say anything that is vacuously circular. Indeed, what has made Wright's (1976) account of functions so intuitively attractive is the fact that it endows functions with just this type of explanatory force: C's having the function of doing X explains why C is there (i.e., why the C-type of thing is prevalent in a given population) *because* C's having the function of doing X is partly constituted by the fact that it is C's doing X that explains why C is there.[4]

However, after acquitting these explanations of the charge of circularity, we should observe several important facts about them. First, the explanans is the claim that a character C has the function of doing X. When understood in the light of an adequate etiological theory of functions, this claim singles out a feature of C, namely, those properties of C in virtue of which it has the disposition to do X (under appropriate conditions), and promises that an explanation of why present tokens of C are prevalent (or why present tokens keep on doing whatever it is that they do to bring about X) can be found in the fact that *past* tokens of C have had that feature. So the explanation consists in the identification of some feature as a feature that is relevant to the answers to (iv)–(vi). Second, an important difference separates a typical causal explanation and a functional explanation. In a typical causal explanation an *individual's* having the causally relevant property is what explains why something happened. For example, this individual pill's being poisonous explains why the pill had the effect of killing the man. In a functional explanation, the individual's having the explanatory property (i.e., its having function of doing X) does not explain anything that the individual does (that is not also explained by its having the disposition to do X); the individual ends up having this functional property owing to the fact that the mere disposition to do X in *other* individuals explains something. As a result, a given individual does not possess any causal powers in virtue of the fact that it has the function of doing something—other than those powers it has virtue of the fact that it has the disposition to do that thing. Hence to attribute a function to an individual is to place the individual in a narrower class than the class of individuals who possess that disposition, namely, a class of individuals with that disposition provided that the disposition has been acquired through a specific causal history that makes the disposition explanatorily relevant to questions (iv)–(vi).

Note that with this preliminary conclusion about the explanatory role of function attributions, we do not mean to diminish the importance of functions. The conclusion merely emphasizes that the force of functional explanations is *not* derived from any causal role that teleological properties may possess. Nonetheless, function attributions do have an important explanatory role. We will return to this point later and discuss how many philosopher's vision of this explanatory role has been obscured by thinking that the only explananda relevant to functions is to be derived from the narrowed class of individuals just mentioned.

In our little skirmish over functions we may appear to be mimicking a major battle in the philosophy of psychology over the causal powers of semantic properties. Although this is not the proper place to try to provide a full argument for this, we think that no mimicry is afoot; the battle is one and the same.[5]

At this point, it may appear that if we abandon etiological theories in favor of the forward-looking propensity account, which locates the function of characters in the survival-enhancing propensity conferred by these characters, we will avoid the peculiarities inherent in etiological accounts. This appearance is deceptive.

III

The fundamental differences between the etiological or backward-looking view of functions and the propensity or forward-looking view are well known. For example, on the etiological approach one cannot grant functionality to new structures that have not undergone selection, or structures that have undergone selection for some other characteristic and which are now put to use to serve a different need (Williams 1966, Ruse 1971, Wright 1976). The forward-looking view suffers similar shortcomings. It attributes functions to structures that might turn out to aid in selection, but where the benefit provided is fortuitous (Boorse 1976, Achinstein 1983, Adams 1979). In this section we will explain why we hold that the etiological account has the upper hand on these matters. However, these basic differences in the accounts are not our main focus. Rather, we wish to pursue the matter of the explanatory power that may or may not be gained by appealing to the fact that a structure has acquired a certain function. We will return to the issue of explanatory power in the final section.

Propensity theories can easily degenerate into eliminative theories of functions, for a propensity is not that different from a plain disposition. To avoid confusion, we here assume that the label "the propensity account of functions" is *not* being used for accounts in which a function is identified with a disposition. The propensities in terms of which functions are analyzed are construed "broadly," so that having a propensity is *not* supervenient on the structural properties of the system which has that propensity.

In what follows, we will take the propensity theory sketched by Bigelow and Pargetter as our working model. Our preliminary remarks are designed to sharpen the model, but our objective is not to develop a propensity account that is defensible against close critical scrutiny; it is rather to formulate a view that has enough texture to make comparisons with the etiological approach meaningful.

The passage in Bigelow and Pargetter's paper that comes closest to an official statement of their account of functions is this: A character has a function if and only if the character confers propensities which are survival enhancing in the creature's natural habitat (1987, 192).[6] This is not a purely eliminative account that equates functions with dispositions because the disposition in question is a special kind of disposition: It is realized only if the natural habitat cooperates. Hence it is not supervenient on the morphological characters of the organisms.

One preliminary point that needs to be made is that, at best, the view yields an answer to the question whether a character has a function or not; it does not tell us what that function is. However, the account could be supplemented to remedy this defect. One could say, "A character of an organism has the function of doing X if and only if the character, in virtue of its activity (or properties) results in X, and its resulting in X confers a survival-enhancing propensity to the character (or the individual, or the population, and so on) in its natural habitat."

The more important point about such a propensity account is that the character that has a survival enhancing propensity has a function *as soon as it appears* in the organism—in contrast with the etiological accounts where the character acquires its function only after it has contributed to the survival of the organism. Now the determinants of a character's contribution to survival derive from two sorts of sources: (i) the features of the character, and (ii) the features of the environment. Hence, as we asserted above, the function of a character cannot, on such an account, supervene solely on the structural features of the character any more than it can on the etiological account. In fact, supervenience is a greater problem on a forward-looking account because if one is not arbitrarily stipulative, purely fortuitous changes in the future of an organism's natural environment may end up legitimizing the wrong function attributions to present characters. In illustration of this point, consider the following example. At time t_1, a butterfly species with a certain wing color pattern exists. This species has as its predator a genus of birds that inhabit the same environment. By sheer chance, at t_2 ($t_2 > t_1$), a second butterfly species develops with the same wing pattern as the first, but the second species is poisonous to the birds. If it were not for this second species, the first would have been driven to extinction by its predators. But the presence of the poisonous second species causes the birds to stop hunting butterflies of the first species. We do not think that the propensity account has any safeguards that would prevent our saying that at t_1, the particular color pattern of the first butterfly species had a function—it certainly conferred a survival-enhancing propensity to that species, that is, it increased the probability of that species' surviving past t_2. This example points to a general problem with the propensity account. It does not have a way of delimiting the class of a character's "mere effects" or "mere propensities" to the class of its propensities which are its functions. Any character will have a large number of propensities and some of them would contribute to survival in appropriate contexts. Some of these contexts may be artificial or "accidental" (see Adams and Enç 1988). The classic example of the soldier who carries his Bible to battle in his vest pocket is a case in point. The Bible's being there has the propensity to contribute to the man's surviving a bullet to the heart. But this is not the function of the Bible, nor of the man's carrying it there.

So the major consequence of a forward-looking account, if one wants to make it distinct from an eliminative account, is that functions of characters end up being supervenient partly on the natural habitat of the organism. The natural habitat, relative to which the probability of survival is to be computed, is constituted by the "relevant" parameters of some "natural" environment for the organism *during* a temporal chunk that starts from the present. The size of this chunk may play an important role in the determination of the function of a character. In fact, different sizes may result in different judgements about the function of one and the same character. On an etiological account, "naturalness," and "relevance" are determined by the past history of the species. As a result, one can, without having to resort to the future changes in the environment of the species, specify a set of conditions that were operative during the period in which the natural selection for the character was taking place. In such an account, one can maintain, for example, that the function of teeth is to pulp food even if the current habitat of the organism is such that all food stuff that was once chewable has now become deep-frozen, for all that is necessary for the function attribution be correct is that the action of teeth result in pulped food under conditions similar to those where, *in the past*, the action of teeth has resulted in advantage conferring consequences.[7] But on the propensity account, unless we take a longish chunk of time into the future under consideration—at least until the time when the food stuff thaws—we would have to maintain that the teeth lost their function as soon as the freeze started. We have to maintain this simply because in a propensity account no non-ad-hoc way of carving out the "relevant" parameters, or of specifying a fixed set of "functionally natural" conditions exists.

Etiological and forward-looking theories entail two different kinds of nonsupervenience: In the etiological accounts, the features of the natural habitat do in fact cause the features of the character to which a function is attributed. When the character executes its function, the execution is implemented *only* by the features of the character that are so caused. Schematically speaking, we can say that the function supervenes on features A (environmental factors), and B (structural properties); features A cause features B, and only B execute the functions. On the other hand, in the forward-looking accounts, the function again supervenes on A and B, and only features B execute the function, but this time, A and B are causally independent of each other.

At this point, note that what we have said in tracing out the consequences of the two types of accounts has not in itself spoken for or against either type. To attempt an evaluation, we need to return to our original problem and seek to identify the scientific (explanatory) role that is played by function attributions. Our earlier discussion of the peculiarity of functions was designed to show that if the explanandum is the behavior of an individual organism (or the possession of some character by the organism), then functions understood from a noneliminative perspective are not going to be causally explanatory—only the disposition component of functions that is supervenient on the structural features of the character will provide the causal explanation. So if we confine ourselves to just this type of explanandum, we maintain that there *is* no real choice between forward-looking and etiological accounts: As far as the issue of the causal explanatory power of the individual is concerned, the difference between them makes no real difference.[8]

IV

We might gain further insight if we compare a choice between an etiological and a forward-looking account of functions with a corresponding choice between two accounts of "purposive" behavior. J. Ringen (1976, 1985) and D. Porpora (1980) conducted an insightful debate over whether goal-directed behavior is best understood as behavior that has been reinforced by past operant conditioning as opposed to understanding it as behavior that increases the probability of the attainment of the goal. In this debate, originally Ringen had maintained that goal-directed behavior is nothing other than behavior that was selected for its consequences. Porpora, in his challenge to Ringen, maintains that teleology, properly understood, needs to advert to the *current* propensity of the behavior to lead to goal-attainment. Following Taylor (1967), Porpora argues that teleological explanations derive their force from requiring that when external circumstances change in such a way as to render the behavior inefficacious for goal-attainment, the teleological system alters its behavior to adapt to the new circumstances. The main point is that when one adopts a Skinnerian paradigm for the analysis of teleology, and thereby grounds the explanation of some behavior on the past consequences of that behavior, one runs the risk of betraying the commitment to the *current* propensities of the behavior. Porpora maintains that Wright's (1972) analysis of goal directedness has an ambiguity:

Organism *S* does *B* in order to *G* means

(i)   *B* tends to bring about *G*,

(ii)   *B* occurs because it tends to bring about *G*.

Here (i) can either be read in the etiological way, "*B* has a history of bringing about *G*," or it can be read in forward-looking, propensity way, "*B* is presently causally connected to *G* in such a way that *B*'s occurrence would increase the probability of *G*'s occurrence." And as Propora points out, if used as a criterion of goal directedness, the two readings would generate incompatible results.

In response to Porpora's challenge, Ringen argues by persuasive examples against Porpora that at least some goal-directed behavior, as ordinarily understood, *is* a function of past successes (1985, 570). The parallels between the logical structure of operant conditioning and natural selection being clear (e.g., Skinner 1984), the close affinity between goal-directed behavior conceived under the operant conditioning model and the etiological account of functions that rests on selection by consequences requires no further elaboration. The more interesting aspect of the debate, though, is that Ringen admits that "instances of behavior guided by precognition, telepathy, or omniscience would surely count as goal-directed, if they were in fact to occur" (1985, 571). He goes on to suggest that the alleged ambiguity in Wright's account is in fact its strength because it reflects an ambiguity inherent in our common sense notion of goal directedness. This debate forms a good parallel to the debate between the etiological and the propensity accounts of functions. Thus, if we pursue the analogy, Ringen's earlier position corresponds to the etiological accounts of functions. Porpora's challenge corresponds to the propensity views of functions. And Ringen's later admission is in conformity with a preference for an account that does not have to choose between them—an account which we also tend to favor.

Suppose an organism behaves in a way that raises the probability of attaining some result. It is clearly *not* a necessary condition for the behavior to be *directed* at that result that the organism's (or its species) history contain episodes in which the behavior achieved that result. If we are puzzled about how the behavior is goal-directed, it would be because in *general* we would not have an *explanation* of how the organism "knew" that the behavior was to lead to the alleged goal. We would want to know how an essential component of the explanation of the behavior could be the fact that the behavior was, under the circumstances, a means to the result. Operant conditioning provides one instance of such an explanation. But telepathy and precognition, were such things to exist, would provide other instances. Thus the choice between a backward-directed and a forward-directed analysis is not the real key to the understanding of goal-directed behavior. The key is rather an explanatory relation between the fact that the behavior has the propensity to lead to the goal and the fact that the organisms of that kind produce that type of behavior. In simple organisms we know that frequently the explanatory relation is realized by something that involves selection by consequences —be it natural selection or operant conditioning. In such selection, some result is correlated with a piece of behavior, and the beneficial consequences of attaining that result when the behavior is manifested under certain conditions increase the probability of the formation or retainment of certain neural connections which code for that piece of behavior. If we understand Porpora and Ringen correctly, we agree that it would be a mistake to freeze the *analysis* of goal-directed behavior into this etiological mold. Goal-directed behavior should be amenable to either of the two types of accounts.

In short, it is not surprising that similarities exist between teleology and functions. These similarities confirm our conviction that an account of functions should not have to choose between the two types of accounts. An account of functions can achieve this kind of impartiality by adopting an explanatory relation similar to the one suggested for goal directedness as a condition. Thus one can say that the function of C is to do X only if the fact that the activity of C leads to X explains why that activity is retained in C-type things. To take this conciliatory position in the debate over the etiological versus the propensity accounts of functions is not of course to solve the "problem" involved in the analysis of functions. The "problem," as it will be remembered, was that involving the explanatory role of functions over and above that of their disposition component. It is still to be shown how, by reverting to the general explanatory relation, one can identify "the scientific usefulness" of functions. We can make the task more explicit by means of an earlier example.

Suppose the function of teeth (C) is to pulp food (X), and they pulp food in virtue of being sharp, hard, and being attached to movable jaws (B). Schematically, the function of C is X, and C executes its function in virtue of doing (being) B. The task is to show *what* the claim, "the function of C is X" explains *that is not explained by* the claim, "C has the disposition to bring about X by doing (or in virtue of being) B." Specifically, now that we endorse an account that does not favor either the etiological or the propensity theory, what does the criterion that we think is essential to functions (i.e., the fact that C's having or doing B, which leads to X, explains why C-type things have or do B) contribute to understanding how functions explain whatever it is that they explain any more than dispositions do?

On the basis of what we have said so far, it does not seem that our proposal is going to fare any better on this count than any we have discussed so far because the explanation we require will most likely be filled out with some etiological account of the disposition of C to do X, and, de facto, the proposal will coincide with a causal theory of functions. Even if the explanation came from elsewhere, the requirement will make it unlikely for functions to be supervenient only on the structural features of the character. It will be unlikely because, given the type of things that are in general acceptable substitutes for X in the schema "the function of C is to do X," whether B leads to X or not will not be decidable purely on considerations *internal* to the character C. Hence on our proposal, too, possession of the teleological property (i.e., the property of having the function to do X) does not endow the individual with a causal power over and above the causal powers it possesses in virtue of having the disposition to do X.

We think this result is basically correct, but we also think that a shift in perspective can save function attributions from the realm of superstition and mythology. It is helpful to express this shift in perspective by reverting to the goal-directed behavior of Ringen's (Skinner's) pigeons.

To set the stage, let us suppose that a pigeon has been operant conditioned to peck at a spot on the wall of the box when some appropriate discriminative stimulus, say a light, is applied. Suppose further that the spot is a shadow that moves very slowly, and that the reinforcers (food pellets) are made contingent on the pigeon's pecking at the *shadow*, as opposed to a fixed point on the wall. To suppose that the pigeon has been operant conditioned, we concur with Ringen, is to suppose that its behavior is goal-directed. Let us assume further, ignoring possible subtleties and complications here, that the goal of the behavior is obtaining food. We still have not specified *what* the behavior is that is goal-directed. Two hypotheses are available. One is that it pecks at a fixed location on the wall, and when the shadow moves a significant distance away from the location, it "relearns" to peck at a new location. The second hypothesis is that it pecks at the moving shadow.

On the first hypothesis, the behavior that is goal-directed at obtaining food keeps *changing* over time; on the second, the behavior remains the *same*. Even if we suppose that the evidence hitherto gathered supports the two hypotheses *equally*, the two hypotheses are distinct. That is, even if the cluster of pecks on the wall over time makes either hypothesis equally likely, whatever one's perspective might be on such issues, it would be irrational to maintain that the hypotheses are empirically equivalent. For example, from a purely behavioristic point of view, other types of evidence may be gathered (e.g., the same pigeon could be confronted with a faster moving spot) which might favor one hypothesis over the other. Also from a point of view we favor, in which one maintains the existence of a neural/psychological reality that underlies and explains the experimental results, the two hypotheses are contraries of each other, and hence cannot be equivalent.

This example brings out the importance of identifying a piece of behavior *as goal-directed*. Our suggestion was that a piece of behavior is directed at a goal G only if the production of that kind of behavior is (partly) explained by the fact that the behavior leads to the attainment of the goal G.[9] On this suggestion, identifying a piece of behavior *as* goal-directed also requires specifying precisely *what* the behavior is. In our example, we should first note that given the proper choice of time scale, *both* pecking at the several stationary spots *and* pecking at the moving shadow are in fact correlated with receiving food pellets. Now, in this example, if the second hypothesis is true, then the fact that pecking at the stationary location was correlated around that time with receiving food pellets would *not* explain why the pigeon is doing what it is doing, because on that hypothesis the neural reality would be such as to have part of the cause of the present packing something like a representation of the correlation between the pigeon's having pecked at the *shadow* (as opposed to the stationary point on the wall) and its receiving food. So as soon as the behavior is diagnosed as goal-directed, the necessary condition we have proposed for goal directedness forces the search for the correct specification of the behavior in question. Each hypothesis favors a different behavior description, whereas, if goal-related considerations are ignored, and the description of the behavior is confined to specifications of the limb movements of the pigeon, the difference between the two hypotheses would become elusive. For, as the example makes clear, the correct descriptions of the various muscular outputs elicited from the pigeon are consistent with both hypotheses.

In short, even if one admits that a piece of behavior's being goal-directed does not cause (or causally explain) anything over and above what the behavior *itself* causes, identifying a piece of behavior as goal-directed introduces a taxonomy of behavior that would otherwise be hard to construct. This taxonomy enables us to formulate predictive hypotheses that would otherwise not be theoretically motivated. In our pigeon example, our classifying the pigeon's pecking as a piece of goal-directed behavior enables us to formulate the hypothesis that it is pecking at the shadow, wherever the shadow may be in the box. This immediately leads to the prediction that if the motion of the shadow is increased, the profile of the peck clusters would change significantly. On the other hand, if the location of the pecks were conceived as theoretically irrelevant, or if they were explained by reference to external causes that did not involve feedback of the kind that characterizes goal-directed behavior, then it would be hard to motivate the contrast between the two hypotheses in question.

We are not claiming here that without the vocabulary of goal directedness such a contrast could never be drawn. A comparable pair of hypotheses could be formulated when one wonders whether the shadow is pulling the pigeon with invisible lines of force or not. What we are rather claiming is that a good portion of the behavior of systems (be they guided missiles, phototropic plants, or intentional agents) *is* goal-directed. And identifying it as goal-directed generates a taxonomy that cuts across those that are obtainable from just the behavioral data (e.g., is the pigeon *tracking* the shadow, *or* is it being pulled along by it?).

It is important to notice that the heuristic need for the attribution of goal directedness is consistent with the claim that the behavior of the individual pigeon is fully explained just by factors that supervene on its physiology. What we are maintaining is just this: Goal directedness contains elements that are not fully supervenient on the features of the pigeon, and these elements help to carve out a classification of behavior that is not derivable from the external causes and the behavioral effects that involve the pigeon. Once the classification is in place, fully supervenient elements take over the explanation of the token behavior so classified. Whether the pigeon is producing random motion that happens to coincide with the trajectory of the spot or it is tracking the moving shadow is a question that can, in principle, be answered by looking inside the pigeon. But it is a question that cannot be *asked* without utilizing the concept of goal directedness.

So far we have emphasized the heuristic role achieved by identifying a piece of behavior as goal-directed. However, it is important to recognize that this role is made possible by setting up a new type of explanandum and thereby changing the explanatory demands on the theory. The change in question may be described in the following way.

When one offers the hypothesis that the behavior of the pigeon is directed at the goal of obtaining food, one commits oneself to, what we may call, a Thesis of Plasticity: (i) under the present circumstances the current behavior of the pigeon has the disposition to result in the attainment of the goal, and (ii) within a certain range, changes in the present circumstances that would tend to diminish the disposition would be accompanied by changes in the behavior so as to restore the disposition. For example, if the behavior *is* goal-directed at food, then clause (ii) of the Thesis requires that when the pigeon's foot is injured, it hop along to follow the spot, or when the pigeon suddenly finds a pile of food pellets at its feet, it stop pecking at the spot and start pecking at the pile; in short, when the goal is obstructed along the current path, it do *whatever is within its means* to attain the goal reasonably efficiently. (Changes in the circumstances have to be restricted to a range because certain types of changes may result in the abandonment of the goal or its substitution by some other goal.)[10]

Thus the hypothesis that the behavior is goal-directed, via the Thesis of Plasticity, which it entails, generates an explanatory structure that applies to a much broader class of limb trajectories than those being actually elicited from the pigeon. In fact, it may not be possible, in principle, to give a closed definition of the class in terms of the mechanically explainable limb movements of the organism.

When we ask, "What is the current behavior that is goal-directed?" we attempt to identify *one member* of this broad class. Here the heuristic value of teleology comes in; it is here that *because* we have identified the behavior as *goal-directed* that we find motivated to ask, "Which of several descriptions, each of which is consistent with the set of limb movements observed, accurately describes what the pigeon is doing?" The answer gives us a description of a type of behavior generalized from the current token. The description makes the behavior fully supervenient on the physiological characteristics of the organism, and as long as the circumstances remain *relevantly similar*, we can assume, the behavior can be characterized in terms of the limb movements of the organism.[11] This characterization, in turn, will enable a fully mechanistic explanation of the type of behavior thus identified. As long as we stay within the parameters that define the "relevant similarities," it would then seem that the pigeon's pecking in such and such a way *in order to get food* is just a *narrowing* of the class of behaviors that consist of the pigeon's pecking just that way. Within those parameters, the accusation that this narrowing has no scientific use because the narrower class yields no generalizations that cannot be formulated in the broader class, is perfectly justified. But the "change in perspective" we spoke of earlier makes it clear that the role of goal attribution is that of *broadening* the class of behaviors to be explained and that of directing the investigation to the prediction and explanation of "novel" behaviors in the face of changes in the circumstances *beyond* the range set by the relevant similarity parameters.

The Thesis of Plasticity, in fact, entails that it is unlikely that a single physical description will capture any type of behavior that is directed to a specified goal. This is fully consistent with the individualistic claim that, for any sufficiently narrowly specified family of behaviors, some physical description will form the full supervenient basis for them. This is why placing a piece of behavior in the broader class can make available predictive generalizations that are not formulable at the level of the narrower classes.

A similar point is true of functions. When we identify the function of a character, we thereby commit ourselves to a thesis comparable to the Thesis of Plasticity—we might call it the Thesis of Multiple Realizability: (i) the type of character in question possesses a set of properties (or has a characteristic activity) which gives it the disposition to execute the function, (and prevalence of those properties among tokens of that type of character is explained by the fact that the properties yield the disposition), and (ii) if the character were significantly different and still had the same function, then it would possess a different set of properties (or have different activities) which would confer the *same disposition* to the character (and the prevalence of the new properties would now be explainable by that same disposition). Thus the Thesis of Multiple Realizability creates a broad class of types of characters all of which have the disposition to execute the function in question, and whatever their diverse properties or activities may be, each type within the class has its own properties or activities *because* those properties or activities have that disposition. As a result, when we explain the prevalence of a character by attributing a function to it, we appeal to a generalization over a much broader class than is constituted by the properties that define the character. The broader class is generated by the fact that there is no one physical description of the type of structures each of which will perform the function—each of which will possess the property of having the function of doing such and such. This property, as we argued above, is definable only nonindividualistically. However, this is fully consistent with there being, for any sufficiently narrowly specified class of structures that possesses the function property, a physical description that forms the full supervenient basis for that property. This is why attributing a function to a structure places the structure in a broader class over which counterfactual predictive generalizations are available that cannot be formulated at the level of the narrower class (e.g., what sort of structure would exist had the conditions under which the function was acquired be different in such and such specific ways).

Again, at this point, we have the motivation, which we mentioned in the discussion of goal directedness, to ask "What is the set of properties (or repertoire of activities) that has the function in question?" The correct answer to this question enables us to carve out a set of properties, or a repertoire of activities that are "function-specific." When we claim, for example, that the function of the governor of a steam engine is to regulate the speed of the engine, and we try to answer the above question relative to this function attribution, we pair up a set of activities of the governor, the lifting of the "wings" due to the increase in rotational momentum, with a goal, the slowing down of the rotational speed of the engine. Here the function consists in the bringing about of a "distal" event, and the function-specific activity is a series of events taking place *within* the character that, under the right conditions, causes the distal event. The distal event is analogous to the goal of goal-directed behavior, and the function-specific activity to the "internally characterized" behavior of the goal-directed system. In this way, knowing the function of a character and knowing its function-specific repertoire enable us to distinguish between a fortuitous occurrence of the events that constitute its repertoire (e.g., the wings of the governor going up as a result of an external force applied to them) and their occurrence in the service of the character's function. This is analogous to the distinction between the pigeon's being pulled along by the shadow and its tracking the shadow. The claim is not that the contrast cannot be specified in terms of the current causal factors; it certainly can. Rather, the claim is that without the concerns introduced by functions and function-specific repertoires, there would not be any theoretical motivation to seek the contrast. The important point is that at the level where the function-specific repertoire is identified, the function attribution to the character seems to be empty of explanatory force; that is, it seems that at that level anything that can be explained about the individual that possesses the character by the function attribution can also be equally well explained by the dispositions of the character fully supervenient on internal features. This collapse of explanatory roles at the individual level has served to conceal from many thinkers both the heuristic role of teleology and its (noncausal) explanatory role at a higher level of generality.

In conclusion, we can say that the solution to the "problem" of function attributions does not lie in the causal explanations of token behaviors yielded by these attributions, for, as we argued, such explanations are fully obtainable from the disposition component of functions. The solutions rather lies in the acknowledgment of a broader taxonomy of properties (or behaviors) classified together by reference to a distal goal, and in the fact that such taxonomy makes it possible to generate noncausal explanations and predictive hypotheses about *types* of behaviors.

Acknowledgments

We have benefitted from the insightful and helpful comments of an anonymous referee. We are also grateful to Elliott Sober, Arda Denkel, Malcolm Forster, and Dennis Stampe for comments on an earlier version of this essay.

Notes

1. A good example of a study in which this contrast is developed is to be found in John Bigelow and Robert Pargetter (1987). Bigelow and Pargetter's paper is especially worthy of attention because it treats the logic of function assignments from a general and metatheoretical viewpoint, and thus brings into focus the choices to be made in any study of functions. They argue for a propensity view, which, they maintain, has many advantages over the etiological views. Our discussion below is partly inspired by their work.

2. These two components of the teleological property are reminiscent of Mayr's (1961) proximate and ultimate causes, or of Dretske's (1988) triggering and structuring causes. That is, dispositions which supervene on the current physical constitution of the organism without reference to the organism's past history explain analogously to the way proximate or triggering causes explain. The historical component, on the other hand, is more like an ultimate or structuring cause in its explanatory role.

3. In the context of Dretske's triggering versus structuring causes, this same point is nicely discussed by Kim (1991) in asking whether explanations in terms of dispositions (triggering causes) *excludes explanation* in terms of the historical component (structuring causes). (This comes out as asking whether explanation in terms of neural states excludes explanation in terms of contents of mental states.) Kim (ibid., 67–68) concludes that the one explanation *does not* exclude

the other. We aim to argue, in a similar vein, that the one explanation in terms of the dispositional component is not in competition with the explanation in terms of the historical component of functions; they are not in competition because the explananda are different.

4. As we previously noted, the explanation involved here is structurally similar to explanation by ultimate causes (Mayr 1961) or to explanation by structuring causes (Dretske 1988).

5. For example, Jerry Fodor (1991) ends his defense of causal-power governed taxonomy, and of his thesis of individualism in psychology, by the following question and answer pair:

—But why can [the] psychologist not allow [twater thoughts and water thoughts] to be *different* states but with the *same* causal powers?
—Because, if he does, his theory misses generalizations, namely, all the generalizations that subsume me and my twin. Good taxonomy is about *not* missing generalizations (p. 25, n. 23).

For further discussion of this point, see Fred Adams (1993).

6. As it stands, "survival enhancing" is incomplete because it ignores comparative judgements about other traits incompatible with the character in question. Are less fit traits to be judged *not* to have a function? Is a character survival enhancing if it merely increases the probability of survival, or only if it maximizes this probability, or something in between? These are questions that need to be addressed, and we assume that they can be satisfactorily answered.

7. Recent philosophical literature (Fodor 1990) contains interesting arguments to the effect that considerations like these at best underdetermine the kind of fine-grained "normal" conditions that are needed for function attributions to do work in the philosophy of psychology. Nothing we say here should be taken to contribute to this issue. What we want to emphasize is that considerations available in etiological accounts for specifying (even if partially) the required set of conditions and for identifying (even if incompletely) the relevance of environmental parameters cannot be found in forward-looking accounts.

8. It might be maintained that the etiological accounts fare better on this issue, for the historical facts causally influence the present activity of the characters, whereas, due to the schematically presented considerations of the previous paragraph, propensity accounts have the net effect of causally insulating the present and future facts about the environment from the present activity of the characters. However, what remains is that on both types of accounts the disposition component of the function, whatever its history or its future effectiveness may be, is all that is needed in the causal explanation of the activity of the characters.

9. We are here ignoring the important but difficult cases of "mis"-direction. So what we say in the text is, strictly speaking, false: As we know all too well, a piece of behavior may be directed at a goal G even when the behavior's occurrence is absolutely neutral to the probability of the attainment of G. Any full account of goal directedness must accommodate such cases.

10. Porpora (1980) makes this point in emphasizing Taylor's thesis that mechanistic explanations cannot exhaust the force of teleological explanations.

11. Even at this level, what we say in the text is not strictly true; the Thesis of Plasticity applies even to descriptions like *pecking at a spot*; that piece of behavior may be accomplished by any one of an indefinite set of muscular movement sequences. We refrain from a lengthy digression on this issue.

References

Achinstein, P. (1983), *The Nature of Explanation*. Oxford: Oxford University Press.

Adams, F. (1979), "A Goal-State Theory of Function Attributions," *Canadian Journal of Philosophy* 9: 493–518.

———. (1986), "Feedback About Feedback: Reply to Ehring." *Southern Journal of Philosophy* 24: 123–131.

———. (1991), "Causal Contents," in B. McLaughlin (ed.), *Dretske and His Critics*. Oxford: Blackwell, pp. 131–156.

———. (1993), "Fodor's Modal Argument," *Philosophical Psychology* 6: 41–56.

Adams, F. and Enç, B. (1988), "Not Quite by Accident," *Dialogue* 27: 287–297.

Bigelow, J. and Pargetter, R. (1987), "Functions," *The Journal of Philosophy* 84: 181–196.

Boorse, C. (1976), "Wright on Functions," *The Philosophical Review* 85: 70–86.

Dretske, F. (1988), *Explaining Behavior*. Cambridge, MA: MIT Press.

Enç, B. (1979), "Function Attributions and Functional Explanation," *Philosophy of Science* 46: 343–365.

Fodor, J. (1990), *A Theory of Content and Other Essays*. Cambridge, MA: MIT Press.

———. (1991), "A Modal Argument for Narrow Content," *The Journal of Philosophy* 88: 5–26.

Kim, J. (1982), "Psychophysical Supervenience," *Philosophical Studies* 41: 51–70.

———. (1991), "Dretske on How Reasons Explain Behavior," in B. McLaughlin (ed.), *Dretske and His Critics*. Oxford: Blackwell, pp. 131–156.

Mayr, E. (1961), "Cause and Effect in Biology," *Science* 134: 1501–1506.

Millikan, R. (1984), *Language, Thought, and Other Biological Categories*. Cambridge, MA: MIT Press.

———. (1989), "In Defense of Proper Functions," *Philosophy of Science* 56: 288–302.

Porpora, D. (1980), "Operant Conditioning and Teleology," *Philosophy of Science* 47: 568–582.

Ringen, J. (1976), "Explanation, Teleology, and Operant Behaviorism: A Study of the Experimental Analysis of Purposive Behavior," *Philosophy of Science* 43: 223–253.

————. (1985), "Operant Conditioning and a Paradox of Teleology," *Philosophy of Science* 52: 565–577.

Ruse, M. (1971), "Functional Statements in Biology," *Philosophical Review* 38: 87–95.

Skinner, B. F. (1984), "Selection by Consequences," *The Behavioral and Brain Sciences* 7: 477–510.

Taylor, C. (1964), *The Explanation of Behaviour.* New York: The Humanities Press.

————. (1967), "Teleological Explanation: A Reply to Denis Noble," *Analysis* 27: 141–143.

Williams, G. (1966), *Adaptation and Natural Selection.* Princeton: Princeton University Press.

Wright, L. (1972), "Explanation and Teleology," *Philosophy of Science* 39: 204–218.

————. (1973), "Functions," *Philosophical Review* 82: 139–168.

————. (1976), *Teleological Explanations.* Berkeley and Los Angeles: University of California Press.

14

Function, Fitness, and Disposition

Sandra D. Mitchell

There has been a recent resurgence of interest in functional explanation, both in the details of the analysis of functions and in the appropriate strategy for carrying out such an analysis (Millikan 1984, 1989; Bigelow and Pargetter 1987; Mitchell 1989, 1993; Horan 1989; Neander 1991; Godfrey-Smith, 1994).[1] Some, including myself, have defended and developed an etiological theory of proper functions whose origin derives, in part, from the work of Larry Wright (1973, 1976). On this theory, a function is a consequence of some component of a system (the pumping of blood consequence of the human heart, the inclusive fitness maximizing effect of the avunculate in societies with low paternity certainty, the keeping open of the door consequence of the brick on the floor). The consequence to be a function, must have played an essential role in the causal history issuing in the presence of that very component. While agreeing on the basic character of functions, the defenders of etiological theories nevertheless disagree with respect to the nature of their enterprise, that is, whether it is an act of conceptual analysis or one of theoretical definition. Clearly the type of argument available to claim success for the etiological theory, and its right to prevail over other contenders in the field depends on what such accounts aim to do. Prior to entering some new battlegrounds in the war between function theories, I will first address this meta-analytic question. I will argue that current disputes may be better understood, if not dissolved, by understanding the explanatory enterprises in which appeal is made to functional ascriptions. In this pursuit I will consider two recent criticisms of the etiological theory, and the relevance of an alternative, dispositional theory of functions.[2]

1   Setting the Stage

Functional explanations are endemic to biology and the social sciences and are clearly intended to do explanatory work. However, the teleological character of functions, namely the fact that a function is identified as a consequence of the very item whose presence it is supposed to explain, has been taken as *prima facie* grounds for denying explanatory status. This methodological criticism was taken as evidence of the gulf between the natural sciences and the social sciences.

Hempel (1959) attempted to restore the unity of sciences under the methodological rubric of the covering-law model of explanation. On his account, which had become the "received view," a function explanation of why a given trait or practice is present in a system consists in its derivation from a set of premises which include a statement of the function of the trait, general laws and initial conditions. However, given the absence of suitable biological, psychological and anthropological laws in the explanans to rule out functional equivalents for the explanandum event, functions logically fail to be explanatory and are allocated a merely heuristic role in scientific discourse (see also Horan 1989, pp. 137–138).

Hempel's negative solution to the philosophical puzzle engendered a new methodological problem. Biologists and anthropologists appeal to functions in ways that assume more than mere heuristic content. They dispute what is the correct function to ascribe to an organ, behavior or social practice. Empirical evidence is brought to bear in identifying functions. Hempel's solution rules that such scientific disputes are misdirected, scientific language and practice must be reformed in light of philosophy. However, some have responded to the mismatch by taking up the challenge to revise the philosophical analysis of functions in order to account for the explanatory character presupposed in scientific practice.[3]

But what type of enterprise is this? What do we want from a theory of explanatory functions, and hence what constitutes evidence that we have achieved our aims? Millikan (1989) defends her etiological theory of proper functions as providing a theoretical definition and rejects Wright's similar account on the grounds that he engages in conceptual analysis. Neander (1991) in response to Millikan's argument, distinguishes weak and strong interpretations of conceptual analysis. In the weak version, the conceptual analysis of the term "function" would identify implicit or explicit criteria of application for its utterance in the appropriate linguistic community. A strong interpretation adds to this the requirement that the analysis provide an account of the meaning of the term, and that it does so by outlining necessary and sufficient conditions for its use. These additional requirements are what Millikan rejects and motivate her promoting a theoretical definition of "function" and at the same time eschewing conceptual analysis. A theoretical definition describes the underlying phenomena that explain the linguistic category.

Neander is correct to defend the weak version of conceptual analysis here, and to suggest that it be used in conjunction with a theoretical definition. However I believe there is a stronger reason for adopting a combined approach than Neander's appeal to the "related aspects of language" which each approach illuminates. That is, "function" does not pick out a material substance like the natural kind terms "gold" or "water" which serve as exemplars for comparing conceptual analyses with theoretical descriptions for Millikan and Neander. It is more like "cause" or "reason" because it is both abstract and picks out an explanatory structure.[4]

Understanding an intended explanatory content of the use of "function" in current science will indeed employ tools of weak conceptual analysis (since it is the scientists' term and not a stipulation outside of science that is at issue), and at the same time provide a theoretical definition which grounds its "correct" use in contemporary science. Rather than ask what scientists mean by "function" or what objects the term denotes in the world, I suggest we explore instead the explanatory import of function ascriptions. That is, we should ask WHAT are they intended to explain and HOW they explain it. This approach will clarify both the content of the etiological theory and its differences with the dispositional alternative offered by Bigelow and Pargetter.

The etiological approach maintains that to explain why something occurs is to describe the causal history which led to the event—i.e., to give its etiology. The plausibility of any posited function derives from the theories in our repertoire. But acceptance of a given ascription is subject to the same evaluation as any other scientific claim.[5] While all theories of scientific explanation might well agree that causes are explanatory, functions are not causal in the usual sense, they are teleological. There have been many accounts of how intentional, goal-directed behavior is appropriately teleological. (Cummins 1975; Nagel 1977; Wimsatt 1972; Wright 1976.) The idea or the representation of the goal of an action in the mind of the actor operates as the proximate cause for that same action. My concerns, however, are with the explanation of systems to which the assignment of goals or intentions may not be appropriate. Can the presence and persistence of properties of biological populations and human societies be explained by appeal to natural and cultural functions? Larry Wright and Ruth Millikan have offered such accounts.

Wright (1976) proposes the following schema:

The function of X is Z if:

i. Z is a consequence (result) of X's being there

ii. X is there because it does (result in) Z.

Satisfying clause (ii) of the formula, that X is there because it does Z, depends on the appropriate causal relationship obtaining between X and Z. A rough version of Millikan's (1989, p. 288) theory is:

A has F as its proper function if one of two conditions hold:

i. A originated as a "reproduction" of some prior item or items that due in part to possession of the properties reproduced, have actually performed F in the past, and A exists (causally, historically because) of this performance or

ii. A has F as a derived proper function, i.e., A originated as the product of some proper device that had performance of F as a proper function and that production of A is the normal means by which F is achieved.

Neither Wright nor Millikan make explicit the specific characteristics of the historical, causal "because" which carries the explanatory burden in their accounts. Clearly not all causal sequences render a consequence a function. To do so two conditions are required: that the organ or behavior has been selected over alternatives on the basis of its consequence, and that it is produced or reproduced as a direct result of that selection process (Mitchell 1989, 1993).[6]

II   Natural Functions and Social Functions

Given this characterization of the two components of an etiological theory, it is clear how the etiological theory applies to natural items when their presence is the result of evolution by natural selection. The presence of traits which are adaptations can be explained by appeal to their past consequences on reproductive success. For example, in primates, males are often larger than females. What function does larger size serve for the males? Two answers were proposed. Darwin (1871) said that larger size in males was a result of sexual selection in that larger size resulted in greater success in male/male competition for mates. Selander (1972) suggested that the dimorphism allowed males and females to exploit different food resources. Clutton-Brock and Harvey (1977) tested the two hypotheses with comparative data. If the function of larger size for males was food acquisition, then one would expect to find this dimorphism in monogamous species of primates where males and females feed together. Here the dimorphism would be due to the consequence of larger size in competition with females for food. If the function was instead, enhancing success in male/male competition, one would expect to find dimorphism in polygamous species. In these situations there is strong competition between males for access to mates where the "winner" acquires a harem of females and the "loser" none. Clutton-Brock and Harvey's study gave evidence supporting Darwin's hypothesis for primates—dimorphism appeared in polygamous species. What is evident from this discussion is that the biological function of a trait is taken to be a real, discoverable property of natural organisms. Determining the correct function requires empirical evidence about conditions of selection and hence the consequences which the trait had in the past.

For an etiological functional relationship to occur, a selection background like natural selection must operate. The selected consequence of an item must furthermore be causally responsible for its replication and, hence, its current presence. Evidence only of a selection background or selection *of* an item will be insufficient to justify ascription of a function. The feature explained, like larger size in male primates, must have been selected *for* the functional consequence, say success in male/male competition, and, secondly, it must have been produced or reproduced as a direct result of that selection process, in this case by genetic transmission.

Function is also used to explain the presence of social practices. This can be understood by an etiological theory if a selection process is operating which takes the alleged functional consequence as causally relevant in the maintenance or transmission of cultural items. Functionalism in anthropology and sociology has been often associated with the program to explain human behavior and social institutions in terms of group level functions. That is, specific beliefs, behaviors, institutions exist because they allow the healthy functioning of the social group. Rituals maintain social cohesion, warfare regulates population size, marriage rules maintain cross-cultural social ties. Social practice X is there because it does Y, i.e. contributes to health or welfare of the social group. A classic contribution to ecological anthropology has been the research by Rappaport and Vayda on the Maring warfare cycle (Rappaport 1968, Vayda 1974). The Marings of Papua New Guinea ritually plant yams and hold pig feasts in a cycle of war and truce-making with neighboring populations. It has been claimed that the function of this complex set of behaviors is to distribute protein (to allies during the pig feasts) and regulate population size (victors confiscate land of the defeated) within the context of the existing modes of production and environmental constraints.

A common criticism of functional explanation in cultural anthropology, the "fallacy of functionalism," is taking the fact that a consequence is beneficial as sufficient to accounting for why the practice exists. The etiological theory of function makes sense of such criticisms. Without evidence of selection and transmission mechanisms operating on that consequence, ascription of etiological function is groundless.

There have been a variety of responses to the recognition of this fallacy. Some, including Vayda, have taken seriously the required causal mechanisms for etiological functions and have re-directed their research toward gathering such evidence.

Contrary to the assumptions made by me in the 1970s and by other anthropologists more recently, the fact that victorious warriors sometimes take enemy land and benefit from doing so has, by itself, no necessary explanatory import. To think that it does is a fallacy, not because it involves putting consequences forth as causes, but rather because it involves putting them forth without due concern for mechanisms, as if the mere fact of their being beneficial automatically conferred causal efficacy upon them. The mechanisms to which I am referring are … intentional action, reinforcement, and natural selection, ontologically grounded in the actions, properties, and experiences of individual human beings. (Vayda, 1989 p. 163)

Just acknowledging the need for selection and replication mechanisms is not in itself sufficient. Vayda cautions against a "ceremonial" invocation of mechanism without any investigation into its operation. Additional evidence is required for identifying the actual selection and replication processes from the set of those that could logically do the job. Vayda now claims that conscious recognition of consequences and intentional action is responsible for the Maring ritual warfare cycle. To support the claim that redistribution of land is indeed the function of at least particular war cycles in the 1950s, Vayda has documented the fact that there was land shortage at the time, and that individuals consciously acted on the basis of this fact.

Individual intentional action is not the only selection and transmission mechanism available to support cultural function Soltis, Boyd and Richerson (1991) raise the question of whether functional accounts of cultural behavior can be justified by mans of cultural group selection. They develop a model which suggests that under certain circumstances, cultural variation can be maintained, and "cultural group selection could cause human societies to exhibit as least some group functional traits" (1991, p. 4). Group dissolution (extinction) and fissioning combined with cultural transmission are thus mechanisms which might explain the presence of practices like the warfare rituals in Papua New Guinea.[7] Groups which have a given practice and which are successful at survival or replication because of it, may pass this practice on through their lineage and to their offshoots by cultural transmission. Boyd, Richerson and Soltis present evidence from New Guinea that support the realism of the assumptions of their model, though it is not sufficient for justifying the claim that group selection actually operated. Just as in the case of natural function, an etiological explanation for cultural function requires detailed, context specific evidence of the actual causal history. Not every beneficial consequence will be explanatory, and not every explanatory consequence will invoke one particular mechanism, be it intentional choice, natural selection or cultural group selection. What the etiological theory accounts for is the structure of cultural functional explanation, whichever selection and replication mechanisms are in evidence.

To summarize, what is crucial to explaining the presence of a trait by its function on the etiological view, is that a specified type of causal background is available and that the functional consequence played the appropriate role in the causal history leading to the trait or social practice. Some practices of biologists and social scientist—theoretical development, empirical investigation and dispute resolution—make sense only if an etiological theory of function is presumed.

III   Malfunction and the Problem of "Doubles"

Two intuition-driven counter-examples have been raised against the etiological approach. The first, the problem of malfunctioning items, like congenitally diseased hearts, has been heralded, curiously, as both a problem for an etiological account by Prior (1985) and an advantage of an etiological account by Millikan (1989). The problem that Prior identifies rests in how stringently one interprets Wright's condition (i) that Z is a consequence (result) of X's being there. If one insists that each individual X must result in Z if Z is a function of X then malfunctioning hearts do not really have the function to pump blood since not only do they fail to, but they cannot succeed under any circumstances. (This distinguishes them from things such as safety devices which may not in any given time result in Z but would if the proper circumstances obtained. The airbag in a car may never get use, but its function is nevertheless to protect the driver from the impact of the steering wheel should a collision occur.)

If one adopts an interpretation of (i) to allow Z to be a typical consequence or result of the presence of things of type X, then the problem with malfunction dissolves. If things identified as Xs typically result in Z and condition (ii) is met, then an individual X which fails to result in Z (the typical consequence responsible for things of types X being present) is malfunctional. Millikan defines biological categories in terms of functions—things are grouped together in virtue of their causal history. An individual member of a category, of course, may be defective and hence unable to produce the consequences which define the category, but that does not diminish their membership. A heart that fails to pump blood, is still a heart.

It is not then the actual constitution, powers, or dispositions of a thing that make it a member of a certain biological category. My claim will be that it is the "proper function" of a thing that puts it in a biological category, and this has to do not with its powers but with its history. (Millikan 1984, p. 17).

The objection that malfunction cannot be identified in an etiological theory is thus met by distinguishing types from tokens and by adopting a non-essentialist interpretation of types. Especially for biology, where traits and groups are both variable and evolving, there will be no set of necessary and sufficient properties which must be expressed by every item classified as a heart, or a Homo Sapiens. What allows the identification of a token as being of a specific type is historical continuity and what is the "normal" or "proper" function of that type is understood in terms of potential consequences whose expression are environmentally conditional.[8] Thus malfunction can be ascribed to items which meet the historical identity conditions for type-classification but fail to have "normal" expression in a given environment.

The second counter-example is the problem of "doubles." Bigelow and Pargetter suggest the following thought experiment to illustrate.[9] "Consider the possible world identical to this one in all matters of laws and particular matters of fact, except that it came into existence by chance (or without cause) five minutes ago." (1987, p. 188). The etiological account would find none of the items we identify as functions in our world to be functions in the parallel world. Clearly in the parallel world, by design of the example, there is no causal history in which to embed them. The problem of "doubles" offers two objects with identical current properties but with different histories. Bigelow and Pargetter claim our intuitions dictate that if the items are identical in structure and in all future consequences and one has function Z, the other must have the same function.

Millikan, in response, objected that they mistake a "mark" of a function (the structure and its current consequences) for the function itself (the causal historical significance of the structure/consequence pair).[10] However, Millikan's argument is not very satisfying. She admits to being brazen in just stating that the cases of doubles are like cases of gold and fool's gold, or H_2O and XYZ—cases which look alike to us, but are simply mistaken identities. This rather begs the question, since for those who take the double case as a persuasive and adequate account of function should construe doubles as functionally identical. For Bigelow and Pargetter it is Millikan who has made the mistake by seeing doubles as functionally different. However, intuitions about fictional worlds, on either side, seem poor grounds for determining allegiance to one or another theory of functional explanation. If intuitions are to play a role, I believe they should be intuitions about our world.

At least for some evolutionary biologists, to ascribe a function to a trait does seem to require a causal history of adaptation. "Evolutionary adaptation is a special and onerous concept that should not be used unnecessarily, and no effect should be called a function unless it is clearly produced by design and not by chance" (Williams 1966, p. vii). This perspective if applied to Bigelow and Pargetter's double world would see traits but no functions—since chance produced those traits.

What intuitions are evoked by considering a "natural double"? In the case of mimicry members of two distinct species display very similar structures or traits, in fact indistinguishable to the appropriate parties, similar current consequences of these traits, and yet the populations have experienced different causal histories. Do we say that the traits of models and mimics—the natural doubles—have the same function?

Take the case of the Monarch butterfly (*Danaus plexippus*). These butterflies, found in North America, are brightly colored, orange and white or black and white, and are abundant. An early puzzle for Darwinism was explaining such conspicuous coloration. After all, one could see that camouflage coloration would be adaptive by helping an organism avoid predators. But conspicuous coloration presents a detectable visual signal to a predator. It turns out that having easily identifiable markings may have evolved for the function of signalling unpalatability. The Monarch caterpillars feed on milkweeds and thereby absorb and store cardenolides, a distasteful substance which makes them inedible both as caterpillars and as butterflies. If predators learn that Monarchs are unpalatable, then they will not eat them. Experiments in the laboratory and in the wild indicate that it is indeed the case that birds have aversive reactions to Monarchs after having tasted one, and they decrease the frequency of biting until completely stopping, and predation rates are thereby substantially reduced.[11] The body of experiments has been taken as evidence that "the function of conspicuous color patterns is aposematic (i.e. warning of unpalatability)."[12] Now consider the Viceroy butterfly (*Nymphalidae, Limenitus archippus*) which is the "natural double" to the Monarch in color pattern, even though it belongs not only to another species, but to a different family. The Viceroy has a diet unlike the Monarch's, and is, indeed, a desirable food source for birds. Nevertheless, the conspicuous coloration in the Viceroy also allows it to avoid predation, by being a mimic of the unpalatable Monarch. In cases of Batesian mimicry in butterflies the operator (predator) cannot visually distinguish individuals from the two unrelated prey species. One—the model—is both conspicuously colored and unpalatable. The other—the mimic—is similar colored, but palatable. The evolution of Batesian mimicry requires that the model organism must be more abundant than the mimic (or the learning process would work against the association of coloration with unpalatability). Studies of models and mimics have shown that mimetic resemblance tracks changes in morphology of the model, disappears in the absence of a model, and is controlled by a complex of genes.[13]

I have presented what I take to be a case of "natural doubles." Two types of organism, the Monarch butterfly and the Viceroy butterfly, are structurally similar or indistinguishable. Parallel to Bigelow and Pargetter's "double world," the morphological structures have the same future consequences, i.e. avoiding predation, but have had different evolutionary histories. Do we want to say that the conspicuous coloration of the Monarch and Viceroy have the same function? No. Mimics and models are not the same. As Wickler (1968, p. 108) says: "In general, we use the term mimicry only when the mimetic characters have been evolved for a specific mimetic function." The function of conspicuous coloration in the Monarch is to warn the predator of its unpalatability. The function of the Viceroy coloration is to mimic the model and deceive the predator into presuming it is unpalatable and thereby avoid predation. The same structure has two functions, one is to warn and the other is to deceive.

Thus, if two butterflies flit into our view and we see virtually identical structures and virtually identical current consequences on predation avoidance, Bigelow and Pargetter might have us conclude that the features of the model and the mimic have the same function. But surely that is a mistake. By knowing the causal history of conspicuous coloration, the models and mimics can be distinguished, and the functions of the traits understood. Perhaps a more compelling argument could be made if two identical structures had more divergent evolutionary histories, i.e. one the product of direct selection of the trait for the resulting consequence, and the other being a case of genetic "hitch-hiking" such that it was not selected directly for any feature save its location on a chromosome. While this is perfectly plausible on current biological theory, I know of no actual case that displays sufficient similarity to be a "natural double." Nevertheless, with the mimicry example I hope to have shown that knowledge of identical current structures is not sufficient to compel assent to identical function. History does matter to biological function, and it is that fact that the etiological account acknowledges.

Bigelow and Pargetter grant that an etiological theory of function renders function explanatory, but at the cost of looking "backward" to causal history. The doubles thought experiment was intended to suggest the need for a theory of function which is not just explanatory, but also "forward-looking." To satisfy this intuition they propose a dispositional theory of function, based on an analogical argument.

IV   Biological Fitness Analogy

Bigelow and Pargetter propose to characterize function not in terms of causal history but rather in terms of future effect, that is, as a disposition. They draw an analogy between biological fitness and biological function and argue that just as "fitness" lacked explanatory status when it was identified with actual reproductive success and that status was restored by construing fitness as a disposition or propensity to have a certain degree of reproductive success—so too with "function." Explanation fails, according to them, when either is characterized solely in terms of actual consequences and is restored when characterized as a disposition to have certain consequences.

There is some initial plausibility for pursuing this line of reasoning. Given that a dispositional interpretation of fitness did solve an explanatory failure problem, it might well apply to other scientific concepts. However, as I have argued (Mitchell 1993) neither the negative nor the positive analogy of function with fitness can be sustained in ways that justify replacing an etiological theory of functions with a dispositional one. In fact, although the dispositional analysis offered by Bigelow and Pargetter may correctly describe the nature of a functional trait, it does not address the problematic of the etiological view, namely explaining why the functional trait with that nature is present. In the remainder of the section I will briefly outline how the strategy of dispositional analysis was successful in the case of biological fitness and show that the correlative move to a dispositional theory of biological function is beside the point.

An often cited criticism of evolutionary theory is that its central principle—the principle of natural selection—is tautological. Since this principle was taken to be empirical and not definitional, so much the worse for the scientific status of evolutionary theory. The principle of natural selection may be represented by the Spencerian motto of "the survival of the fittest." Or, to put it in a more acceptable form: "Those organisms which are more fit will reproduce more successfully" (you can substitute genes, groups, species, etc. for organisms). If, however, fitness if defined as actual reproductive success, an interpretation found in population genetics, then the principle of natural selection becomes the tautology: "Those organisms which have greater actual reproductive success will reproduce more successfully." Dispositional theories of fitness argue that actual reproductive success is a mistaken interpretation (Brandon 1978; Mills an Beatty 1979; Brandon and Beatty 1984). Identical twins in the same environment are equally fit to that environment. Nevertheless, by chance, one may perish before it reproduces while the other does not. Thus while the actual reproductive success of the twins differs, their fitness does not. If fitness is construed instead as a propensity to have a certain reproductive success, then the alleged tautology is transformed into an empirical explanatory principle. The twins have identical dispositions, but only one of them actualizes or manifests its potential.

How does the appeal to disposition work in an explanation, and what is thereby explained? Dispositional properties are commonly characterized as subjunctive conditionals describing the resultant behavior (the manifestation of the disposition) when certain antecedent conditions are realized. Furthermore, dispositional properties are explanatory, an implicit assumption of Bigelow and Pargetter's argument, when associated with a causal basis.[14] The basis may be a physical structure that in conjunction with the antecedent circumstances, *ceteris paribus*, will cause the manifestation described by the subjunctive conditional. For example, fragility is a real dispositional property characterized by the conditional

"If this object should be dropped, *ceteris paribus*, then it will break." The manifestation (breaking) is caused by one of a number of molecular structures $(M, N, . . .)$ given the object is dropped from a sufficient height onto a hard surface, etc. The dispositional property and the particular molecular structure that forms its basis in a specific object are distinct properties. It may well be the case that in different objects, different molecular structures will all be such that the object will break when dropped—hence fragility is multiply realised and supervenes on its bases. Which set of molecular structures are co-extensive with fragility is an empirical matter. That fragile objects break when dropped is a matter of definition.

When we ask, why did this egg break, we might explain it by its disposition, i.e. because it is frgaile. Or we might explain it by its causal basis, because it has molecular structure M. The first answer abstracts away from the particular material of the causal basis and picks out a class of causally efficacious properties. Hence, fragility can explain why, say, chicken eggs break when dropped but marble eggs do not. The second explanation appeals to the basis, thus giving the specific cause of the manifestation in this instance.[15]

Biological fitness as a dispositional property or propensity is characterized by the subjunctive: "if the organism (or organism type) should be environment E, then it will leave N offspring." Fitness is probabilistic in that its causal bases determine probability distributions of a range of offspring outcomes in contrast to an all-or-nothing disposition like fragility. Thus a specific causal basis—like camouflage coloration in peppered moths or large male size in primates—given a set of environmental circumstances, confers probabilities onto the individual (or type) to display specific levels of reproductive success. Fitness thus may be expressed as the disposition to have an expected number of offspring.[16]

As we have seen, the disposition "fragility" explains why a certain object breaks. Either abstractly or by appeal to a causal basis in an object, a disposition explains the occurrence of its manifestation under the appropriate circumstances. Similarly with fitness. Differential fitness between organisms or organism types explains why one individual or type of individual reproduces more successfully, that is, manifests a specific reproductive success. Why did primate A produce more offspring than primate B? Answer, because primate A was more fit than primate B. The abstract, general disposition appeals to overall fitness specified only as a probabilistic disposition or propensity and is explanatory in the same way a general law is explanatory.[17] It explains the similarity with respect to reproductive success of two *prima facie* dissimilar phenomena. Darwin's original insight was that the features of Galapagos finches and the English pigeons were the results of the same causal process. Selection, whether by nature or by pigeon fanciers, "preserves" any advantageous character through time via differential reproduction of its bearers, given the heritability of the character. Fitness, the disposition to leave N offspring, describes a property which is often important to evolutionary change. Of course, like any other disposition, there may be reasons why it is not expressed (and hence could not effect evolutionary change). The unfortunate twin who, by chance, failed to reproduce at all, failed to manifest the dispositional fitness endowed by its properties.

At a more concrete level, fitness or the disposition to leave N offspring identifies the causal basis, say, large size, and thus explains in a given circumstance why some individuals or types are more reproductively successful than others. Large size in conjunction with triggering conditions of male-male competition causes the manifestation of a reproductive success of N. In either case the dispositional property—fitness N—explains actual reproductive success. By construing fitness as a disposition, we see what fitness is—both at the concrete level by attaching it to a specific causal basis in a specific population and at the abstract level by seeing how whatever the particulars of the causal basis—be it large size or small number of eggs in the nest—having such a disposition can have consequences in the processes of natural selection and evolution.[18]

Thus the fitness of an organism, being a disposition or propensity to leave an expected number of offspring, can explain why the organism realizes a certain reproductive success, just as the disposition of fragility can explain why an object breaks. But does biological function operate like biological fitness in explanations?

V Functions as Dispositions

The classic puzzle regarding functions is how the consequence of a trait could explain why the trait is, in fact, present. As we have seen the etiological account identifies function with that consequence in the past which was responsible for the current presence of the item in question. Both selection *for* the functional consequence over alternatives and replication *of* the structure with that consequence as a direct result of selection, are required to warrant etiological function ascription.

Bigelow and Pargetter objected to the causal history strategy for being too "backward looking." "Fitness is forward-looking. Functions should be forward-looking in the same way and, hence, are explanatory in the same way ... something has a function just when it confers a survival-enhancing propensity on a creature that possesses it" (Bigelow and Pargetter 1987, pp. 191–2). But what do these sorts of properties—dispositional functions—allow us to explain? Bigelow and Pargetter say (1987, p. 193):

... functions will be explanatory of survival, just as dispositions are explanatory of their manifestations; for they will explain survival by positing the existence of a character or structure in virtue of which the creature has a propensity to survive.

Applied to the case of sexual dimorphism discussed above, we get the following claims. Let us say that the function of large body size in male primates in a specified environment is to contribute to success in male/male competition for mates. What does that mean? On the etiological view, it is clear that identifying the function explains why the males are larger by appealing to the causal history of natural selection and evolution which took the differential effect on male/male competition as a cause for differential female acquisition and hence differential reproductive success. Given the heritability of the trait, this selection process would propagate larger body size over smaller size through a series of generations. On the dispositional view, identifying the function as contribution to success in conflict is to say that this is a survival-enhancing propensity (although the conflicts usually don't have such direct effects on physical survival of the winners and losers, but rather on their acquisition to mates in this case) which accrues to the trait in question. Hence when the right conditions obtain (two males meet in competition), the function will be manifested in actual domination and female acquisition and thereby enhance the survival of the larger male. What function explains on the dispositional view is why a large male primate reproduces more successfully than a small one. What function explains on the etiological view is why male primates are large.

Another way to put this distinction is as follows: the dispositional account tells us how having a trait with a function (a disposition to have a certain consequence) contributes to the survival and reproductive success of individuals with that trait, i.e. how they fare in the struggle for existence that constitutes natural selection. In contrast, the etiological view tells us how having a trait with a function (a consequence which has played a certain role in its causal history) contributes to the presence of the trait, i.e. how the trait has evolved by means of natural selection.

This is not to say that Bigelow and Pargetter's dispositional analysis renders function non-explanatory. Rather, scientific explanations are developed to do different things, are directed at different targets and give explanations at different levels of abstraction or concreteness. As Cartwright (1986, p. 203) puts it: "Explanations give answers not only to *why* questions, but also to *what* questions. They say *of* something, what it really is." If one asks what is this property of conspicuous coloration? One can answer by outlining the role it might play in contributing to survival. On the dispositional theory, a function identifies the nature of such a trait—its function in the current system. However, it does not account for the explanatory use of function to answer the original question of why the trait is there.[19]

Thus, the dispositional account does not offer a competing analysis to the etiological theory of functional explanation. Function is used in different scientific projects, there is not a single, univocal explanatory task for which such language is employed in scientific practice. The two theories discussed each focuses on a different use.[20] One might argue about which project is most significant, or most common in biological discourse. However, I believe such arguments are fruitless. Rather, the philosophical task is to recognize the plurality of explanatory projects, to clarify their relationships, and to explicate their structures. By attempting to do so in this paper, I hope to have shown that the etiological theory of function accounts for the explanatory structures implicit in at least some scientific practices, and that a dispositional theory does not offer a competing alternative account.

Notes

1. An early version of this paper was presented at the International Society for History, Philosophy and Social Studies of Biology, in London, Ontario, July 1989. See also Mitchell (1993) for related arguments against Bigelow and Pargetter (1987). I wish to thank the anonymous reviewer for detailed, helpful comments.

2. Bigelow and Pargetter (1987).

3. See Bigelow and Pargetter (1987) for an account of the range of such views in the current market.

4. Thus it is a "methodological" concept, as Nordmann suggests, which needs to be "worked through again and again" (1990, p. 380).

5. Or, if you prefer, which function ascription is accepted by the scientific community will depend on the constitutive standards for evidence and confirmation in that community.

6. Wright suggests this: "functional explanation depends essentially on a selection background." (Wright 1976, p. 101)

7. See also Boyd and Richerson (1991).

8. See Hull (1987), Millikan (1989), p. 300.

9. See also similar cases given by Prior (1985).

10. Similar objections to adaptation claims are made by Gould who criticizes biologists who mistake current consequences for traits as sufficient evidence for the ascription of adaptive function. For discussion of this argument see Mitchell (1992), Gould (1987a, 1987b) and Alcock (1987).

11. Brower (1958), L. Brower (1988) for experimental results.

12. Guilford (1988), p. 9. Guilford however argues that this is insufficient evidence for the conclusion, and suggests various evolutionary explanations for the association between conspicuous coloration and unpalatability.

13. Wickler (1968).

14. Armstrong (1969), Prior (1985), Prior, Pargetter and Jackson (1982) defend the necessity of a causal basis, but seem to allow for the disposition, i.e., conditional behavior, to supervene over a variety of different causal bases. Sober, on the other hand, defends the reality of dispositions as similarly independent of the behavioral subjunctive conditional description, but requires something stronger, namely that "A scientifically respectable dispositional property must be a univocal characteristic that underlies all the instances in which the subjective conditional displays itself." (Sober 1984b, p. 47).

15. General issues concerning the explanatory status of dispositions arise in the particular case of biological fitness. Namely, here has been discussion about how "fitness" is to be understood dispositionally, namely whether (i) dispositional language itself is non explanatory and hence that fitness be taken as an undefined primitive (Rosenberg 1986, 1985), (ii) "fitness" as a dispositional term merely

marks a temporary stage in scientific development, and should discarded since biology has made sufficient advances to fill out the place-holder for which it was used (Waters, 1986), or (iii) the dispositional character of "fitness" is still required to be open-ended (Resnik 1988 and Nordmann 1990). In the argument that follows, I defend an explanatory role for the abstract place-holder and thus in this respect share the views of Resnik and Nordmann.

16. This is the view presented by Mills and Beatty (1979), 270–275.

17. See Cartwright (1986) for similar argument regarding teleological explanations in general, and Kitcher 1981 on the unifying role of schematic explanations.

18. Ayala (1970) had something like this in mind when he spoke of two levels of teleology in organisms, the proximate end of features of organisms, as well as the ultimate goal of contribution to increase in reproductive success.

19. For an argument that the two types of function are not always compatible, see Mitchell (1993).

20. See Wouters (1995) for other types of explanatory projects associated with function.

References

Alcock, J.: 1987, "Ardent Adaptationist." *Natural History* 96, 4.

Armstrong, D.: 1969, "Dispositions are Causes," *Analysis* 30, 23–26.

Ayala, F.: 1970, "Teleological Explanations in Evolutionary Biology," *Philosophy of Science*, 37, 1–15.

Bigelow, J. and Pargetter, R.: 1987, "Functions," *Journal of Philosophy*, LXXXIV, 4, 181–196.

Boyd, R. and Richerson, P. J.: 1991, "Culture and Co-operation," in R. A. Hinde and J. Groebel (eds,); *Cooperation and Prosocial Behavior*, Cambridge University Press, pp. 27–48.

Brandon, R.: 1978, "Adaptation and Evolutionary Theory," *Studies in the History and Philosophy of Science* IX, 3: 181–206.

Brandon, R. and Beatty, J.: 1984, "Discussion: The Propensity Interpretation of 'Fitness'—No Interpretation is No Substitute," *Philosophy of Science* 51, 342–347.

Brower, J. 1960 "Experimental Studies of Mimicry," *American Naturalist* 44, 271–83.

Brower, L. 1988 "Avian Predation on the Monarch Butterfly and its Implications for Mimicry Theory," in L. Brower (ed.), *Mimicry and the Evolutionary Process*, University of Chicago Press, Chicago, pp. 4–6.

Cartwright, N.: 1986, "Two Kinds of Teleological Explanation," in A. Donagan, A. N. Perovich, Jr. and M. V. Wedin (eds.), *Human Nature and Natural Knowledge*, D. Reidel, Dordrecht, pp. 201–210.

Clutton-Brock, T. H. and Harvey, P. H.: 1977, "Primate Ecology and Social Organization," *Journal of Zoology London* 183, 1–39.

Clutton-Brock, T. H., Harvey, P. H. and Rudder, B.: 1977, "Sexual Dimorphism, Socioeconomic Sex Ratio and Body Weight in Primates," *Nature* 269, 797–800.

Cummins, R.: 1975, "Functional Analysis," *Journal of Philosophy* 72, 741–765.

Darwin, C.: 1871, *The Descent of Man and Selection in Relation to Sex*, Princeton University Press, Princeton (1981).

Godfrey-Smith, P: 1994, "A Modern History Theory of Function," *Noûs* XXVIII: 3, 344–62.

Gould, S. J.: 1987a, "Freudian Slip," *Natural History* 92, 14–21.

Gould, S. J.: 1987b, "Stephen Jay Gould Replies," *Natural History* 96, 4–6.

Guildford, T.: 1988, "The Evolution of Conspicuous Coloration," in L. Brower (ed.), pp. 7–21.

Hempel, C. G.: 1959, "The Logic of Functional Analysis," in L. Gross (ed.), *Symposium on Sociological Theory*, Harper and Row, New York. Reprinted in C. G. Hempel, 1965, *Aspects of Scientific Explanation*, The Free Press, New York, pp. 297–330.

Horan, B. L.: 1989, "Functional Explanations in Sociobiology," *Biology and Philosophy* 4, 131–158.

Hull, D.: 1987, "On Human Nature," *PSA 1986, Volume 2, Philosophy of Science Association*, 3–13.

Kitcher, P.: 1981, "Explanatory Unification," *Philosophy of Science*, 48, 507–531.

Millikan, R. G.: 1984, *Language, Thought, and Other Biological Categories*, The MIT Press, Cambridge.

Millikan, R. G.: 1989, "In Defense of Proper Functions," *Philosophy of Science* 56, 288–302.

Mills, S. and Beatty J.: 1979, "The Propensity Interpretation of Fitness," *Philosophy of Science* 46, 263–286.

Mitchell, S. D.: 1987, "*Why Functions (in Evolutionary Biology and Cultural Anthropology)*," unpublished Ph.D. thesis, University of Pittsburgh, microfilm no. 8808338, U. M. I., Ann Arbor, Michigan.

Mitchell, S. D.: 1989, "The Causal background for Functional Explanations," *International Studies in the Philosophy of Science* 3, 213–230.

Mitchell, S. D.: 1992, "On Pluralism and Competition in Evolutionary Explanations," *American Zoologist* 32.

Mitchell, S. D.: 1993, "Comments and Criticism: Dispositions or Etiologies? A Comment on Bigelow and Pargetter" *The Journal of Philosophy* XC, 249–259.

Nagel, E.: 1977, "Teleology Revisited," *Journal of Philosophy* 74, 261–301.

Neander, K.: 1991, "Functions as Selected Effects: The Conceptual Analyst's Defense," *Philosophy of Science*, 58, 168–184.

Nordmann, A.: 1990, "Persistent Propensities: Portrait of a Familiar Controversy," *Biology and Philosophy* 5, 379–399.

Prior, E.: 1985, *Dispositions*, Humanities Press, Atlantic Highlands, N. J.

Prior, E., Pargetter, R. and Jackson, F.: 1982, "Three Theses about Dispositions," *American Philosophical Quarterly* 19, 252.

Rappaport, R.: 1968, *Pigs for the Ancestors*, Yale University Press.

Resnik, D.: 1988, "Survival of the Fittest: Law of Evolution or Law of Probability?" *Biology and Philosophy* 3, 349–362.

Rosenberg, A.: 1982, "Discussion: On the Propensity Interpretation of Fitness," *Philosophy of Science* 49, 268–273.

Rosenberg, A.: 1985, *The Structure of Biological Science*, Cambridge University Press, Cambridge.

Selander, R. K.: (1972), "Sexual Selection and Dimorphism in Birds," in B. G. Campbell (ed.), *Sexual Selection and the Descent of Man 1871–1971*, Aldine, Chicago, pp. 180–230.

Sober, E.: 1984a, *The Nature of Selection*. The MIT Press, Cambridge.

Sober, E.: 1984b, "Force and Disposition in Evolutionary Theory," in C. Hookway (ed.), *Minds, Machines and Evolution*, Cambridge University Press, Cambridge, pp. 43–61.

Soltis, J., Boyd, R. and Richerson, P. J.: 1991, "Can Group-Functional Behaviors Evolve by Cultural Group Selection? An Empirical Test," *Preprint Series of the Research Group on Biological Foundations of Human Culture* (1991/92), at the Center for Interdisciplinary Research, University of Bielefeld, Germany.

Thornhill, R.: 1979, "Review of *Insect Behavior*, by R. W. Matthews and J. R. Matthews," *Quarterly Review of Biology* 54, 365–366.

Vayda, A.: 1974, "Warfare in Ecological Perspective," *Annual Review of Ecology and Systematics* 5, 183–193.

Vayda, A.: 1989, "Explaining Why Marings Fought," *Journal of Anthropological Research*, 45, No. 2, 159–177.

Waters, K.: 1986, "Natural Selection without Survival of the Fittest," *Biology and Philosophy*, 1, 207–225.

Wickler, W.: 1968, *Mimicry in Plants and Animals*. McGraw Hill.

Williams, G. C.: 1966, *Adaptation and Natural Selection*, Princeton University Press, Princeton.

Wimsatt, W. C.: 1972, "Teleology and the Logical Structure of Function Statements," *Studies in History and Philosophy of Science* 3, 1–80.

Wouters, A (1995) "Viability Explanation," *Biology and Philosophy* 10: 435–457.

Wright, L.: 1973, "Functions," *Philosophical Review* LXXXII, 2, 139–168.

Wright, L.: 1976, *Teleological Explanations*, University of California Press, Los Angeles.

IV

Synthesis or Pluralism?

15

The Concept of Function

R. A. Hinde

1 Introduction

Niko Tinbergen, more than any other ethologist, has stressed the importance of the problem of function. He has shown not only that its understanding contributes to the distinct but yet related questions of causation and evolution, but also, and more fundamentally, that in its own right it is an important, interesting, and too frequently neglected question. In particular, the paper Tinbergen presented at the 1966 Ornithological Congress (Tinbergen 1967) contained a vivid demonstration of the increased understanding of the biology of a species which studies of function provide, an evaluation of the types of evidence rele ant to it, and a demonstration of the effectiveness of the experimental analysis of function which he had pioneered. But in spite of his hard-headed example, the problem of function is still too often treated as a matter for speculative asides (Tinbergen 1965). Therefore, although it means some retreading of old ground, it seems worthwhile to give further consideration to the nature of the concept, and to the sorts of evidence used in its discussion.

2 The Behavioural Nexus

Each item of behaviour forms a link in a nexus of events which precede and follow it. If we focus on the behaviour, we refer to some of the former as causes and some of the latter as consequences. The consequences can in principle be classified according to their effects on the organism's subsequent reproductive success: some are detrimental, some neutral, and some beneficial.[1] For example, when a gull incubates its eggs it (1) perhaps exposes itself to predation and loses time that it could spend feeding (harmful consequences); (2) slightly expands the egg shell (presumably a neutral consequence);[2] and (3) keeps the embryos warm and thus enhances its chance of reproductive success (beneficial consequence) (Tinbergen 1942). In a broad way, beneficial consequences are referred to as "functions." A character of structure or behaviour is said to be adaptive if its beneficial consequences outweigh the deleterious ones.

Now according to this formulation, the function of all adaptive characters is ultimately the same, namely, contributing to eventual reproductive success. In practice, of course, the interesting question is how they so contribute. In the case of behaviour each act has ramifying consequences: if the behaviour is adaptive, one or more of the interlinked chains of consequences contributes to reproductive success. The term "function" is often applied to items on such chains following fairly soon after the behaviour itself, and preceding the subsequent enhancement of reproductive success. For instance, if territorial behaviour spaces out a breeding population, and spacing out reduces predation, and thus improves reproductive success (Tinbergen 1956), either spacing-out or the reduction in predation may be called the function of the territorial behaviour.[3]

3   Weak and Strong Meanings of the Term "Function"

Some characters of structure or behaviour are present in some individuals of a population but not others.[4] More usually, however, a character is present in all, but to varying degrees. In such a case, its mere presence could (in theory at least) be sufficient to ensure that a given beneficial consequence occurs with full efficiency in each individual. That consequence could not then provide material on which natural selection could act to maintain the character in the population. Natural selection can operate only through consequences that are achieved to a different extent in different individuals. (It is, of course also necessary that the differences have a genetic basis.) For example, a bird's body feathers provide insulation, and also carry patterns and colours that are biologically advantageous. But if the natural variation in the number of feathers is within the range which would affect insulation but not coloration, natural selection cannot act through the latter to maintain the number of feathers at its present level. Thus the effectiveness of the colour pattern may not be the consequence through which natural selection is acting to maintain the present density of plumage. The coloration of the feathers could also affect insulation, but if the variation in coloration is such that insulation is virtually unaffected, natural selection cannot affect coloration through its heat-radiating consequences. Again, the preservation of biological requirements and the prevention of epidemic diseases have both been suggested as beneficial consequences of avian territorial behaviour. If the variability in territory size is sufficient to affect the availability of necessities, but not the spread of diseases, then it could be said that the former is a function and the latter is not (Hinde 1956).

Thus not all beneficial consequences provide material for the action of natural selection. In practice, of course, the dichotomy just implied between beneficial consequences of a character through which natural selection does and does not act is an oversimplification. The extent to which the various beneficial consequences provide material for natural selection will vary along a continuum, some consequences being achieved so nearly fully in all individuals that the effectiveness of natural selection operating through them is overridden by its effects through others. But the argument does indicate that a concept of function equated with "any beneficial consequence" does little more than answer the question, "What is the character good for?" and need tell us nothing about the dynamics of the evolutionary process. How then can the concept be hardened up?

First, if function is to have empirical relevance it must refer ultimately to the consequences of a *difference*. Thus if we speak of the function of flocking, we refer to the beneficial consequences of associating with others not enjoyed by individuals living alone. If we speak of the function of "the cliff-nesting behaviour of kittiwakes," we imply "as compared with the ground-nesting behaviour of gulls." Sometimes the comparison is with a hypothetical organism: when we speak of "the function of bird song," we refer to events consequential upon the birds' singing which would not occur if birds did not sing, and thus imply a hypothetical population which does not sing.

Second, we must recognize that the concept of function carries a spectrum of meanings. A weak meaning answers the question "what is it good for?" In so far as a comparison is implied, it is with a hypothetical organism similar in every respect except that it lacks the character in question. Furthermore, a weak use of function may contain no indication of precisely how the character in question contributes to reproductive success. The statement that the function of birds' wings is flight is a weak one in so far as it implies at most comparison with a hypothetical bird without wings, or an avian ancestor not yet possessing them, and because it contains no indication of how flight contributes to reproductive success. (This is not so obvious as it seems: on oceanic islands birds tend to become flightless, presumably because they are liable to be blown out to sea [Lack 1947].) Discussion of function at this level has little meaning in terms of the dynamic processes of evolution: only when the differences between this wing and that are discussed does the problem begin to relate to the operation of natural selection.

By contrast, "function" in a strong sense attempts to answer the question "through what consequences does natural selection act to maintain this character?" Now, to answer this question, it would strictly be necessary to assess reproductive success with and without the character. Cross-species and even cross-population studies are seldom strictly adequate for this, simply because the different characters of each form a co-adapted complex. For example, the young of hole-nesting passerine birds tend to spend longer in the nest than those of open-nesting species, presumably because they are less vulnerable to predation. This means that they need less food (weight for weight) each day, and the parent can thus rear more individuals. Furthermore, there is less premium on female crypticity, and thus the sexes are more likely to be coloured similarly. Thus nest-site, fledgling period, clutch size, and sexual dimorphism are adaptively related characters. This argument could be continued almost indefinitely: the point at issue is that, from the point of view of natural selection, characters cannot be considered in isolation, but only in relation to each other.[5] Thus it is usually impossible to find two populations differing in only one character relevant to the consequence under examination. Sometimes, however, it is possible to use existing variations within a population, assessing the consequences of that variation. Such a procedure can lead to hard evidence on function in this strong sense—provided, of course, that it can be assumed that the variations in the character in question are not correlated with variations in other characters which affect its consequences.

All this seems rather pedantic. Indeed it leads at first sight to the improbable view that it is meaningless to ask about the function (in the strong sense) of stable characters equally present in all members of the population. But it is in fact the case that as we move from gross characters (for example, bird's wings) to finer ones (for example, long narrow wings in this species as opposed to short broad ones in that) to yet finer ones (for example, longer wings in this population than that) the questions that we can ask about beneficial consequences move from gross ones (what is it good for?) to fine ones about how natural selection actually acts. We shall return in a moment to the nature of the data available for answering such questions: the issues here are that when we ask "what is its function?" "it" implies a difference and "function" can have a spectrum of meanings.

4   The Concept of Function: Summary

The argument so far is summarized in figure 15.1. A and P represent two items of behaviour. A leads to biologically deleterious, neutral and advantageous chains of consequence a, b, c, and d, only the last leading to consequences through which natural selection acts. P leads to comparable chains q, r, s, and t. Consequences c and s are functional in a weak sense, d and t in a strong sense. The biologically advantageous chain c also leads to a neutral consequence e, and s to u. The adaptive nature of the further consequences of c_2 (itself a consequence of A) are affected by t_1 (itself a consequence of P). Consequences c, d would represent functions of possessing the character A in the weak sense but only d in the strong sense.

5   Is All Behaviour Functional?

Only characters which confer a positive biological advantage can survive for long against the forces of mutation and selection. Biologists are thus

Characters   Chains of consequences

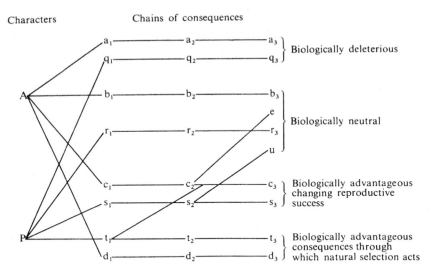

Figure 15.1

prone to consider that characters of behaviour are biologically advantageous unless proved otherwise: this assumption is one reason why the problem of function has not been studied more assiduously in the past. In any case, it is a natural point of view when to prove function in a strict sense is a task of such complexity. The climatic, vegetational, and zoogeographic circumstances in which a species lives are not constant. To the extent that the variations are orderly, the patterning will exert a selective influence. To assess the function of a character subject to shifting selective forces could be an impossible task.

Nevertheless, certain categories of exceptions are recognized as accounting for the presence of characters that appear not to be biologically advantageous.

1. The character in question may be a by-product of a character adaptive in another context. Thus Tinbergen (1969) regards the aggressiveness of a male Black-headed Gull to its mate as a by-product of aggressiveness adaptive in the context of territorial disputes. Similarly, Kruuk (1972) argues that the phenomenon of surplus killing by carnivores is a relatively trivial disadvantageous consequence of behavioural characteristics otherwise adaptive in obtaining food.

2. Just as a structure may be biologically advantageous but not used all the time, so every expression of a character of behaviour does not have to be functional. Thus when Blue Tits (*Parus caeruleus*) enter houses and tear paper, this may be a non-functional expression of a motor pattern biologically adaptive for feeding, and normally used in that way (Hinde 1953).

3. A structure or item of behaviour may be a relict of a formerly adaptive character which has not yet been lost. For example, many birds extend their ipsilateral wing downwards when scratching their head; this is often held to be a phylogenetic remnant from tetrapod ancestors (but see discussion in Wickler (1961)). The persistence of neutral characters is perhaps particularly likely in view of the co-adapted nature of the character complex.

In addition a character may be adapted but not perfect because of conflicting requirements. Thus the extent to which a bird will display at or attack a predator near its nest is the result of the conflicting requirements of protecting its young and protecting itself (Tinbergen 1965, 1967).

6 Observational Evidence for "Functions"

We may now consider some of the types of evidence by which "functions" are identified, bearing in mind the distinction between weak and strong meanings of "function."

The first indication of function often comes from contextual information. For example, it was primarily (though not solely) because Hooker and Hooker (1969) found that, in the shrikes they studied, "duetting occurs throughout the year between members of an established pair occupying and defending a territory," that they concluded that duetting functions "in maintaining the bond between family groups in dense vegetation and in joint aggressive vocal display during territorial disputes."

In addition to contextual information, statements about function are usually based on observations that the behaviour in question has a certain consequence, and that that consequence appears to be a "good thing." Thus the observation that an intruder onto a robin's territory flies away when it hears the owner sing (Lack 1939) is evidence that song functions in maintaining territorial ownership, additional to the contextual evidence that a robin sings only when on its territory. The hypothesis that gulls roost on open beaches up to the time when incubation starts because they are safer from predators there was confirmed by data showing that the number of gulls killed by foxes is much greater in the breeding areas than on the beaches (Kruuk 1964; Tinbergen 1965).

A further type of evidence, perhaps to be regarded as a form of contextual evidence, concerns correlations between characters of structure or behaviour and environmental factors. For example, if body size in a number of distantly related groups tends to become larger the nearer the poles that they breed, this suggests that body size is related to some environmental factors that change with latitude. If conspicuous coloration and behaviour are common in male birds living in areas where the ranges of closely related species overlap, whereas where only one species is present both sexes tend to be cryptic, this suggests that the conspicuous features have a function related to the presence of closely related species. A wealth of further examples of this sort has been given by Lack (1966, 1968). When evidence of this sort is added to evidence about consequences that appear to be advantageous—for instance, that large animals withstand cold better, or conspicuous colours are used in pair formation —the evidence for function even in the strong sense becomes powerful.

It would probably be true to say that most statements about functions of behaviour are based on evidence of these types. Often further evidence is impossible to obtain. However, it is as well to recognize not only that they can provide at most indirect evidence about the action of natural selection, but also that a number of other difficulties may arise.

One of these is that even the characters chosen for discussion may have little biological relevance. Thus there has recently been considerable speculation about the functions of the different types of social structure to be found among non-human primates, and attempts have been made to relate the size or composition of groups to ecological variables (Crook and Gartlan 1966). As more information has become available, the inadequacy of these attempts has become increasingly apparent (for example, Struhsaker 1969; Crook 1970). This is not only because of the complexity of the problems, but because social structure depends on the behaviour of individuals, and it is the behaviour of the individuals for which functions must be identified. Characters of group size or structure may be by-products of behaviour adaptive through other consequences. Of course if such characters were adaptively disadvantageous, they might be selected against; the point here is that they may be near-neutral.

Another type of difficulty can arise if the observer focuses his attention on the wrong consequence, which may be a mere by-product of a nexus of consequences leading also to a quite different event that is adaptively important. For example, it is often said that the function of threat displays is to reduce actual physical combat, and it is in fact the case that physical contact often does not occur while animals are displaying. However, it seems reasonable to suppose that an animal that is definitely going to attack or to flee would do best to do so immediately, without giving warning by displaying first. On this view, threat displays would be useful only in moments of indecision. If what one individual will do depends in part on the probable behaviour of the other, threatening by one or both individuals may convey the information necessary for a decision. Such a view is indeed prompted by Stokes' (1962) finding that the threat displays of Blue Tits do not give an absolute prediction of whether the bird will subsequently attack, stay, or flee; and by Simpson's (1968) description of the fighting of the Siamese Fighting Fish (*Betta splendens*), which involves reciprocal displays of mutually similar and gradually increasing intensity, often until one individual suddenly capitulates before either has given a bite. On this view, then, the function of a display lies in communication with the opponent, which maximizes the chances of either a successful attack or escape; a reduction in the actual amount of combat is a by-product. This view, of course, requires substantiation; the issue here is that it is a reasonable alternative to that often advanced. The observation that display has a consequence which appears to be advantageous is not proof of function in a strong sense.

A related example concerns the formation of dominance hierarchies. This is accompanied by a reduction in aggressive behaviour, and this is often said to be its function. But examination of the behaviour of individuals suggests that the subordinate individuals behave as they do because under natural conditions (and perhaps in the long run) they stand a better chance of obtaining access to the resources they need if they do not attempt to obtain them in the presence of the dominants (for example, Rowell 1966). A reduction in aggressive behaviour is one consequence of the interactions between individuals, a pattern of relationships which can be described as a dominance hierarchy is another, but to say that the former is a function of the latter in a biological sense is both imprecise and misleading. It would be nearer the truth to conjecture that the reduction in fighting has consequences which are beneficial in different ways for dominants (for example, access to food without risk of injury) and subordinates (for example, reduction in risk of injury by dominant and possibility of finding food elsewhere); and that the hierarchy is a consequence of the reduction in fighting, rather than vice versa.

Furthermore, there are often difficulties in proving that the supposed function really is biologically advantageous, and really is a consequence of the behaviour in question. An example is provided by a point in Marler's (1968) important review of the mechanisms of group spacing in primates. He suggests that the signals used can be classified as distance-increasing, distance-maintaining, distance-reducing, and proximity-maintaining signals, and that their properties conform with these functions. Marler himself points out that the distinction between distance-increasing and distance-maintaining functions is not always clear, but there are also further difficulties with the latter category. First, one may ask whether there is indeed evidence that it is advantageous for a primate group to have another group just so far away but no further; one can speculate that it may be, but that is all, and unless distance-maintenance has biologically advantageous ramifications, it cannot be designated as a function. Second, when a particular signal does result in the maintenance of distance, this may be because it either attracts or repels other groups, the effect being balanced by a tendency to withdraw or approach with a quite different basis. The consequence of maintaining distance may thus be a joint consequence of the signal and some other factor and cannot be designated as a function of the signal *per se*. Third, Marler himself points out that distance-maintaining calls may sometimes either increase or decrease the distance between troops, depending on their spatial relations. For example, Ellefson (1968) reports that if a gibbon group locates the source of a call outside the area it occupies, it may reply in kind, but if the call comes from a point within or near that area, it may approach and then give distance-increasing calls. This suggests that the important consequence of such calls is that of facilitating localization and that, depending on this, other mechanisms are called into play which promote over-dispersal.

Another type of difficulty which sometimes arises involves the confusion of consequences at different points in the chains ramifying from the behaviour in question. For example, Barnett (1969), generalizing about the functions of agonistic behaviour says that "(1) it influences population density; (2) it determines the structure of the groups of colonial species, and (3) it regulates interactions between groups, or, in solitary species, between individuals." But (1) is consequence of (2) and (3); and (2), in so far as agonistic behaviour also regulates interactions between individuals of colonial species, of (3).

7 Experimental Evidence

So far we have considered evidence derived from field observations. Can experimentation help? It can, but again its limitations must be recognized. Van den Assem (1967) demonstrated in the laboratory that the territorial behaviour of Three-spined Sticklebacks could limit density, that the possession of an adequate territory increased reproductive success in a number of ways, and that owners of large territories were more successful in rearing young than owners of small ones. From these experiments he concluded that the reduction of interference by conspecifics is the main function of territorial behaviour. But, as Tinbergen (1968) points out, the interesting results obtained in this study cannot be extrapolated beyond the experimental situation; other variables, such as food availability and predator pressure, in relation to which territorial behaviour might have important consequences in nature, were not manipulated. Although the validity of deductions from laboratory experiments about selection in nature varies from case to case, it must be acknowledged that, in the absence of evidence of other sorts, laboratory experiments of this sort can provide evidence about beneficial consequences only in the circumstances in which they were conducted, and thus lead only to suggestions about function in a fairly weak sense.

To a lesser degree, of course, the same is true of field studies; the selective forces that impinge on a species are not uniform throughout its range.

However, much stronger evidence for function in the strong sense can be obtained in the natural situation. Tinbergen and his pupils have pioneered this work, and we may consider three of their experiments.

1. Patterson (1965) was concerned to show whether the aggressive displays of territory-owning Black-headed Gulls (*Larus ridibundus*) do or do not have a deterrent effect on others. By comparing the behaviour of intruders onto territories in which the owners were present but immobilized by a stupefying drug with that of intruders on territories in which the owners were behaving normally, it was shown that the displays do have a deterring effect. This experiment by itself (and Patterson, of course, used also other types of evidence) showed that the difference between territory-owning gulls that display normally and individuals (anaesthetized or hypothetical) that occupy territories but do not display has consequences for territorial ownership. Accepting that the latter is adaptively important (see below) this demonstrates a function in the weak sense for territorial display, but does not demonstrate that natural selection acts through this consequence of the display to maintain it at its present level.

2. Black-headed Gulls remove fragments of egg shells from the nest soon after the young have hatched. Tinbergen and his colleagues (Tinbergen, Broekhuysen, Feekes, Houghton, Kruuk, and Szulc 1962; Tinbergen, Kruuk, and Paillette 1962) suspected that this functions in reducing predation on unhatched eggs or chicks. This was tested by laying out eggs in the colony with or without a broken egg shell nearby. The former were taken by predators more rapidly than the latter. This demonstrates that the difference between gulls that do and do not remove egg shells, or that remove them with different latencies, is adaptive. Furthermore, since individual gulls remove egg shells with different latencies, so that the difference between eggs with a broken shell nearby and eggs without is a difference that actually occurs in the colony, and since it can be assumed that the difference has some genetic basis, the experiment provides compelling evidence for function in the strong sense at least in this colony.

3. As indicated above, Patterson (1965) showed that territorial behaviour promotes the spacing out of nest. By taking advantage of the natural variation in spacing in the colony he showed that spacing out reduced predation on pipped eggs and newly hatched chicks by other gulls and by other species (see also Kruuk 1964; Tinbergen 1967). Again, since Patterson demonstrated differential success in relation to natural variation within the species, and since it can be assumed that this variation has some genetic basis, this comes as near as possible to showing that natural selection is operating to maintain spacing out at its present level.

In his discussion of the adaptive features of the Black-headed Gull, Tinbergen (1967) is careful not to overrate the value of his experimental approach. In some cases it is not necessary, sometimes it is not possible, and often it is not sufficient. Furthermore, description and interpretation must always precede experimentation. However, the descriptive-comparative method and the experimental method can together carry the study of function far beyond the reach of armchair speculation.

8 Conclusion

I have argued that the concept of function can properly be applied only to the consequences of differences between characters, and that it is advisable to recognize that it covers spectrum of meanings. At the weak end it is equivalent to beneficial consequences, while at its strongest it has reference to the action of selective forces.

In discussing the sources of evidence adduced for function, I have emphasized the difficulties of proof. Just because hard evidence is so difficult to obtain, it has become respectable to speculate about the function of behaviour in a manner that would never be permissible in studies of causation. This is counterproductive. But I hope that this emphasis on its dangers will not tend to inhibit those types of speculation that may suggest new avenues to explore or new experiments to conduct—or indeed speculation that feeds the sense of wonder that is the mainspring of biological research.

Acknowledgements

I am grateful to Professor N. Tinbergen for the help and advice which he has given me on many occasions, and in particular when I was first starting in research. The present essay has been improved as the result of critical comments from P. P. B. Bateson, T. Clutton-Brock, C. Goodhart, and N. W. Humphrey.

Notes

1. Strictly, success should be assessed in terms of gene survival rather than progeny survival (Hamilton 1964), but that does not affect the present argument.

2. Characters are said never to be neutral (Cain 1964), though that depends in part on what is meant by a character. But the present discussion is concerned not with characters (for example, incubation behaviour) but with their consequences. That some of these can be neutral is self-evident.

3. This usage of function by evolutionarily minded biologists to refer forward in time must be distinguished from two others. One closely related usage implies the question "How does it work?" For instance, the bastard wing of a bird has the function of smoothing the airflow over the wing when stalling conditions are approached. This usage is closely related to that under discussion in that it also refers to a beneficial consequence of the character, but differs in that that consequence is more or less immediate. A quite different usage is the more mathematical one implying merely the existence of a relationship between two events. "Behaviour Y is a function of X" or "Y is a function of behaviour X" can mean merely that a mathematical relation exists between X and Y, without implication that the causal relations, if any, are direct or indirect.

4. The range of meanings of the term "character" will not be pursued here: in the present context it is used both for the properties of a behaviour pattern (for example, the precise form or frequency of a signal movement) and for the pattern itself.

5. Well worked-out examples are provided by the studies of Tinbergen and his pupils on gulls (for example, Cullen 1957) or Crook (1964) on reproduction in weaver-birds.

References

Van den Assem, J. (1967). Territory in the three-spined stickleback, *Gasterosteus aculeatus*. *Behaviour, Suppl.* 16, 1–164.

Barnett, S. A. (1969). Grouping and dispersive behaviour among wild rats. In *Aggressive behaviour* (eds. S. Garattini and E. B. Sigg). Excerpta Medica, Amsterdam.

Cain, A. J. (1964). The perfection of animals. In *Viewpoints in biology* (eds. J. Carthy and R. Duddington), Vol. 3, pp. 36–62. Butterworth, London.

Crook, J. H. (1964). The evolution of social organisation and visual communication in the weaver birds (Ploceinae). *Behaviour, Suppl.* 10, 1–178.

———— (1970). (ed.). The socio-ecology of primates. In *Social behaviour in birds and mammals*. Academic Press, London.

———— and Gartlan, J. S. (1966). Evolution of primate societies. *Nature, Lond.* 210, 1200–1203.

Cullen, E. (1957). Adaptations in the kittiwake to cliff-nesting. *Ibis* 99, 275–302.

Ellefson, J. O. (1968). Territorial behaviour in the common white-handed gibbon, *Hylobates lar*. In *Primates* (ed. P. C. Jay). Holt, Rinehart, and Winston, New York.

Hamilton, W. D. (1964). The genetical evolution of social behaviour. *J. Theor. Biol.* 7, 1–52.

Hinde, R. A. (1953). A possible explanation of paper-tearing behaviour in birds. *Br. Birds* 46, 21–3.

———— (1956). The biological significance of the territories of birds. *Ibis* 98, 340–69.

Hooker, T. and Hooker, B. I. (1969). Duetting. In *Bird vocalizations* (ed. R. A. Hinde). Cambridge University Press.

Kruuk, H. (1964), Predators and ant-predator behaviour of the Black-headed Gull (*Larus ridibundus* L.). *Behaviour, Suppl.* 11, 1–130.

———— (1972). Surplus killing by carnivores. *J. Zool. Lond.* 166, 233–44.

Lack, D. (1939). The behaviour of the robin: I and II. *Proc. zool. Soc. Lond.* A109, 169–78.

———— (1947). *Darwin's finches.* Cambridge University Press.

———— (1966). *Population studies of birds.* Clarendon Press, Oxford.

———— (1968). *Ecological adaptations for breeding in birds.* Methuen, London.

Marler, P. (1968). Aggregation and dispersal: two functions in primate communication. In *Primates* (ed. P. C. Jay). Holt, Rinehart, and Winston, New York.

Patterson, I. J. (1965). Timing and spacing of broods in the black-headed gull, *Larus ridibundus. Ibis* 107, 433–459.

Rowell, T. E. (1966). Hierarchy in the organization of a captive baboon group. *Anim. Behav.* 14, 430–443.

Simpson, M. J. A. (1968). The display of the Siamese fighting fish, *Betta splendens. Anim. Behav. Monogr.* 1, No. 1.

Stokes, A. W. (1962). Agonistic behaviour among blue tits at a winter feeding station. *Behaviour.* 19, 208–218.

Struhsaker, T. T. (1969). Correlates of ecology and social organization among African cercopithecines. *Folia. Primat.* 11, 80–118.

Tinbergen, N. (1942). An objective study of the innate behaviour of animals. *Biblthca Biotheor.* 1, 39–98.

Tinbergen, N. (1956). On the functions of territory in gulls. *Ibis* 98, 401–411.

———— (1959). Comparative studies of the behaviour of gulls (Laridae): a progress report. *Behaviour* 15, 1–70.

———— (1965). Behaviour and Natural Selection. In *Ideas in modern biology* (ed. J. A. Moore). *Proc. int. Congr. Zool.* 6, 521–42. New York.

———— (1967). Adaptive features of the black-headed gull, *Larus ridibundus. L. Proc. Int. orn. Congr.* 14, 43–59.

———— (1968). Book review. *Anim. Behav.* 16, 398–399.

————— Broekhuysen, G. J., Feekes, F., Houghton, J. C. W., Kruuk, H., and Szulc, E. (1962). Egg shell removed by the black-headed gull, *Larus ridibundus*, L.; a behaviour component of camouflage. *Behaviour* 19, 74–117.

————— Kruuk, H., and Paillette, M. (1962). Egg shell removed by the black-headed gull *Larus r. ridibundus*: II. *Bird Study* 9, 123–131.

Wickler, W. (1961). Über die Stammesgeschichte und den taxonomischen Wert einiger Verhaltensweisen der Vögel. *Z. Tierpsychol.* 18, 320–342.

16

Functional Analysis and Proper Functions

Paul E. Griffiths

1 Introduction

Etiological theories of "proper functions" have become increasingly popular in recent years. Neander (1991) has effectively defended the approach against the criticisms directed by Boorse (1976), Prior (1985) and Bigelow and Pargetter (1978) against the early etiological theory of Wright (1973), (1976). Millikan (1984), (1989b) has developed a complex and ingenious semantic theory using an etiological account of function. The analysis of proper function offered here differs from other etiological theories in two ways. First, it relates the etiological approach to Cummins' (1975) general account of function ascription. Second, it is sensitive to the real form of selective explanations in biology.

In section 2, I discuss Cummins' account of the nature and purpose of function ascriptions. Cummins is primarily concerned with the role of functions in the explanations of complex capacities. He is not concerned to distinguish teleological or "proper" functions from other, non-teleological functions. This has led authors such as Millikan (1989a, pp. 293–5) to think that Cummins' analysis does not contribute to the understanding of the "proper functions" of biological items and human artifacts. Millikan notes that many "Cummins-functions" are not proper functions. (Any function which derives from a functional analysis of a system is a Cummins-function. The Cummins-function of an item is always relative to the overall capacity under analysis.) Conversely, some proper functions are not Cummins-functions. These objections are correct, but they are not to the point. I show in section 3 that the proper functions of a biological trait are the functions it is assigned in a Cummins-style functional explanation of the fitness of ancestral bearers of the trait. The adequacy of teleological explanations given using proper functions depends on the validity of these earlier functional explanations.

In section 4 I discuss some shortcomings of Neander's (1990) analysis of proper function. The analysis is not adequate as it stands because of a mistake concerning the forms of selective explanation that back ascriptions of biological function. My own analysis, presented in section 5, is closer to some versions of Millikan's. It differs, however, because my central concern is to define a notion adequate to the purposes of biology; Millikan's concern is primarily with semantics. Thus, for example, Millikan does not distinguish vestigial traits from those which currently have functions. I discuss the distinction between functional traits and vestiges in section 6.

In the final sections of the chapter I give an account of design processes which allows me to extend my analysis to the proper functions of artifacts, and suggest a general account of naturalistic teleology.

2   Functional Analysis

Cummins argues that the practice of assigning functions to parts of systems derives from an explanatory strategy which he calls "functional analysis." Functions are assigned when analysing a complex capacity into a set of simpler capacities that are to be explained by subsumption under laws. The function of an item is its contribution to the overall capacity. The overall capacity is explained in terms of the contributing capacities, or functions, of parts of the system.

Cummins (1983) is concerned to show that functional analysis is the basic form of explanation in psychology (the form of explanation described by Lycan 1981, 1987 as the "Homuncular" strategy). But functional analysis has application wherever the aim of scientific investigation is to explain the overall capacities of a complex system. One complex capacity which might be explained by functional analysis is the ability of an animal to survive and reproduce. This can be analysed into a set of simpler capacities, such as the capacity to move about, to feed, to escape predation, to mate, and so on. Each of these can in turn be analysed into even simpler capacities. In the case of feeding, the ability to ingest food, masticate it, break it down into simple nutrients, to absorb these, and so forth. These capacities in turn can be analysed into still simpler capacities, arriving eventually at such simple capacities as that of a membrane to permit diffusion of some substance. These base level capacities are directly explicable by physical laws. Each capacity at each level can be attached to some subsystem of the organism. The function of this subsystem is the capacity which it realizes, and which contributes to the overall capacity of the organism.

A similar strategy can be applied to artifacts. Artifacts have capacities, often very complex ones, which suit them for their intended uses. The various parts of the artifact have functions which contribute to this overall capacity.

The overall capacity explained by functional analysis need not be one that will yield so called "proper" functions. The human body has capacities to die of various diseases. Each of these complex capacities can be analysed in the way Cummins suggests, yielding some very strange functions for the various body-parts. Similarly, Cummins' strategy can be applied to a bit of dirt which has become stuck in a pipe. We can regard this as a one-way valve, and use this assignment of function to explain the overall capacity of the pipe to control flow.

3   Proper Functions

Proper functions are the sorts of functions that biologists assign to the organs of animals, and the sorts of functions that human artifacts have. The notion is sometimes introduced by pointing out that proper functions are what things are *for*, whilst other functions are not. It is also possible to distinguish *having the function* F from merely *functioning as an* F. Both these linguistic distinctions can be used to mark a rough boundary for the class of proper functions.

Proper functions differ from other functions in that they can be cited to explain the presence of a functional item. The presence of the liver can be partially explained by its capacity to store glycogen and secrete bile. These functions enter into an evolutionary explanation of the presence of the liver. But the presence of the liver cannot be explained by its capacity to accommodate liver flukes. This is the Cummins-function of the liver relative to the capacity to die of fluke infestation, but it is not a proper function of the liver.

In the same way, the presence of a bit of dirt in a pipe cannot be explained by the fact that it functions as a one-way valve. This is not a proper function of the piece of dirt. The presence of one-way valves in human veins and in man-made pumps can be explained by the fact that they perform this function. It is a proper function of these features to act as one-way valves.

Explanations which cite an item's proper functions are philosophically interesting because they are teleological. The existence and form of an item seems to be explained by its goal, or purpose, rather than its antecedent causes. The etiological approach to proper functions is an attempt to demystify these teleological explanations. On this view to ascribe a proper function to an item is to claim that earlier items of the same type had the effect which we now label a proper function and that their having had that effect helps explain the presence of later items of the type.

The etiological approach was inspired by the attempts of evolutionary biologists to explain their use of teleology. In evolutionary biology it is natural to interpret the claim that a trait of an animal has the proper function F as the claim that F is the property in virtue of which the trait evolved. Lorenz, for example, states explicitly that he intends his use of function locutions in ethology to be interpreted etiologically:

If we ask "What does a cat have sharp, curved claws for?" and answer simply "To catch mice with," this does not imply a profession of any mythical teleology, but the plain statement that catching mice is the function whose survival value, by the process of natural selection, has bred cats with this particular form of claw. Unless selection is at work, the question "What for?" cannot receive an answer with any real meaning (Lorenz 1963, p. 9).

We can incorporate the etiological approach into the Cumminsesque picture of function ascription. The proper functions of a biological trait are the functions it is ascribed in a functional analysis of the capacity to survive and reproduce (fitness) which has been displayed by animals with that feature. This means that a feature will have a proper function only if it is an *adaptation* for that function. The trait must have been selected because it performs that function.

This picture of proper functions gives them a role in two kinds of biological explanation. First, the biological fitness of a type of organism can be explained by Cumminsesque functional analysis. An organism's fitness is a measure of its overall capacity to survive and reproduce, relative to the capacities of competing types in the population. (The classical fitness of a trait is a measure of the average expected number of offspring of systems with that trait. The trait may be genotypic or phenotypic. It is relative fitnesses within a population that are of interest to population genetics, and fitness values are usually normalized so that the fittest trait in a population has the value 1.) The analysis of this capacity will reveal a number of "fitness components." Fitness components are those effects of traits which enhance the fitness of their bearers. They are the Cummins-functions of those traits relative to the overall capacity of the animal to survive and reproduce (fitness). The proper functions of a trait are those effects of the trait which were components of the fitness of ancestor. They are the effects in virtue of which the trait was selected, the effects for which it is an adaptation.

I have already noted that proper functions are characterized by their capacity to enter into a second kind of biological explanation, the teleological explanation of the presence of certain traits. Proper functions can be used in this second kind of explanation precisely because they figure in the first kind of explanation (though they need not figure there as *proper* functions). The proper functions of traits are those effects for which they are adaptations. To explain a trait by alluding to its proper function is to explain it as the result of natural selection, in the way with which we are all familiar.

4 Current Etiological Theories

The main etiological accounts currently on offer are those of Millikan (1984), (1989a), (1989b) and Neander (1991). (Neander's work is well known to those working in the field. See Prior 1985, and Lycan 1987.) Millikan's main aim is to give an account of intentionality in terms of the proper functions of systems, such as people, which contain and produce representations, and the derivative functions of those representations themselves. It is impossible to do justice here to the elaborate system that Millikan devises to this end. She herself gives a sketch of the central element of her theory as follows:

very roughly, for an item A to have a function F as a "proper function," it is necessary (and close to sufficient) that ...[1] A originated as a "reproduction" (to give one example, as a copy, or a copy of a copy) of some prior item or items that, due in part to possession of the properties reproduced, have actually performed F in the past, and A exists because (causally, historically because) of this or these performances.

This captures the central element of the etiological approach, the idea that the effects of past tokens of a type provide an explanation for the existence of current tokens of that type, but it does not spell out exactly which explanations in biology support function ascriptions. For example, this analysis and the more formal analysis of "direct" proper function given elsewhere (Millikan 1984, p. 28) make no distinction between currently functional traits and vestiges of past adaptations. I am inclined to take Millikan at her word when she describes her analysis as a tool forged for a specific job in the philosophy of language (1984, p. 18). My own interest is in deriving a notion of proper function adequate to the purposes of biology. The two accounts are certainly in sympathy, but there are distinctions, such as that between functional traits and vestiges, which are important to biology, and which Millikan has only minimally gestured at (e.g. 1984, p. 32).

Neander's account, on the other hand, is explicitly intended to capture the use of function ascriptions in biology. It is conceptual analysis, but of the concepts of current biologists, not of ordinary people. According to Neander:

It is a/the proper function of an item (X) of an organism (O) to do that which items of X's type did to contribute to the inclusive fitness of O's ancestors and which caused the genotype, of which X is the phenotypic expression (or which may be X itself where X is the genotype) to increase proportionally in the gene pool (Neander 1991).

This analysis fails to capture the notion of function in use in modern biology because of the requirement for proportional increase. It is commonplace for a population to contain several competing traits, whose proportions vary. In such circumstances there are many legitimate selective explanations that do not explain the proportional increase of a trait in the population. They explain the current level, which is often the result of a recent decrease. So the function of some traits is to do that which has led to their representation at a reduced level in the population. I shall give detailed examples of this in a moment, after dismissing some possible replies to the general point.

Neander might reply that whatever proportion of the population currently display a trait, that trait must have originated in one or a few individuals, and spread because of its adaptive value. So her requirement for proportional increase may be taken to refer to the spread of the trait after its initial introduction. But this reply would leave her unable to distinguish vestigial traits, such as the appendix, which have lost their functions. In fact, Neander makes it quite clear that "the etiological theory looks back to the recent evolutionary past" and that functional traits are distinguished from vestiges by the fact that selection for them has occurred fairly recently.

Neander might say that the function of a trait is the effect which figured in the most recent episode in which it did proportionally increase. But this will not do either. Traits arise and spread for non-adaptive reasons, perhaps as side-effects of adaptive traits, and only later acquire a function. It is also common for traits to lose old functions and acquire new ones. Neither of these cases could be accommodated on the current proposal unless each acquisition of function were accompanied by an increase in the proportion of the trait in the population. It can easily be shown that this need not occur.

Darwin (1892) describes a scenario in which the utilitarian function of certain facial expressions declines, whilst their importance in intraspecific communication increases. Tooth-baring in certain primates evolved as a preparation for attack. It later acquired the function of expressing anger. Finally, in man, it has become purely expressive and is vestigial with respect to its original function. (The notion of vestigiality relative to a particular function is formally defined in section 6). There is no reason to assume that the acquisition of the new function must have increased its prevalence in the population.

There are also standard scenarios in evolutionary theory where a trait acquires a new function whilst the proportion of the population with the trait actually decreases. This happens in versions of the hawk/dove game (Maynard-Smith 1982). Imagine a population in which all resource disputes are settled by conventional posturing. Any animal subject to a serious attack retreats. This is the "Dove" trait. A new trait is introduced by migration or mutation. Animals with the new trait, "Hawks," attack and defend resources to an extent where serious injury can occur. Selection between these two traits is frequency dependent. Under certain assumptions about costs and pay-offs the following picture holds. When Doves are common, Hawks are fitter than Doves, as they can capture a large share of resources with little risk. However, as the proportion of Hawks increases, the cost of being a Hawk also increases. If the proportion of Hawks rises above a certain level, Doves become fitter than Hawks, as their reduced risk of injury outweighs any loss of resources. Under certain assumptions, the proportion of Hawks to Doves will settle at an equilibrium level such that any proportional increase in Hawks would reduce the fitness of Hawks below that of Doves.

The persistence of the Dove trait in the population is explained by the selective advantage that accrues to Doves in virtue of their avoiding injury. The exact proportion of Doves at any generation can be explained by the relative fitness of Hawks and Doves in past generations. Given the underlying rationale for the etiological approach, the conclusion is clear. In organisms whose ancestors competed with Hawks, one of the proper functions of Dove behaviour is to reduce the chance of injury. But Dove behaviour acquired this function whilst the proportion of Doves in the population decreased. Neander's requirement for proportional increase is inappropriate.

5   A New Etiological Theory

The etiological theory can avoid these pitfalls if reformulated:

Where i is a trait of systems of type S, a proper function of i in Ss is F iff a selective explanation of the current non-zero proportion of Ss with i must cite F as a component in the fitness conferred by i.

This analysis avoids the assumption of proportional increase and differs from Neander's analysis in two other, relatively minor ways. First, it is phrased in terms of classical fitness. It is not always realized that classical fitness includes kinship effects (Grafen 1982). The considerably more complex notion of inclusive fitness simply makes these effects more perspicacious. Second, it leaves implicit various points about the gene/phenotype relation. The analysis indexes any function to a class of systems. In many cases the system will be the organism which bears the trait in question, but there are two alternative cases.

First, there are cases where genes have evolved adaptations which benefit only the gene itself. Certain genes have the capacity to subvert the normal mechanisms of cell division so as to ensure their over-representation in the sex cells. These genes are more highly prevalent than they would otherwise be. In one of the best documented cases (Lewontin and Dunn 1960), this allows an allele to survive despite disastrous effects on organisms which are homozygous for the allele. In such cases the advantageous trait of the gene must be assigned functions relative to the gene itself, not the organism at whose expense it survives. Speculations on "intragenomic conflect" between nuclear and mitochondrial DNA (Cosmides and Tooby 1981) suggest that phenotypic features may also have functions relative to a particular segment of DNA, rather than the entire organism.

Second, there are cases where a trait of one phenotypic individual has a function for another individual. Dawkins (1982) has drawn attention to this phenomenon under the slogan of the "extended phenotype." The galls which grow on oak trees are the product of selection acting on the genes of gall-wasps, not the genes of oak trees. They have a function for the wasp but are functionless for the tree. Both these cases and the genic cases can be accommodated on my account, simply by choosing the right class of systems S.

Neander has pointed out to me that allowing extended phenotypic traits to have functions lets in some very strange cases. The items of food which come into an animal's possession are assigned the function of feeding the animal, for example, because there is a selective explanation of why current animals have these items of food which cites feeding the animal as the effect by which the bits of food possessed by ancestors enhanced the ancestors fitness. Although my analysis could be reformulated so as to exclude these cases, I do not wish to do so because the selective stories which back them seem sound, and because I can see no way of excluding them which would not also exclude such highly plausible claims as the claim that the function of the discarded mollusc shells possessed by hermit crabs is to protect them from predators.

6   Vestigial Traits

Any theory of proper functions needs to distinguish currently functional traits from vestigial traits. The idea of vestigiality is linked to the notion of regressive evolution. This is the process whereby useless traits tend to atrophy because of the costs of producing them. Perhaps the most striking example is the evolution of subterranean forms "troglodytes." Cave-dwelling populations from many different taxonomic groups display a characteristic pattern of evolution. As well as losing pigmentation, such animals characteristically possess vestigial eyes, reduced in size, complexity and effectiveness.

Whilst the classic examples of vestigiality are atrophied in this way, not all vestiges need be so. There are two obvious cases. First, a trait may change its function but be preserved intact in virtue of its performing a new function. Darwin regards facial expressions as vestiges relative to their original, utilitarian functions, although they are currently selected for their value in intraspecific communication. Second, a useless trait may subsist for an extended period of time simply because there is no genetic variation. Evolutionary forces are not platonic entities, and where there is no variation there are no selective forces.

A successful account must allow non-atrophied vestiges. But it will not do to make every trait that cannot perform its functions vestigial. First, as Neander has pointed out, a malfunctioning trait cannot perform its proper function. But it must still have the function in order to count as malfunctioning.[2] Second, it is important for traits to be able to have proper functions they cannot perform in order to prevent functions fluctuating wildly in response to temporary environmental changes.

It is possible to allow for non-atrophied vestiges without objectionably reifying selective forces, and without classifying every trait which ceases to function as vestigial. Define an evolutionarily significant time period for a trait T as a period such that, given the mutation rate at the loci controlling T and the population size, we would expect sufficient variants for T to have occurred to allow significant regressive evolution if the trait was making no contribution to fitness. A trait is a vestige relative to some past function F if it has not contributed to fitness by performing F for an evolutionarily significant period. A trait is a vestige simpliciter if it is a vestige relative to all its past functions. This account allows a trait to become a vestige relative to one function whilst remaining intact in virtue of its other functions. It also allows a trait to become vestigial whilst remaining intact because of a lucky absence of mutations. Interestingly, it makes it possible for the trait to be a vestige in one population but not in another. This seems to accord with biological usage.

The functional trait/vestige distinction is not made explicit in the analysis given above. The current prevalence of the human appendix, for example, can be given an expanation that involves selection. There were, presumably, distant ancestors for whom the appendix was an adaptation. It is possible to express it by saying that it *had* this function. If it is to have a function at the ancestors. But it is more natural and sits better with the notion of vestigiality to express it by saying that it had this function. If it is to have a function at the present time, selection for it must have occurred in the last evolutionarily significant time period. We might incorporate this into the definition by defining a *proximal* selective explanation as one that involves the action of selective forces during the last evolutionarily significant period, or would have involved such action during that period had the mutation rate not fallen below expectation.[3] The functional trait/vestige distinction can then be made explicit:

Where i is a trait of systems of type S, a proper function of i in Ss is F iff a proximal selective explanation of the current non-zero proportion of Ss with i must cite F as a component in the fitness conferred by i.

7   Artifact Functions

The other main class of objects that have proper functions are human artifacts. A corkscrew is for removing corks. Damaged, malformed and badly designed corkscrews retain this function although they cannot perform it. Most importantly, the existence and form of artifacts can be explained by alluding to their functions. Why are there so many cork-screws? They are for removing corks. Why are they shaped thus and so? In order to better remove those corks.

The teleological properties of artifacts are often though to be unproblematic. Before presenting my own analysis of artifact functions, I want to show that this is a mistake. Providing an adequate account of artifact teleology is not a trivial task.

A first suggestion might be that the functions of artifacts are their intended uses. In support of this it might be said that artifacts can be explained by their functions because they have been designed to fulfil their intended uses. But it is not just artifacts as wholes that have proper functions. Their parts and features have proper functions too. Nearly every detail of a car, down to the last shim, has a function. Something has to be said about how to determine these functions from the overall function.

One obvious way would be to perform a Cummins-style functional analysis of the car's ability to fulfil its intended uses. The parts of the car contribute to its capacity to fulfil its intended use. These contributions are their functions. But not all functions which come out of this analysis will be proper functions. Some items may make additional, accidental contributions to the car's capacity to perform its intended use. These contributions cannot be used to explain the items, and are not proper functions.

This problem can be avoided by saying that the function of every part is its *intended* contribution to the overall use. But this solution leads straight into another difficulty. Many features of artifacts make no intended contribution and yet have proper functions. In societies with low-level technologies, artifacts are often designed by trial and error over periods of many generations. The contribution that a feature makes to the performance of overall function may never be appreciated. Such features do not have an intended use but they do have functions, and they can be explained by their functions. Various shapes of a tool are tried out. One is more effective than the others because, for example, it is better balanced. This shape is copied more than the others and eventually becomes the norm in that culture. The function of the shape is to balance the tool and the occurrence of the shape can be explained by the fact that it balances it. The shape retains this function even when it cannot perform it. A badly made or broken tool still has the shape, and the functional explanation can still be given. The shape can even become vestigial, when the tool is reproduced in jade as a votive object. The shape's function is, I contend, a proper function.

The account of artifact function I offer here is able to avoid the problems just discussed. It does not assign proper functions to artifact traits which accidentally contribute to intended uses. Conversely, it does assign proper functions to artifact traits that arise as a result of trial and error.

8   Etiology and Artifacts

Artifact functions can be handled in a manner analogous, although not identical, to my treatment of biological functions. The etiological account can be extended to artifacts because human selection does for artifacts what natural selection does for organisms. The prevalence of an artifact, or an artifact trait, can be explained by selective processes in which people meet their needs, sometimes by conscious design, sometimes by trial and error, and sometimes by an amalgam of the two.

The extension of my account requires a "selection type theory" (Darden and Cain 1989) of the processes that give rise to artifacts. I exploit the fact that these processes share certain features with natural selection to construct selective explanations for the features of artifacts. It turns out that this can be done in such a way that the formal analysis given above can be applied directly to artifact functions.

Although artifacts are not in actual competition with other artifacts during the design process, they are in hypothetical competition. The designer conceives a range of alternatives and chooses amongst these in virtue of their perceived possession of certain capacities. Some are more capable of performing the intended use than others. As in biological competition, there is only a certain range of available alternatives. Maori canoe designers did not have to consider the idea of the winged keel. It was not part of the "population" of designs which were in competition at that time.

The "fitness" of an artifact or artifact trait is a rather vaguer notion than the fitness of a biological system or trait. It is still the propensity of the system or trait to be reproduced, relative to the alternatives. (Note that the fitness of an artifact is its propensity to be reproduced, not its efficiency in fulfilling its intended use.) But this class of alternatives will be hard to specify once hypothetical selection is admitted, and accurate measures of fitness will not be possible. However, as I remarked above, some artifacts and artifact traits are not consciously designed but occur through trial and error. In such cases the selection process which has produced the current design has involved selection between actual alternatives in virtue of actual performances. (An explanation of the spread of iron weapons and farming gear might take this form.) In such cases the class of alternatives will be easier to specify, and a more accurate assessment of fitness may be possible.

The other major element of my account of biological function, the theory of vestiges, also has its artifact analogue. There are many vestigial artifact traits, from those elements of classical stone architecture derived from wood construction to Maori "fish-hook" pendants. They are vestiges, not merely because they cannot perform their original function but because they have not been selected in virtue of their original function for so long that they would have been eliminated if they had not acquired a second, decorative function.

There is one major superficial difference between biological functions and artifact functions. The overall capacity to which biological functions contribute is the animal's fitness—its relative capacity to survive and reproduce. The overall capacity to which artifact functions contribute is usually thought to be the capacity to perform the intended use. If the design of artifacts is to be assimilated to natural selection in the way described, this difference will have to be overcome. I suggest that the function of an artifact is its intended use only because its ability to fulfil its intended use gives it a propensity to be reproduced. The overall capacity to which the proper functions of artifacts contribute is the capacity to be reproduced, but they contribute to this capacity via the capacity to fulfil the intended use.

When we say of a whole artifact "this is what it is for," we refer to a *penultimate* level capacity, realized by the general configuration of the artifact. There may be several such capacities, as an artifact can have more than one intended use. The ultimate capacity which gives rise to functions is the capacity to be reproduced, just as it is with biological systems. An analogue to the overall function of an artifact might be the ability of an animal to occupy its niche. What is a wolf for? To be an effective predator in temperate climates.

The subordination of intended use to this secondary role might seem to give rise to a difficulty. If an artifact's function is its intended use, it can have a function which neither it nor any ancestor has ever performed. Consider the tapered tail of an old racing car. This feature is intended to streamline the car, to reduce its drag coefficient. But it does not do this, and nor have any of the other designs it has "evolved" from. They are all based on a false theory about drag. It is hard to see how there could be a selective explanation of a trait of this kind. There have, it seems, been no episodes when it has performed well, and hence been selected.

This type of situation cannot arise with biological functions. Natural selection can only operate on a trait in virtue of something it actually does. A trait may cease to perform its function, but it must at one time have performed it.

The solution to this difficulty is actually implicit in what has already been said. I have shown how to create a more abstract notion of "selective process" by allowing selection amongst hypothetical alternatives. This selection amongst hypothetical alternatives occurs in a hypothetical environment constituted by the beliefs of the designer. When the designer has false beliefs about the real world this results in artifacts functioning well in his hypothetical environment when they do not function in the real environment. The tail of the facing car did perform its function, but only in the mind of its designer.

9 Conclusion

There are considerable differences between the kinds of selective processes that give rise to artifact teleology and those that give rise to biological teleology. Artifact functions are not just the biological functions of inanimate objects! This can be seen clearly by considering the contrasting artifact and biological functions of traits of selectively bred animals. Suppose pigeon breeders select a long tail because they falsely believe that it will make the pigeon fly faster. Its artifact function will be to make the pigeon fly faster. Its biological function, however, must be a property it actually has, since only actual properties can be subject to natural selection. In this case, its biological function will be to fool people into thinking it is useful, just as the biological function of the yellow stripes on a harmless insect is to fool other organisms into thinking it dangerous!

So I cannot offer a unified theory of proper functions. But I can offer a unified theory of naturalistic teleology. Both kinds of proper function derive their teleological force form the contribution of past performances of function to reproduction. In the case of artifact teleology, however, these performances are frequently hypothetical, and sometimes occur in an unrealistic hypothetical environment.

There may be other types of process that give rise to teleology supporting functions. Darden and Cain (1989) have pointed out analogies between biological evolution and other "selection type theories," such as the clonal selection theory of antibody production. I believe it is to be a general characteristic of such theories that they allow the prevalence and form of selected traits to be explained by the effects of these traits on survival and reproduction. They will, therefore, give rise to proper functions. But there should be nothing surprising in the fact that wherever there is selection, there is teleology.

Notes

1. Millikan offers a pair of disjoint conditions, but the second condition is satisfied by "derived proper functions" and need not trouble us here.

2. Non-etiological accounts of proper function can try to incorporate this feature by appealing to "normal" circumstances. Millikan (1989a) has suggested that this involves an implicit reference to evolutionary history. Neander (1991) has pointed out that it fails to handle pandemic diseases, such as the viral infections of some plants.

3. I am grateful to Karen Neander for pointing out the necessity of this last clause. It excludes counterexamples parallel to those which led me to include a probabilistic element in my definition of a vestige. A trait might be thought to be currently contributing to fitness although it is not being selected because of an improbable absence of mutations.

References

Bigelow, J. and Pargetter, R. 1987: "Functions," *Journal of Philosophy*, LXXXIV, pp. 181–96.

Boorse, C. 1976: "Wright on Functions," *Philosophical Review*, LXXXV, pp. 70–86.

Cosmides, L. M. and Tooby, J. 1981: "Cytoplasmic Inheritance and Intragenomic Conflict," *Journal of Theoretical Biology*, 89, pp. 83–129.

Cummins, R. 1975: "Functional Analysis," *Journal of Philosophy*, LXXII, pp. 741–765.

Cummins, R. 1983: *The Nature of Psychological Explanation*. Cambridge, MA: Bradford Books/MIT Press.

Darden, L. and Cain, J. A. 1989: "Selection Type Theories," *Philosophy of Science*, 56, pp. 106–129.

Darwin, C. 1872: *The Expression of Emotions in Man and Animals*. New York. Philosophical Library, 1955.

Dawkins, R. 1982: *The Extended Phenotype*. San Francisco, CA: Freeman.

Gould, S. J. and Lewontin, R. 1979: "The Spandrels of San Marco and the Panglossian Paradigm: A Critique of the Adaptationism Programme," *Proceedings of the Royal Society of London*, 205, pp. 581–598. Reprinted in Elliott Sober (ed.) 1984: *Conceptual Issues in Evolutionary Theory: An Anthology*. Cambridge, MA: Bradford Books/MIT Press.

Grafen, A. 1982: "How not to Measure Inclusive Fitness," *Nature*, 298, pp. 425–426.

Lewontin, R. and Dunn, L. C. 1960: "The Evolutionary Dynamics of a Polymorphism in the House Mouse," *Genetics*, 45, pp. 702–722.

Lorenz, K. 1963: *On Aggression*. London: Methuen, 1968.

Lycan, W. G. 1981: "Towards a Homuncular Theory of Believing," *Cognition and Brain Theory*, 4, pp. 139–157.

Lycan, W. G. 1987: *Consciousness*. Cambridge, MA: Bradford Books/MIT Press.

Maynard-Smith, J. 1982: *Evolution and the Theory of Games*. Cambridge: Cambridge University Press.

Millikan, R. 1984: *Language, Thought, and Other Biological Categories: New Foundations for Realism*. Cambridge, MA: Bradford Books/MIT Press.

Millikan, R. 1989a: "In Defense of Proper Functions," *Philosophy of Science*, 56, pp. 288–302.

Millikan, R. 1989b: "Biosemantics," *Journal of Philosophy*, LXXXVI, pp. 281–297.

Neander, K. 1991, "Functions as Selected Effects: The Conceptual Analyst's Defense." *Philosophy of Science*, 58, pp. 168–184.

Prior, E. W. 1985: "What is Wrong with Etiological Accounts of Biological Function?" *Pacific Philosophical Quarterly*, LXVI, pp. 310–328.

Wright, L. 1973: "Functions," *Philosophical Review*, 82, pp. 139–168.

Wright, L. 1976: *Teleological Explorations*. Berkeley, CA: University of Calif. Press (Berkeley) 1976.

17

A Modern History Theory of Functions

Peter Godfrey-Smith

I Introduction

Biological functions are dispositions or effects a trait has which explain the recent maintenance of the trait under natural selection. This is the "modern history" approach to functions. The approach is historical because to ascribe a function is to make a claim about the past, but the relevant past is the recent past; modern history rather than ancient.

The modern history view is not new. It is a point upon which much of the functions literature has been converging for the best part of two decades, and there are implicit or partial statements of the view to be found in many writers. This paper aims to make the position entirely explicit, to show how it emerges from the work of other authors, and to claim that it is the right approach to biological functions.

Adopting a modern history position does not solve all the philosophical problems about functions. It deals with a family of questions concerning time and explanation, but there are other difficulties which are quite distinct. The most important of these concern the extent to which functional characterization requires a commitment to some form of adaptationism (Gould and Lewontin 1978). These issues will not be addressed here. Further, as many writers note, "function" is a highly ambiguous term. It is used in a variety of scientific and philosophical theories, several domains of everyday discourse, and there is probably even a plurality of senses current within biology. This paper is concerned with one core biological sense of the term, which is associated with a particular kind of explanation. In this sense a function has some link to an explanation of why the functionally characterized thing exists, in the form it does.

Cummins (1975) argued that functions are properly associated with a different explanatory project, that of explaining how a component in a larger system contributes to the system exhibiting some more complex capacity. Following Millikan (1989b) I suggest that both kinds of functions should be recognized, each associated with a different explanatory project. If it is claimed, for instance, that the function of the myelin sheaths round some brain cells is to make possible efficient long distance conduction of signals, it may not be obvious which explanatory project is involved—that of explaining why the sheath is there, or that of explaining how the brain manages to perform certain tasks. Often the same functions will be assigned by both approaches, but that does not mean the questions are the same.

The aim of this chapter is to analyze an existing concept of function, which plays a certain theoretical role in biological science. So the aim is a certain sort of conceptual analysis, a conceptual analysis guided more by the demands imposed by the role the concept of function plays in science, the real weight it bears, than by informal intuitions about the term's application. Also, though I will defend the modern history view within the context of a particular theory of functions which draws on the work of Larry Wright and Ruth Millikan, the overall value of the modern history approach stands independently of many of the details of my theory.[1]

II The Wright Line

Our point of departure is a simple formula proposed by Larry Wright in 1973 and 1976: "The function of X is that particular consequence of its being where it is which explains why it is there" (1976 p. 78). That is:

The function of X is Z iff:

i. Z is a consequence (result) of X's being there,

ii. X is there because it does (results) in Z. (1976 p. 81)

Wright argued that his theory dealt with a broad range of cases, handling both the functions of artifacts and biological entities without significant modification. The function of spider webs is catching prey, because that's the thing they do that explains why they are there; the function of tyre tread is improving traction because that's also the thing it does that explains why its there; and the function of the newspaper under the door is to prevent a draft, for the same reason.

However, Wright's analysis covers more cases than these. Boorse (1976) notes that when a scientist sees a leak in a gas hose, but is rendered unconscious before it can be fixed, on Wright's schema the break has the function of releasing gas. The break is there because it releases gas, keeping the scientist immobilized, and the leaking gas is a consequence of the break in the hose. Similar cases take us even further from the plausible realm of purpose. One might see a small, smooth rock supporting a larger rock in a fast-flowing creek, and note that if it did not hold up that larger rock, it would be washed away, and no longer "be there." But it is not the function of the small rock to support the larger one. The problem here is with the broad range of "X" and "Z", with the need to restrict the kinds of things to which the schema can be applied. A restriction of this kind is a key component of Ruth Millikan's theory (1984, 1989a).

Before moving on however, it is important to recognize Wright's aims. Wright's strategy is to avoid convoluted analysis by trusting many details to pragmatic factors which will apply case by case. For Wright, function hinges directly on explanation, and explanation is pragmatically sensitive in a multitude of ways. There is a sense in which Wright's theory is not an "analysis" of function in the sense that earlier accounts are. Earlier writers were largely concerned with how it can ever be that something's existence can be inferred from its function, given that other things could often have done the same job (Hempel 1965). Without this inference, it was thought there could be no functional explanation. Wright simply insists that with a less demanding, more realistic picture of explanation, it becomes clear that people do explain the presence of things in terms of what they do, and a function is any effect that operates in such an explanation.

Wright also hopes, I suspect, that some natural slack in the notion of function will be mirrored and explained by corresponding slack in the notion of explanation, that the analysis will bend where the concept analyzed naturally bends. Wright's vague formulation of the relevant explanandum—"why its there"—is intended to wrap unsystematically around a variety of explanatory projects, in biology, engineering and everyday life. Nonetheless, counterexamples such as Boorse's do suggest that Wright has backed off too early, and a sensitivity to pragmatics should not prevent us from pushing an analysis as far as we profitably can.

Millikan's analysis, like Wright's, is historical. It locates functions in actual selective histories. The most important sophistication of the historical approach in Millikan (1984) is her detailed treatment of functional *categories*. The first concept she defines is that of a "reproductively established family." A reproductively established family is a group of things generated by a sort of copying. Family members can be copied one off the other, or be common copies off some template, or be generated in the performance of functions by members of another family. These different kinds of copying are all distinguished by Millikan, but the finer divisions are not important here. Call any entities which can be grouped as tokens of a type by these lines of descent by copying, members of a "family." Understand "copy" as a causal matter involving common properties and counterfactuals. The copy is like the copied in certain respects, though it is physically distinct, and if the copied had been different in certain ways, then, as a consequence of causal links from copied *to* copy, the copy would have been different in those ways too (1984 p. 20).[2] So two human hearts are members of the same family, as are two frill-necked lizard aggressive displays, two AIDS viruses, and two instances of the acronym "AIDS," assuming that acronym was hit upon only once. But two planets, and two time-slices of a rock or hose are not, as one was not copied off the other, nor are they produced off a common template, and so forth. Functions are only had by family members, and the performance of a function must involve the action of one of the properties copied, one of those properties defining the family.

This restriction deals with many of Boorse's counterexamples, such as the gas hose case. It also removes from the realm of function some cases Wright was concerned to capture, such as the newspaper under the door. However, our project here is to capture the biological usage. Preserving a continuity between biological cases and other domains can be sacrificed.[3]

The next step is to add to this an explanation-schema in the style of Wright. The explanandum is the existence of current members of the family. The explanans is a fact about prior members.

(F1) The function of *m* is to *F* iff:

i. *m* is a member of family *T*, and

ii. among the properties copied between members of *T* is property or property cluster *C*, and

iii. one reason members of *T* such as *m* exist now is the fact that past members of *T* performed *F*, through having *C*.

Most simply, a family member's function is whatever prior members did that explains why current members exist (see also Brandon 1990 p. 188).

It is one of the strengths of the historical approach combined with an appeal to "families" that it can say without strain that some particular thing which is in principle unable to do *F* now, nonetheless has the function to do *F*. It has this function in virtue of its membership in a family which has that function. Whether this member can do *F* is irrelevant to its family membership, as long as it was produced by lines of copying that are generally normal enough. A genetic defect may produce a heart unable to ever pump blood, but if this token was produced in more-or-less the same way as others, it has the function characteristic of the family.

At this point we must confront an issue unrelated to history. It is striking that while analyses such as Wright's and Millikan's permit any activity or power explaining survival to qualify as a function, biologists apparently reserve "function" for activities or powers which are, in some intuitive sense, helpful and constructive. If being inconspicuous and avoiding attention by doing nothing is itself "doing something," then pieces of junk DNA, which sit idly on chromosomes and are never used to direct protein synthesis, have the function to do nothing. That is the thing past tokens of junk DNA types have done, which explains the survival of present tokens. If doing absolutely nothing is a behavior when an animal does it for concealment, why is it not something that junk DNA "does"? Perhaps the function of junk DNA is, alternatively, to be more expensive to get rid of than to retain. But biologists do not describe junk DNA like this; it is the paradigm of something with no function. Similarly, characters which hitchhike genetically on useful traits or persist through developmental inevitability (like male nipples) might, in extended senses, be "doing" things which lead to their survival. So we might consider making some restriction on the selective processes relevant to functional status.

This will not be easy. A simple requirement that the trait do something positive, that the null power is not a power, will not suffice. Beside the cases where biological entities persist through doing nothing, there are positive and selectively salient powers which seem unlikely candidates for functions. As well as junk DNA, which does nothing, there is "selfish DNA" (Orgel and Crick 1980). Selfish DNA can move around within the genome, replicating itself as it goes, and proliferate in a population despite having deleterious effects on individuals carrying it.

Similarly, segregation distorter genes disrupt the special form of cell division (meiosis) which produces eggs and sperm (gametes). Meiosis usually results in a cell with two sets of chromosomes giving rise to four gametes with one set each, and on average a particular type of chromosome will be carried by half the gametes produced. Segregation distorters lever their way into more than their fair half share of gametes, by inducing sperm carrying the rival chromosome to self-destruct as they are formed (Crow 1979). Fruit flies, house mice, grasshoppers, mosquitoes and a variety of plants are known to have segregation distorters in their gene pools. Now, disrupting meiosis is something that segregation distorter genes do, that explains their survival (Lewontin 1962). Further, this explanation appeals to natural selection, at the gametic level; the problem can not be solved by disqualifying traits that survive for non-selective reasons. Disrupting meiosis is not generally claimed to be the genes' *function* though. Should we restrict the powers which can become functions, to exclude these subversive cases?[4]

There are two attitudes we might have to this issue. First, as a question of conceptual analysis, there is not much doubt that biologists typically restrict the powers that can qualify as functions. Many might say we should then change the selective theory of functions to include this factor. An obvious move is to bring in some reference to the goals of some larger system. Disrupting meiosis makes no contribution to the goals of individuals bearing segregation distorter genes, so this is not a function.

An appeal to goals is certainly a step backwards however. So we might consider a more aggressive attitude to the problem. It may be that many biologists reserve "function" for powers with some intuitively benign nature, and withhold it from more subversive activities, with there being no theoretically principled reason for this distinction. Some hold that biology since the 1960s has produced, for better or worse, an increasingly cynical view of the coalitions that make up organisms (Dawkins 1982, Buss 1987), families (Trivers 1974), and larger groups (Williams 1966, Hamilton 1971). The feeling that functions must involve harmonious interactions may, from this point of view, be a holdover from an earlier, more truly teleological view of nature. It might be claimed that the theoretically important category of properties, the category our concept of function should be tailored to, is simply the category of selectively salient powers and dispositions.[5] If so, we should remain with the simpler analysis that allows any survival-enhancing power, however subversive, to qualify as a function.

Although some may favor this more heartless approach I will adopt a third, intermediate position. Consider first another counterintuitive consequence of an unembellished selective account: whole organisms, like people, have functions. Past tokens of people did things—survived and reproduced—that explain why current tokens are here. Hence, we have the function to survive and reproduce. This usage seems odd—note that these are not functions people might have with respect to some social group, they are functions people just have, individually. One way to exclude both people as bearers of functions and also exclude disruption of meiosis as a function of segregation distorters is to stipulate that (i) the functionally characterized structure must reside within a larger biologically real system, and (ii) the explanation of the selection of the functionally characterized structure must go via a positive contribution to the fitness of the larger system. My account here resembles that of Brandon, who requires that a functional trait increase the "relative adaptedness of [its] possessors" (1990 p. 188). Brandon requires not just selective salience, but selective salience which goes via the fitness of a larger system "possessing" the trait.

The catalog of "real systems" is taken from biology, and clarifying the catalog is part of the units of selection problem.[6] Individuals, kin groups and perhaps populations and species might be examples of these systems. Thus hearts reside within people, and survive by aiding people's fitness. But people, considered individually, reside within no such systems. There may, however, be groups within which people do things which contribute to the selection of the group, and then people would have functions.

Similarly, segregation distorter genes do not have the function of disrupting meiosis, because their proliferation under selection does not occur through a positive contribution to the fitness of individuals bearing these genes. Indeed, many segregation distorters, when present in two copies, greatly impede the fitness of their carriers. One the other hand, as some readers may have felt earlier, there could well be functional characterization of *parts* of segregation distorter genes or gene combinations. Some part of the gene or combination might have its current presence explained by the fact that it has been selected for carrying out some part of the segregation distortion project. Crow (1979) distinguishes two genes which cooperate to produce segregation distortion in fruit flies. The "S" gene produces sabotage in sperm, and the "R gene stops the chromosome that the S and R are on from sabotaging itself. So a chromosome with S but no R sabotages itself, and a chromosome with R but no S does not distort, but is immune to distortion by its rival. Here the segregation distorting chromosome is the larger system, and the selective explanation of S goes via the explanation of the success of the whole chromosome. S has the function of sperm sabotage, and it has this function with reference to the segregation distortion gene complex. The selection of R is only partly an explanation in terms of the selection of the distorter chromosome, as R is useful without S, once the population contains some chromosomes with S. So R has the function of preventing sabotage, and it has this function with reference to two larger units, the segregation distorter complex and the individual.

It is important that not all failures on the part of evolution to produce intuitively well-engineered animals disqualify selective episodes from bestowing functions. A question sometimes arises concerning the status of traits which are explained in terms of some forms of sexual selection. If it is true that sexual selection can operate through females favoring characteristics in males which have no other benefit or use to the male (Fisher 1930, Lande 1981), then the explanation of a bird's long tail is not an explanation in terms of anything intuitively useful the tail does. The explanation is simply that females prefer long tails (Andersson

1982). Once a female preference gets established, for any reason, it can be sustained and made stronger through the association of the gene for the preference in females (unexpressed in males) and the gene for the preferred trait (unexpressed in females). The preference leads to the selection of long tails, and the selection of long tails leads to the strengthening of the associated preference. The long tail could be a hindrance elsewhere in life. Consequently, some biologists hesitate to describe the tail as an adaptation, and functional in the ordinary sense: "Runaway sexual selection is a fascinating example of how selection may proceed without adaptation" (Futuyama 1986 p. 278). On the present account however the tail has the function to attract females. It has been selected because of that power, and this explanation goes via the augmentation of the individual's fitness.[7]

Here is an amended definition:

(F2) The function of *m* is to *F* iff:

i. *m* is a member of family *T*,

ii. members of family *T* are components of biologically real systems of type *S*,

iii. among the properties copied between members of *T* is property or property cluster *C*,

iv. one reason members of *T* such as *m* exist now is the fact that past members of *T* were successful under selection, through positively contributing to the fitness of systems of type *S*, and

v. members of *T* were selected because they did *F*, through having *C*.

III Looking Forward

Although philosophers have discussed a variety of intuitive problems with the view that functions derive from a selective history (Boorse 1976), the most damaging charge against this view derives from the biological literature, from the wide acceptance of the distinctions made in "Tinbergen's Four Questions."

It is common in ethology and behavioral ecology to distinguish four questions "why?" we can ask about behavior. Someone who asks why frill-necked lizards extend the skin around their necks so spectacularly might want an answer:

1. In terms of the physiological *mechanisms* and the physical stimuli that lead to the behavior.
2. In terms of the current *functions* of the behavior.
3. In terms of the evolutionary *history* of the behavior.
4. In terms of the *development* of the behavior in the life of the individual lizard.

This four-way distinction is usually attributed to Tinbergen 1963. Tinbergen in turn credits Julian Huxley with distinguishing questions 1–3, and adds question 4. Tinbergen, it must be admitted, uses the term "survival value" rather than "function" in the official formulation of question 2. But generally he uses these two expressions interchangeably (1963 p. 417, 420).

Tinbergen's distinctions are often endorsed in the opening pages of books about animal behavior (Krebs and Davies 1987 p. 5, Halliday and Slater 1983 p. vii, and see also Horan 1989). This is clearly an embarrassment for any historical theory of function which seeks to capture biological usage: on the historical view there should be three questions, not four, as the functional question *is* a question about evolutionary history, as long as the rest of (F2) above is satisfied. Related distinctions with this separation between function and history are found elsewhere in evolutionary writings as well. Mayr 1961 distinguishes "functional" from "evolutionary" biology, and Futuyma's widely used textbook echoes Mayr in dividing the study of biology into functional and historical "modes" (1986 p. 286).[8]

There are various ways to respond to this problem. Many ahistorical usages of "function" are probably best understood as referring to Cummins' functions. However, it is common for writers to both regard functions as ahistorical *and* regard them as intrinsically tied to natural selection, sometimes via the expression "survival value." This supports the proposal of a number of writers that functions involve not actual selective histories, but probable futures of selective success, or atemporal dispositions to succeed. Tinbergen may have accepted such a view: "the student of survival value, so-to-speak, looks 'forward in time'" (1963 p. 418). Tinbergen (p. 428) also casts the question about a structure's function as a question about how deviations from the actual structure *would* lower the fitness of the bearer. John Staddon concurs (1987 p. 195). One way to develop this approach is with an appeal to propensities.

Bigelow and Pargetter (1987) develop a theory of functions modelled explicitly on the widely accepted propensity view of fitness (Mills and Beatty 1979). The propensity view of fitness claims that the fitness of an individual is not the actual fact of its reproductive success, but its propensity to have a certain degree of reproductive success. Similarly, Bigelow and Pargetter claim, functions should be understood as dispositions or propensities to succeed under natural selection. "Something has a (biological) function just when it confers a survival-enhancing propensity on a creature that possesses it" (1987 p. 192).

The propensity view is not satisfactory, though its failure performs the valuable service of narrowing the discussion down, along with Tinbergen's Four Questions, to a point where the modern history view will become compelling. I will discuss first some internal difficulties with the propensity view and then argue that the whole forward-looking approach is on the wrong track.[9]

The central internal problem is that as one tries to fill in some more details, the theory tends to go in one or other of three undesirable directions. It can become enmeshed in strong counterfactual commitments. Alternately, it draws on the historical facts it sought to avoid. Or thirdly it makes the wrong kinds of demands on the future. Putting it briefly: propensities to be selected and survive bestow functions, but, the questions swarm: survive where? be selected over *what*? Bigelow and Pargetter address the first question, admitting that their account "must be relativized to an environment" (p. 192). The context assumed is the creature's "natural habitat." "Natural habitat," it appears, is understood historically by Bigelow and Pargetter. The statistically most common context for a trait now might be odd and unnatural (Neander 1991b).

More worrying is the question of the competitors that have a propensity to be ousted from the population by the trait we are interested in. Bigelow and Pargetter make no mention of the fact that claims about propensities to do well under natural selection are surely always comparative claims. A trait does not have a propensity to be selected and survive simpliciter, but always a propensity to be selected over some range of alternatives. Evolution is driven by differences in *relative* fitness.

Bigelow and Pargetter cannot claim that current useful traits would triumph over any possible alternatives. Which are the relevant ones? Those alternatives genetically attainable (given mutation rates, population structure, other constraints ...) now? Those that could enter the fray during the next thousand years? Those that could enter the fray if the ozone layer goes and mutation rates are elevated? If Bigelow and Pargetter think there is a range of alternatives, and circumstances of selection, appropriate to the trait in question independently of history, they are making strong modal commitments. These might be avoided with an appeal to what is most likely to happen in the actual future, but then problems are created by (what appear to be) irrelevant contingent features of this future. If a trait is adaptive, but doomed because of linkage to something bad, then it is not likely to survive. But this should not make a trait itself non-functional.

So, though the propensity theory is tailored to avoid dragging up the past, the propensities involved must either make tacit reference to millennia gone by, inappropriate predictions about the future, or questionable modal commitments about relevant ranges of alternatives and circumstances of selection. These internal problems are important, because it is easy to think that propensity views are somehow more economical than analyses appealing to the past. Still, the propensity view has recommendations. It does seem to be a way to accommodate the intuition that functions derive from selection with the observation that many biologists keep functional and historical questions separated. In addition, I am often told that no matter how questionable philosophers may find the modal commitments outlined above, many biologists constantly talk as if these facts are quite unproblematic and accessible. It is difficult to work our the right attitude to such a datum. Further, one principled way to deal with these internal problems is to fashion a mixed theory, using the basic propensity format with an appeal to history to answer the objections raised above. (This mixing was suggested to me by Elisabeth Lloyd).

The mixed theory claims that functions derive from propensities to be selected, but all the factors that Bigelow and Pargetter left vague are understood historically. The relevant ecological conditions are the actual ones that obtained during the development of the trait. The range of alternatives the trait has a propensity to be selected over are the ones it actually triumphed over, and continues to be selected over. The propensity that bestows functions is strictly atemporal; a trait is held to have a certain advantage under certain conditions over certain rivals. But these conditions and rivals are determined by the actual world. So it does seem likely that the propensity approach can be developed in a coherent way, at the price of narrowing the gap between it and the historical view. This is the general form of the contemporary functions debate: each theory is made more plausible by setting it on a course of convergence with its rivals.

There is, however, a more important problem with propensity theories, and other forward-looking views. These theories inevitably distort our understanding of functional explanation. In the first section I claimed that the sense of function under discussion is a sense linked in some way to explanations of why the functionally characterized entity exists, or exists in the form it does. The most straightforward way to envisage this link, which I have been assuming, is to say that functions are used in explanations of why the functionally characterized thing exists now. If this is granted, and the explanation is understood causally, then there is a simple argument against propensity views. The only events that can explain why a trait is around now are events in the past. Forward-looking accounts claim that functions are not bestowed by facts about the past, but rather by how things are in the present. But then appealing to a function cannot itself explain the fact that the trait exists now. If the environment is uniform, then present propensities to do well under selection may be a good guide to actual prior episodes of selection. But this epistemological point does not alter the fact that it is not the present propensities, but the prior episodes, that are causally responsible for how things are now (see also Millikan 1989b, Neander 1991a).[10]

I do not claim that Bigelow and Pargetter have missed this straightforward point. On their view, there is a problem with the background assumptions I have made about the explanatory role of functions, and which the argument above assumes. Bigelow and Pargetter claim that if the fact that some effect is a function itself depends on the fact that this effect explains the survival of the trait in question, if the assignment of a function is always retrospective in this way, "then it is no longer possible to explain why a character has persisted by saying that the character has persisted because it serves a given function" (1987 p. 190). This vacuity problem can be solved, according to Bigelow and Pargetter, if functions are understood as propensities. These propensities can be used to explain the existence of a trait in the present if we claim, in addition, that the propensities in question did exist in the past, and were causally active in the past. This postulation of the past action of the propensities is an extra claim; it is not guaranteed by the mere fact that the effects in question are functions.

Bigelow and Pargetter's claims about explanatory vacuity and the historical view have been criticized effectively by Sandra Mitchell (1993). She points out that if we say "Trait X persisted because it had a consequence responsible for its selection and consequent evolution," this is only vacuous if we read "persisted" as meaning "evolved by natural selection." That is, it is only vacuous if we assume that the *only* mechanism which could explain some trait being around today is natural selection, though in fact there are alternative evolutionary forces which could play this explanatory role (1993 p. 253–54). This is correct, and it shows that Bigelow and Pargetter's argument about the vacuity of the historical view assumes an implausible adaptationism. There is also another objection to Bigelow and Pargetter's claim, which is compatible with even the strongest adaptationism. On the historical view and with the assumption of adaptationism, it will be truly vacuous to say that X persisted because it serves *some* function, because we are assuming that this is the only possible type of explanation. But even against this background it will of course not be vacuous to say that X persisted because it provided effective camouflage, or because it attracted mates, or because it conserved heat. Neither is it vacuous to say that the trait persisted because some specific effect was its *function*. If the historical theorist says "X persisted because its function was to conserve heat," this is to be translated into something which is ungainly, and contains a redundancy—"X persisted because its actually-selected effect was that of conserving heat." But this is not vacuous; it does contain a real explanation, though to express it this way mentions the explanatoriness of the effect twice. So this is not the most natural mode of expression for the historical view; on that view the sentence "The function of X is to conserve heat" is itself explanatory, and if someone is asked "Why is X there?" they can reply by simply citing the function. This is not possible at all on the propensity view. On the propensity view, a functional explanation must give a function and also make an additional claim that the function was causally active in the past.

So despite what Bigelow and Pargetter claim, as long as "a given function" is understood to refer to some specific task or benefit, it is not trivial to say that "the character has persisted because it serves a given function," even assuming adaptationism. This, along with Mitchell's argument, shows that there is no vacuity problem with the background assumptions about explanation that proponents of the historical view make. It is possible to retain the explanatory force of function ascriptions, along with the philosophically attractive view, argued by Wright, that actual explanatory salience is exactly what *distinguishes* functions from mere effects.

A "forward-looking" approach to functions has also been endorsed by Barbara Horan (1989), but the claims she makes about explanation are more problematic than those of Bigelow and Pargetter. Horan says "questions about the function of a given pattern of social behavior are a way of asking how that behavior enhances the fitness of an individual who engages in it" (1989 p. 135). Nevertheless, she claims soon after that the presence of a trait like a social behavior can be explained by an attribution of a function to that behavior. The model of explanation she applies, citing G. A. Cohen, is called a "consequence explanation." Consequence explanations use laws of the form: "If (if C then E), then C." In the present context: "if a behavior pattern would increase individual fitness, individuals will come to display that behavior" (1989 p. 136).

This is trying to have it both ways. It is true that useful things a behavior does now can lead to its prevalence in the future. So forward-looking functions may predict and explain the future prevalence of a trait. But if the explanandum is how things are now, nothing present or future can be the explanans. Only the past will do. Of course, traits that are useful now were often useful then, so we can often infer that a propensity existing now was also causally active then. But if so, it is explanatory with respect to the present *because* it was causally active then. To claim that present usefulness in itself explains the morphologies and behaviors organisms presently display, and to build this into an account of functions, is to distort the explanatory structure of evolutionary theory.

IV   The Modern History Theory

It might appear that we are painting ourselves into an analytical corner. Historical analyses are unacceptable because they fail to respect an apparently important distinction in biology between functional and evolutionary explanation. Forward-looking analyses are unacceptable because they distort our understanding of functions' explanatory role. In fact there are several options available at this point. Bechtel (1989) suggests that we retain a forward-looking account of functions while giving up our prior conception of functional explanation. We might, alternately, claim that functional explanation just is evolutionary explanation, and banish other notions of function (except for Cummins') as creatures of teleological darkness. A third option is to analyze functional explanation as a particular *kind* of evolutionary explanation. One alternative here is to regard a functional explanation as a selective explanation which satisfies (F2) above, hence a subset of evolutionary explanation. The option I prefer, however, is to construe functional explanation more narrowly still.

This brings us, at last, to the modern history view: functions are dispositions and powers which explain the recent maintenance of a trait in a selective context. Several people have already said, in effect, that this is the answer, but these people either make the suggestion in passing (Kitcher 1990), or more often, they only say it some of the time. Horan says "to explain the maintenance of a trait in a species, one gives a functional explanation" (1989 p. 135), but insists on an atemporal construal of this explanation. And consider this remark of Millikan's, in response to Horan:

If natural selection accounts for a trait, that is something that happened in the past, but the past may have been, as it were, "only yesterday." Indeed, *usually* the relevant past is only yesterday: the *main business* of natural selection is steady maintenance of useful traits against new intruders in the gene pool. But only yesterday is not outside of time. (1989b p. 173)

We need not endorse the claim about the "main business" of natural selection; whether or not maintaining traits is the main business of selection, it is one important kind of selection. It might be important enough to make this a constitutive part of the concept of function. Millikan does not take this step; her historical account does not *build into* functions the historically recent nature of the relevant selective episodes. Indeed, in her 1984 treatment she explicitly allows powers which were important in ancient history, but not in modern history, to be functions (1984 p. 32). In the 1989b treatment her emphasis is different, and she claims the relevant past is *"usually"* only yesterday. But perhaps, as far as functions go, it *must* be only yesterday.

The modern history view does not respect the letter of Tinbergen's Four Questions, but it is faithful to their spirit. Tinbergen makes the modern/ancient history distinction himself (1963 pp. 428–29), but he regards *both* these explanations as "evolutionary" rather than functional. This puts two distinct questions under one head, however, as well as leaving the explanatory significance of functions in the dark. From the present viewpoint, the "evolutionary" question is the question about the forces which originally built the structure or trait in question. This may or may not be a selective explanation, and this explanation might be different from the explanation of why the trait has recently been maintained in the population.

Some might wonder how recent the selective episodes relevant to functional status have to be. The answer is not in terms of a fixed time—a week, or a thousand years. Relevance fades. Episodes of selection become increasingly irrelevant to an assignment of functions at some time, the further away we get. The modern history view does, we must recognize, involve substantial biological commitments. Perhaps traits are, as a matter of biological fact, retained largely through various kinds of inertia. Perhaps there is not constant phenotypic variation in many characters, or new variants are eliminated primarily for non-selective reasons. That is, perhaps many traits around now are not around because of things they have been doing. Then many modern-historical function statements will be false. If functions are to be understood as explanatory, in Wright's sense, there is no avoiding risks of this sort.

One way to support the modern history view of function is to demonstrate that the category of explanation it distinguishes is a theoretically principled one. This can be done by focusing on traits for which the modern historical explanation and the ancient historical explanation diverge, so the selective forces salient in the origin of the trait are different from those salient in the recent maintenance of the trait. Here is where a distinctively functional style of characterization—in the modern history sense—can be seen to be useful.

The importance of the distinction between modern and ancient evolutionary explanations is discussed, in support of an analysis of function quite opposed to mine, in Gould and Vrba 1982. The central concern of Gould and Vrba is a distinction between adaptations and "exaptations" (their coinage). They understand adaptations as characters shaped by natural selection for the role they perform now. Exaptations are characters built originally by selection for one job, or characters with no direct selective explanation at all, which have since been coopted for a new use. This analysis has consequences for their concept of function; only adaptations have functions, and exaptations have "effects." Gould and Vrba do not discuss the recent past, as distinct from the present, so I am uncertain how they would classify modern-historical functions. Generally they seem to understand effects-of-exaptations as propensities (1982 p. 6). Their effects-of-exaptations correspond to the functions of Bigelow and Pargetter, and Horan. It should be clear why I think their way of dividing the cases is inadequate: modern history and ancient history can *both* furnish genuine explanations, which we should distinguish, for why something exists now, while present propensities cannot themselves furnish such explanations.

Gould and Vrba's central point is the importance of cases where a trait's original and current uses diverge. But these are also cases where the selective forces that built a character and those maintaining it in the recent past diverge, so they also illustrate the origin/function distinction as I understand it. Gould and Vrba make two claims about such cases. Firstly, there are many of them, and secondly, the cases are theoretically significant. The co-opting of existing traits for new uses is important in the development of complex and novel adaptive characters.

Feathers, it has been argued, did not originate as adaptations for flight. The earliest known bird *Archaeopteryx* did not have the skeleton for anything beyond very rudimentary flight, but was well-covered with feathers. It has been claimed that feathers originated as insulation, and only later were coopted for flight (Gould and Vrba 1982 p. 7 cite Ostrom 1979). Thus the question about the evolutionary origin of feathers is answered in terms of selection for effective insulation, but if we ask today about the function of feathers, in a sub-tropical bird for instance, the answer appeals to the reason feathers have recently been maintained— their facilitating flight.

A similar story can be told about the development of bone. Bone is essential as a support for land-dwelling vertebrates, but it developed in sea animals well before it could be put to its modern use. Gould and Vrba discuss the hypothesis that bone was developed as store of phosphates needed for metabolic activity (Halstead 1969). In this case, the original use continues, and bone functions in modern vertebrates as storage for mineral ions, including phosphate ions, as well as support.

Gould and Vrba's examples can be augmented easily. The electric eel's ability to kill prey and defend itself with electric shocks is a development of the weaker electric abilities of other fish, which generate electric fields as part of a perceptual system, used in orientation and communication (Futayama 1986 pp. 423–24). Shepherd (1988 p. 67) discusses a suggestion made by J. B. S. Haldane about the origin of neurotransmitters, the chemicals whose function now is passing signals between neurons in the brain. Haldane suggested that these chemicals may have developed originally as chemical messengers between individuals. There are a number of neurotransmitters which can induce effects on other organisms.

A final illustration of the importance of the distinction between originating and maintaining selection is found in some of the literature applying game theory to animal behavior (Maynard Smith 1982).[11] An ESS, or evolutionarily stable strategy, is a strategy which, once prevalent in a population, cannot be invaded by rival strategies. However, an ESS need not be a strategy that can evolve from scratch in any situation. Often a critical mass of like-minded individuals is needed before a strategy becomes stable. Thus to explain a behavior by showing it to be an ESS is not necessarily to explain how that behavior originally became established. Rather, it is to point to the selective pressures responsible for the recent maintenance of the strategy in population.[12]

The point is not just the apparent commonality of a divergence between modern and ancient history, but the fact that this distinction has sufficient theoretical importance to justify its place in an analysis of functions.

One final problem must be discussed, which can be introduced with a feature of Wright's analysis. It is initially perplexing that Wright uses the present tense in the expression: "*X* is there because it does (results in) *Z*" (1976 p. 81). If his account is historical ("etiological"), why does he not make it explicit that the performances of *Z* that explain the presence of *X*'s are in the past?

> In general, when we explain something by appeal to a causal principle, the tense of the operative verb is determined by whether or not the principle still holds at the time the explanation is given.... We might say, for example, "The Titanic sank because when you tear a hole that size in the bow of a ship it sinks," using the verb "to sink" in the present tense even though the sinking in question took place in the past.... If we were to throw the statement into the past tense it would imply that nowadays one could get away with tearing a hole that size in the bow of a ship without it sinking. (1976 pp. 89–90)

Wright requires that the effects appealed to in a functional explanation still exist at the time of the functional ascription, and these effects must still have the same causal efficacy that they have had in the past. If this means that the structure in question must now have a propensity to continue to be selected for the same reason that it was selected in the past, Wright's account converges with that of Gould and Vrba, who demand that functions presently "promote fitness" (1982 p. 6).

Should the modern history view include these extra requirements? In my view, there may be good reason to require that the trait still be able to do now what it was selected for doing, but we should not require that the trait also have the same propensity to succeed under selection that it has had in the past. This problem is less pressing for the modern history view than for other historical views. If a trait has very recently been selected for doing *F*, it will tend to still be able to do *F* now. As it is *possible* for it (the type) to be unable to do *F* now, no matter how recently it has been selected for doing *F*, it is probably reasonable to add an extra clause requiring the continuation of the disposition into the strict present.[13] Whichever way one goes here, it is an advantage of the modern history view that these uncooperative cases should be made very rare.

Here is my final attempt at a definition of function.

(F3) The function of *m* is to *F* iff:

i. *m* if a member of family *T*,

ii. members of family *T* are components of biologically real systems of type *S*,

iii. among the properties copied between members of *T* is property or property cluster *C*, which can do *F*,

iv. one reason members of *T* such as *m* exist now is the fact that past members of *T* were successful under selection in the recent past, through positively contributing to the fitness of systems of type *S*, and

v. members of *T* were selected because they did *F*, through having *C*.

Much of this definition is proposed tentatively. The most important part is the appeal to modern history, which can also be incorporated in other theories of functions. The central recommendation of the modern history view is the fact that it accounts for the explanatory force of function ascriptions, but does this while making sense of the biological distinction between "functional" and "historical" explanation. It is a theory which steers a principled middle course.

Acknowledgment

This work developed largely out of a series of discussions with Phillip Kitcher. Along with many of the ideas, the term "modern history theory" was his, though he should not be taken to endorse (or reject) the modern history view. I have also benefitted from discussions with Elisabeth Lloyd, Ruth Millikan, Sandra Mitchell and everyone at Kathleen Akins' Functions Reading Group. An anonymous referee for *Noûs* made a number of valuable criticisms of earlier drafts. I would also like to thank the University of Sydney for generous financial support during the period when most of this work was done.

Notes

1. Neander (1991a), Mitchell (1989, 1993) and Brandon (1990) have defended theories of functions running along similar lines. Sober's (1984) analysis of adaptation is also a relative.

2. Those familiar with some units of selection debates in philosophy of biology will note that family members need not be replicators: see Dawkins 1982, Hull 1981.

My definition of copying is not supposed to be airtight, and may be too inclusive. Kim Sterelny suggested that it lets in molecular structures in a crystal lattice, for instance, though it is not so certain that this case should be kept out. See Millikan 1984 for more details.

3. Millikan presents her 1984 account as a stipulative definition, not an analysis of an existing concept, so this is not a problem for her. It is also important that Millikan's restrictions do not prevent the analysis being applied to artifacts generated by copying in the right ways.

4. The treatment in Millikan 1984 fudges here. Millikan's official definition of function begins with a stipulated function F, and explains why something has this function F. Can *any* activity or power qualify as function F, as long as it promotes survival? If not, Millikan owes us an account of what sorts of properties can be functions. If on the other hand she allows any power to be a function, then why does she take the indirect route, of starting with a function to be fulfilled and then explaining why one structure, rather than a rival, has this function as its own?

5. Most philosophical commentators on an earlier draft of this material inclined towards the heartless line on this question.

6. In the terms of the units of selection debate, the larger system needs to be a real *interactor* (Hull 1981, Lloyd 1988, Brandon 1990).

7. Wright 1976 discusses the possibility of an appeal to the broader system (p. 106). He dismisses it firmly (though this fails to prevent other writers from attributing such an appeal to him: Nagel 1977 p. 283, Hampe and Morgan 1988 p. 123). Wright however does not discuss examples like those causing trouble in the present discussion.

8. A puzzling case is George Williams (1966) Williams is usually regarded as an advocate of a Wright-style account of functions (Boorse 1976 p. 85, Wright 1976 pp. 92–93), as suggested by this well-known passage: "One should never imply that an effect is a function unless he can show that it is produced by design [natural selection] and not by happenstance" (1966 p. 261). But when Williams lays down principles for the general study of adaptation, he seems to imply that the basic fact of something's having a function is not a historical fact. It appears that the "prime" question asked about a character in such a study—"What is its function?"—is answered in terms of contributions to goals (1966 p. 258, citing Pittendrigh 1958). The *second* question asked is the historical one about selec-

tion (p. 259, see also p. 264). Williams does go on to say that an activity is not a function unless it was produced by design rather than chance (p. 261, quoted above). So the ahistorical nature of the "prime" question might be merely epistemological.

9. The version of the propensity view I am discussing is based on the survival propensities of *character types*. The propensity is possessed by human hearts as a type, not by individual hearts, and not by individual people. Bigelow and Pargetter are not consistent here. Sometimes they talk about the survival of the individual bearing the functionally characterized trait (1987 p. 192). But late, when speaking more strictly, they focus explicitly on the character type (p. 195, see also p. 194). On my reading, their talk of the "survival" of individuals is really talk of individuals' inclusive fitness (in the biological cases at least). Sandra Mitchell pointed out to me that if their propensities are read as belonging to individual trait-bearers, their theory is more like a classical goal theory. Admittedly, they do regard their account as a "cousin" of goal theories (p. 182). Neither interpretation squares with everything they say, but this exegetical question is less important than the theoretical issue of the viability of a propensity-based selective account.

10. Focusing on causal explanation in this way also makes it clear why the selective advantage relevant to functional status cannot be understood with reference to a range of counterfactual alternative traits, as opposed to actual ones, as some propensity views might maintain. Only competition with actual, past rivals is causally relevant in explaining why a trait exists today.

11. I am indebted to Philip Kitcher for this point.

12. The distinction between the original establishment and the maintenance of a strategy is stressed, for instance, in Axelrod and Hamilton's well-known discussion of the properties of tit-for-tat in the iterated prisoner's dilemma (1981).

13. This suggestion is made cautiously—perhaps all these additional requirements are ill-advised (Neander 1991b p. 183).

References

Andersson, M. (1982) Female Choice Selects for Extreme Tail Length in a Widowbird. *Nature* 299: 818–820.

Axelrod, R. and W. Hamilton (1981) The Evolution of Cooperation. *Science* 211: 1390–1396.

Bechtel, W. (1989) Functional Analyses and their Justification. *Biology and Philosophy* 4: 159–162.

Bigelow, J. and R. Pargetter (1987) Functions. *Journal of Philosophy* 84: 181–197.

Boorse, C. (1976) Wright on Functions. *Philosophical Review* 85: 70–86.

Brandon, R. (1990) *Adaptation and Environment*. Princeton: Princeton University Press.

Brandon, R. and R. Burian., eds. (1984) *Genes, Organisms, Populations: Controversies over the Units of Selection.* Cambridge, MA: MIT Press.

Buss, L. (1987) *The Evolution of Individuality.* Princeton: Princeton University Press.

Crow, J. (1979) Genes that Violate Mendel's Rules. *Scientific American* 240: 134–146.

Cummins, R. (1975) Functional Analysis. *Journal of Philosophy* 72: 741–765.

Dawkins, R. (1982) *The Extended Phenotype.* Oxford: Oxford University Press.

Fisher, R. A. (1930) *The Genetical Theory of Natural Selection.* Oxford: Clarendon.

Futuyama, D. (1986) *Evolutionary Biology.* 2nd edition. Sunderland: Sinauer.

Gould, S. J. and R. C. Lewontin (1978) The Spandrels of San Marco and the Panglossian Paradigm: A Critique of the Adaptationist Program. *Proceedings of the Royal Society, London* 205: 581–598.

Gould, S. J. and E. Vrba. (1982) Exaptation—A Missing Term in the Science of Form. *Paleobiology* 8: 4–15.

Halliday, T. R. and P. J. B. Slater, eds. (1983) *Animal Behavior, vol. 2: Communication.* New York: Freeman.

Halstead, L. B. (1969) *The Pattern of Vertebrate Evolution.* Edinburgh: Oliver and Boyd.

Hamilton, W. D. (1971) Geometry for the Selfish Herd. *Journal of Theoretical Biology* 31: 295–311.

Hampe, M. and S. R. Morgan. (1988) Two Consequences of Richard Dawkins' View of Genes and Organisms. *Studies in the History and Philosophy of Science* 19: 119–138.

Hampel, C. G. (1965) The Logic of Functional Analysis. In *Aspects of Scientific Explanation.* New York: Free Press.

Horan, B. (1989) Functional Explanations in Sociobiology. *Biology and Philosophy* 4: 131–158.

Hull, D. (1981) Units of Evolution: a Metaphysical Essay. Reprinted in Brandon and Burian 1984.

Kitcher, P. S. (1990) Developmental Decomposition and the Future of Human Behavioral Ecology. *Philosophy of Science* 57: 96–117.

Krebs J. and N. Davies (1987) *An Introduction to Behavioural Ecology,* 2nd edition. Oxford: Blackwell.

Lewontin, R. C. (1962) Interdeme Selection Controlling a Polymorphism in the House Mouse. *American Naturalist* 96: 65–78.

Lloyd, E. A. (1988) *The Structure and Confirmation of Evolutionary Theory.* New York: Greenwood Press.

Maynard Smith, J. (1982) *Evolution and the Theory of Games.* Cambridge: Cambridge University Press.

Mayr, E. (1961) Cause and effect in biology. *Science* 134: 1501–1506.

Millikan, R. G. (1984) *Language, Thought, and Other Biological Categories.* Cambridge, MA: MIT Press.

Millikan, R. G. (1989a) In Defence of Proper Functions. *Philosophy of Science* 56: 288–302.

Millikan, R. G. (1989b) An Ambiguity in the Notion "Function." *Biology and Philosophy* 4: 172–176.

Mills, S. and J. Beatty (1979) The Propensity Interpretation of Fitness. *Philosophy of Science* 46: 263–286.

Mitchell, S. (1989) The Causal Background of Functional Explanation. *International Studies in the Philosophy of Science* 3: 213–229.

Mitchell, S. (1993) Dispositions or Etiologies? A Comment on Bigelow and Pargetter. *Journal of Philosophy* 90: 249–259.

Nagel, E. (1977) Teleology Revisited: The Dewey Lectures 1977. (1) Goal-directed Processes in Biology. (2) Functional Explanations in Biology. *Journal of Philosophy* 74: 261–301.

Neander, K. (1991a) The Teleological Notion of "Function." *Australasian Journal of Philosophy* 69: 454–468.

Neander, K. (1991b) Functions as Selected Effects: The Conceptual Analyst's Defence. *Philosophy of Science* 58: 168–184.

Orgel, L. E. and F. H. C. Crick (1980) Selfish DNA; the Ultimate Parasite. *Nature* 284: 604–606.

Ostrom, J. H. (1979) Bird flight: How Did it Begin? *American Scientist* 67: 46–56.

Pittendrigh, C. S. (1958) Adaptation, Natural Selection, and Behavior. In A. Roe and G. G. Simpson, eds, *Behavior and Evolution.* New Haven: Yale University Press.

Shepherd, G. M. (1988) *Neurobiology,* 2nd edition. Oxford: Oxford University Press.

Sober, E. (1984) *The Nature of Selection.* Cambridge, MA: MIT Press.

Staddon, J. E. R. (1987) Optimality Theory and Behavior. In J. Dupré, ed. *The Latest on the Best: Essays on Evolution and Optimality.* Cambridge, MA: MIT Press.

Tinbergen, N. (1963) On the Aims and Methods of Ethology. *Zeitschrift für Tierpsychologie* 20: 410–433.

Trivers, R. (1974) Parent-offspring Conflict. *American Zoologist* 14: 249–264.

Williams, G. C. (1966) *Adaptation and Natural Selection.* Princeton: Princeton University Press.

Wright, L. (1973) Functions. *Philosophical Review* 82: 139–168.

Wright, L. (1976) *Teleological Explanations.* Berkeley: University of California Press.

18

Function and Design

Philip Kitcher

I

The organic world is full of functions, and biologists' descriptions of that world abound in functional talk. Organs, traits, and behavioral strategies all have functions.[1] Thus the function of the *bicoid* protein is to establish anteriorposterior polarity in the *Drosophila* embryo; the function of the length of jackrabbits' ears is to assist in thermoregulation in desert environments; and the function of a male baboon's picking up a juvenile in the presence of a strange male may be to appease the stranger, or to protect the juvenile, or to impress surrounding females. Ascriptions of function have worried many philosophers. Do they presuppose some kind of supernatural purposiveness that ought to be rejected? Do they fulfil any explanatory role? Despite a long, and increasingly sophisticated, literature addressing these questions, I believe that we still lack a clear and complete account of function-ascriptions. My aim in what follows is to take some further steps towards dissolving the mysteries that surround functional discourse.

I shall start with the idea that there is some unity of conception that spans attributions of functions across the history of biology and across contemporary ascriptions in biological and non-biological contexts. This unity is founded on the notion that the function of an entity *S* is *what S is designed to do*. The fundamental connection between function and design is readily seen in our everyday references to the functions of parts of artifacts: the function of the little lever in the mousetrap is to release the metal bar when the end of the lever is depressed (when the mouse takes the cheese) for that is what the lever is designed to do (it was put there to do just that). I believe that we can also recognize it in pre-Darwinian perspectives on the organic world, specifically in the ways in which the organization of living things is taken to reflect the intentions of the Creator: Harvey's claim that the function of the heart is to pump the blood can be understood as proposing that the wise and beneficent designer foresaw the need for a circulation of blood and assigned to the heart the job of pumping.

Now examples like these are precisely those that either provoke suspicion of functional talk or else prompt us to think that the concept of function has been altered in the course of the history of science. Even though we may retain the idea of the "job" that an entity is supposed to perform in contexts where we can sensibly speak of systems fashioned and/or used with definite intentions—paradigmatically machines and other artifacts—it appears that the link between function and design must be broken in ascribing functions to parts, traits, and behaviors of organisms. But this conclusion is, I think, mistaken. On the view I shall propose, the central common feature of usages of function—across the history of inquiry, and across contexts involving both organic and inorganic entities—is that the function of S is what S is designed to do; design is not always to be understood in terms of background intentions, however; one of Darwin's important discoveries is that we an think of design without a designer.[2]

Contemporary attributions of function recognize two sources of design, one in the intentions of agents and one in the action of natural selection. The latter is the source of functions throughout *most* of the organic realm —there are occasional exceptions as in cases in which the function of a recombinant DNA plasmid is to produce the substance that the designing molecular biologist intended. But, as I shall now suggest, the links to intentions and to selection can be more or less direct.

II

Imagine that you are making a machine. You intend that the machine should do something, and that is the machine's function. Recognizing that the machine will only be able to perform as intended if some small part does a particular job you design a part that is able to do the job. Doing the job is the function of the part. Here, as with the function of the whole machine there is a direct link between function and intention: the function of X is what X is designed to do, and the design stems from an explicit intention that X do just that.

It is possible that you do not know everything about the conditions of operation of your machine. Unbeknownst to you, there is a connection that has to be made between two parts if the whole machine is to do its intended job. Luckily, as you were working, you dropped a small screw into the incomplete machine and it lodged between the two pieces, setting up the required connection. I claim that the screw has a function, the function of making the connection. But its having that function cannot be grounded in your explicit intention that it do that, for you have no intentions with respect to the screw. Rather, the link between function and intention is much less direct. The machine has a function grounded in your explicit intention, and its fulfilling that function poses various demands on the parts of which it is composed. You recognize some of these demands and explicitly design parts that can satisfy them. But in other cases, as with the luckily placed screw, you do not see that a demand of a particular type has to be met. Nevertheless, whatever satisfies that demand has the function of so doing. The function here is grounded in the contribution that is made towards the performance of the whole machine and in the link between the performance and the explicit intentions of the designer.

Pre-Darwinians may have tacitly relied on a similar distinction in ascribing functions to traits and organs. Perhaps the Creator foresaw all the details of the grand design and explicitly intended that all the minutest parts should do particular things. Or perhaps the design was achieved through secondary causes: organisms were equipped with abilities to respond to their needs, and the particular lines along which their responses would develop were not explicitly identified in advance. So the Creator intended that jackrabbits should have the ability to thrive in desert environments, and explicitly intended that they should have certain kinds of structures. However, it may be that there was no explicit intention about the length of jackrabbits' ears. Yet, because the length of the ears contributes to the maintenance of roughly constant body temperature, and because this is a necessary condition of the organism's flourishing (which is an explicitly intended effect), the length of the ears has the function of helping in thermoregulation.

Understanding this distinction enables us to see how earlier physiologists could identify functions without engaging in theological speculation.[3] Operating on the presupposition that organisms were designed to thrive in the environments in which they are found, physiologists could ask after the necessary conditions for organisms of the pertinent types to survive and multiply. When they found such necessary conditions, they could recognize the structures, traits, and behaviors of the organisms that contributed to satisfaction of such conditions as having precisely such functions—without assuming that the Creator explicitly intended that those structures, traits, and behaviors perform just those tasks.

I have introduced this distinction in the context of machine design and of pre-Darwinian biology because it is more easily grasped in such contexts. I shall now try to show how a similar distinction can be drawn when natural selection is conceived as the source of design, and how this distinction enables us to resolve important questions about functional ascriptions.

III

We can consider natural selection from either of two perspectives. The first, the organism-centered perspective, is familiar. Holding the principal traits of members of a group of organisms fixed, we investigate the ways in which, in a particular environment or class of environments, variation with respect to a focal trait, or cluster of focal traits, would affect reproductive success. Equally, we can adopt an environment-centered perspective on selection. Holding the principal features of the environment fixed, we can ask what selective pressures are imposed on members of a group of organisms. In posing such questions we suppose that some of the general properties of the organisms do not vary, and consider the obstacles that must be overcome if organisms with those general properties are to survive and reproduce in environments of the type that interests us.

So, for example, we might consider the selection pressures on mammals whose digestive systems are capable of processing vegetation but not meat (or carrion) in an environment in which the accessible plants have tough cellulose outer layers. Holding fixed the very general properties of the animals that determine their need to take in food and the more particular features of their digestive systems, we recognize that they will not be able to survive to maturity (and hence not able to reproduce) unless they have some means of breaking down the cellulose layers of the plants in their environments. Thus the environments impose selection pressure to develop some means of breaking down cellulose. Organisms might respond to that pressure in various ways: by harboring bacteria that can break down cellulose or by having molars that are capable of grinding tough plant material. If our mammals do not have an appropriate colony of intestinal bacteria, but do have broad molars that break down cellulose, we may recognize the molars as their particular response to the selection pressure and ascribe them the function of processing the available plants in a way that suits the operation of their digestive systems. At a more fine-grained level, we may hold fixed features of the dentition, and identify properties of particular teeth as having functions in terms of their contributions to the breakdown of cellulose.

This illustration can serve as the prototype of a style of functional analysis that is prominent in physiology and in general zoological and botanical studies. One starts from the most general evolutionary pressures, stemming from the competition to reproduce and concomitant needs to survive to sexual maturity, to produce gametes, to identify and attract mates, and to forth. In the context of general features of the organisms in question and of the environments they inhabit, we can specify selection pressures more narrowly, recognizing needs to process certain types of food, to evade certain kinds of predators, to produce particular types of signals, and so forth. We now appreciate that certain types of complex structures, traits, and behaviors enable the organisms to satisfy these more specific needs. *Their* functions are specified by noting the selection pressures to which they respond. The functions of their constituents are understood in terms of the contributions made to the functioning of the whole. Here, I suggest, we have a mixture of evolutionary and mechanistic analysis. There is a link to selection through the environment-centered perspective from which we generate the selection pressures that determine the functions of complex entities, and there is a mechanistic analysis of these complex entities that displays the ways in which the constituent parts contribute to total performance.

I claim that understanding the environment-centered perspective on selection enables us to draw an analogous distinction to that introduced in section II, and thus to map the diversity of ways in which biologists understand functions. However, before offering an extended defense of this claim, two important points deserve to be made.

First, the environment-centered perspective has obvious affinities with the idea that organisms face selective "problems," posed by the environment, an idea that Richard Lewontin has recently criticized.[4] According to Lewontin, there is a "dialectical relationship" between organism and environment that renders senseless the notion of an environment prior to and independent of the organism to which "problems" are posed. Lewontin's critique rests on the correct idea that there is no specifying which parts of the universe are constituents of an organism's environment, without taking into account properties of the organism. In identifying the environment-centered perspective, I have explicitly responded to this point, by proposing that the selection pressures on organisms arise only when we have held fixed important features of those organisms, features that specify limits on those parts of nature with which they causally interact. Quite evidently, if we were to hold fixed properties that could easily be modified through mutation (or in development), we would obtain an inadequate picture of the organism's environment and, consequently, of the selection pressures to which it is subject. If, however, we start from those characteristics of an organism that would require large genetic changes to modify—as when we hold fixed the inability of rabbits to fight foxes—then our picture of the environment takes into account the evolutionary possibilities for the organism and offers a realistic view of the selection pressures imposed.

Second, as we shall see in more detail below, recognizing a trait, structure, or behavior of an organism as responding to a selection pressure imposed by the environment (in the context of other features of the organism that are viewed as inaccessible to modification without severe loss of fitness) we do not necessarily commit ourselves to claiming that the entity in question originated by selection or that it is maintained by selection. For it may be that genetic variation in the population allows for alternatives that would be selectively advantageous but are fortuitously absent. Thus the entity is a response to a genuine demand imposed on the organism by the environment even though selection cannot be invoked to explain why it, rather than the alternative, is present. In effect, it is the analogue of the luckily placed screw, answering to a real need, but not itself the product of design. I shall be exploring the consequences of this point below.

IV

The simplest way of developing a post-Darwinian account of functions is to insist on a direct link between the design of biological entities and the operation of natural selection. The function of X is what X is designed to do, and what X is designed to do is that for which X was selected. Since the publication of a seminal article by Larry Wright, etiological accounts of function have become extremely popular.[5] Wright claimed that the function of an entity is what explains why that entity is there. This simple account proved vulnerable to counterexamples: if a scientist conducting an experiment becomes unconscious because gas escapes from a leaky valve, then the presence of the gas in the room is explained by the fact that the scientist is unconscious (for otherwise she would have turned off the supply), but the function of the gas is not to asphyxiate scientists.[6] Such objections can be avoided by restricting the form of explanations to explanations in terms of selection, so that identifying the function of X as that for which X was selected enables us to preserve Wright's idea that functions play a role in explaining the presence of their bearers without admitting those forms of nonselective explanation that generate counterexamples.[7] However, this move forfeits one of the virtues of Wright's analysis, to wit, its recognition of a common feature in attributions of functions to artifacts and to organic entities.

There are other issues that etiological analyses of functional ascriptions must confront, issues that arise from the character of evolutionary explanations. First is the question of the *time* at which the envisaged selection regime is supposed to act. Second we must consider the *alternatives* to the entity whose presence is to be explained and the extent of the role that selection played in the singling out of that entity.[8] If these issues are neglected—as they frequently are—the consequence will be either to engage in highly ambiguous attributions of function or else to fail to recognize the demands placed on functional ascription.

Selection for a particular property may be responsible for the original presence of an entity in an organism or for the maintenance of that entity.[9] In many instances, selection for P explains the initial presence of a trait *and* the subsequent maintenance of that trait: the initial benefit that led to the trait's increase with respect to its rivals also accounts for its superiority over alternatives that arose after the original process of fixation. But as a host of well-known examples reveals, this is by no means always the case. To cite one of the most celebrated instances, feathers were apparently originally selected in early birds (or their dinosaur ancestors) for their role in thermoregulation; after the development of appropriate musculature (and other adaptations for flight) the primary selective significance of feathers became one of making a causal contribution to efficient flying.

Faced with examples in which the properties for which selection initially occurs are different from those for which there is selection in maintaining a trait, behavior, or structure, the etiological analysis must decide which of the following conditions is to govern functional attributions:

(1)   The function of X is Y only if the initial presence of X is to be explained through selection for Y,

(2)   The function of X is Y only if the maintenance of X is to be explained through selection for Y,

(3)   The function of X is Y only if both the initial presence of X and the maintenance of X are to be explained through selection for Y.

But deciding among these three conditions is only the beginning of the enterprise of disambiguating the etiological analysis of function. Just as the properties important in initiating selection may not be those that figure in maintaining selection, it is possible that an entity may be *maintained* by selection for different properties at different times. Hence, both (2) and (3) require us to specify the appropriate period at which the maintenance of X is to be considered. I believe that there are two plausible candidates with respect to (2), namely the present and the recent past, and that the most well-motivated version of (3) requires that the character of the selective regime is constant across all times. Thus we obtain:

(2a)   The function of X is Y only if selection of Y has been responsible for maintaining X in the recent past,

(2b)   The function of X is Y only if selection for Y is currently responsible for maintaining X,

(3)   The function of X is Y only if selection for Y was responsible for the initial presence of X and for maintaining X at all subsequent times up to and including the present.

A consequence of adopting (1)—which effectively takes functions to be *original* functions—is that two of Tinbergen's famous four why-questions are conflated: there is now no distinction between the "why" of evolutionary origins and the "why" of functional attribution.[10] In those biological discussions in which an etiological conception of function is most apparent (ecology, and especially behavioral ecology), Tinbergen's distinction seems to play an important role. Thus I doubt that an etiological analysis based on (1) reflects much that is significant in biological practice.

Etiological analyses clearly based on (3) can sometimes be found in the writings of those who are critical of unrigorous employment of the notion of function. So, for example, Stephen Jay Gould's and Elisabeth Vrba's contrast between functions and "exaptations" seems to me to thrive on the idea that specification of functions must rest on the presupposition that selection has been operating in the same way in originating and maintaining traits (and, indeed, that traits maintained by selection were originally fashioned by selection).[11] Because there is frequently no available evidence for this presupposition, adoption of etiological conception based on (3) can easily fuel skepticism about ascriptions of function.

I suspect that some biologists do tacitly adopt an etiological conception of function founded on (3), and that their practice of ascribing functions is subject to Gould's strictures. Others plainly do not. Thus, Ernst Mayr explicitly recognizes the possibility of change of function over evolutionary time, suggesting that he acknowledges *two* notions of function, one ("original function") founded upon (1) and another ("present function") based on some version of (2).[12] For biologists who draw such distinctions, Gould's criticisms will seem to claim novelty for a point that is already widely appreciated. (Of course, one of the most prominent features of the debates about adaptationism is the opposition between those who believe that the criticisms tiresomely remind the evolutionary community of what is already well known and those who contend that what is professed under attack is ignored in biological practice.)[13]

The most prevalent concept of function among contemporary ecologists is, I believe, an etiological concept founded on some version of (2). Claims about functions are founded on measurements or calculations of fitness, and the measurements and calculations are made on *present* populations. Faced with the question, "Do you believe that the properties for which selection is now occurring are those that originally figured in the fixation of the trait (structure, behavior)?" sophisticated ecologists would often plead agnosticism. Their concern is with what is currently occurring, and they are happy to confess that things may have been different in a remote past that is beyond their ability to observe and analyze in the requisite detail. Hence the concept of function they employ is founded on the link between functions and contemporary processes of selection that maintain the entities in question, a link recorded in (2).

But which version of (2) should they endorse? Here, I believe, philosophical analyses reveal unresolved ambiguities in biological practice. An account of functions that effectively endorses (2b) has been proposed by John Bigelow and Robert Pargetter (who, idiosyncratically it seems to me, attempt to distance themselves from Wright and other etiological theorists).[14] My own prior discussions of functional ascriptions presuppose a concept based on (2a), and this notion of function has been thoroughly articulated by Peter Godfrey-Smith.[15] On what basis can we decide among these accounts?

As Godfrey-Smith rightly notes, a "recent history" notion of function, committed to (2a), gives functional ascriptions an explanatory role. Identifying the function of an entity outlines an explanation of why the entity is now present by indicating the selection pressures that have maintained it in the recent past. Arguing that philosophers ought to identify a concept that does some explanatory work, he concludes that (2a) represents the right choice. But this seems to me to be too quick. The conception of function defended by Bigelow and Pargetter, founded on (2b), is perhaps most evident in those biological discussions in which the recognition that a trait is functional supports a prediction about its future presence in the population. Yet the "forward-looking" conception also allows ascriptions of function to serve as explanations of why the trait will continue to be present. There is still an explanatory project, but the *explanandum* has been shifted from current presence to future presence.

Biological practice seems to me to be too various for definitive resolution of these differences. Sometimes attributions of function outline explanations of current presence, sometimes offer predictions about the course of selection in the immediate future, sometimes sketch explanations of the presence of traits in succeeding generations. Moreover, since it is often reasonable to think that the environmental and genetic conditions are sufficiently constant to ensure that the operation of selection in the recent past was the same as the selection seen in the present, it will be justifiable to combine the main features of the "recent past" and "forward-looking" accounts to found a notion of function on a combination of (2a) and (2b)

(2c)   The function of X is Y only if selection of Y is responsible for maintaining X both in the recent past and in the present.

In situations in which there is reason to think that the action of selection has been constant across the relatively short time periods under consideration, use of a notion of function founded on (2c) will allow functional attributions to play a role in all the explanatory and predictive projects I have considered.

If biological practice overlooks potential ambiguities with respect to the timing of the selection processes that underlie attributions of function, it is even more silent on issues about the competition involved in such processes. What are the alternatives to the biological entity whose presence is due to selection? And to what extent is selection the *complete* explanation of the presence of that entity?

Ecologists working on pheromones in insects or on territory size in birds can sometimes specify rather exactly the set of alternatives they consider. Holding fixed certain features of the organisms they study, features that would, they suppose, only be modifiable by enormous genetic changes that render rivals effectively inaccessible, they can impose necessary conditions that define a set of rival possibilities: pheromones must have such-and-such diffusion properties, territories must be able to supply such-and-such an amount of food, and so forth. In light of these constraints, they may be able to construct a mathematical model showing that the entity actually found in the population is optimal (or, more realistically, "sufficiently close" to the optimum).[16] A different strategy is to consider alternatives that arise by mutation in populations that can be observed and to measure the pertinent fitness values. Either of these approaches will support claims about selection processes that have occurred/are occurring in the recent past or the present. In both instances there may be legitimate concern that unconsidered alternatives might have figured in historically more remote selection processes, either because the organisms were not always subject to the constraints built into the mathematical model or because the genetic context in which mutations are now considered is quite different from the genetic contexts experienced by organisms earlier in their evolutionary histories. So far this simply underscores our previous conclusions about the greater plausibility of analyses based on some version of (2).

But now let us ask how exactly selection is supposed to winnow the alternatives. Suppose we ascribe a function to an entity X, basing that function on a selection process with alternatives $X_1, \ldots, X_n$. Must it be the case that organisms with X have higher fitness than organisms with any of the X_i? On a strict etiological analysis of functional discourse, this question should be answered affirmatively: where selection is the *complete* foundation of the design that underlies X's function, X is favored by selection over *all* its rivals. Thus on the strongest version of an etiological conception, functional ascriptions should be based either on recognition that X has greater fitness than all the alternatives arising by mutation in current populations, or on an analysis that shows X to be strictly optimal. I believe that some biologists—particularly in ecology and behavioral ecology—make functional claims in this strong sense and attempt to back them up with careful and ingenious observations and calculations.[17] Nonetheless, there is surely room for a less demanding account of biological function.

Consider two possibilities. First, our optimality analysis shows that, while X is reasonably close to the optimum, it is theoretically suboptimal. We do not know enough about the genetics and developmental biology of the organisms under study to know whether mutations providing a genetic basis for superior rivals could arise in the population. Under these circumstances, one cannot claim that the presence of X is entirely due to the operation of selection. It may be that X is present because theoretically possible mutants have not (recently) arisen, and selection, acting on a limited set of alternatives, has fixed X. Second, we may be able to identify actual rivals to X that are indeed superior in fitness but that have fortuitously been eliminated from the population. During the period that concerns us (present or recent past) organisms bearing some entity X_i have arisen, and these have had greater fitness than organisms bearing X. By chance, however, such organisms have perished. Here, we can go further than simply recognizing an inability to support the strong claim about optimality—we recognize that X is definitely suboptimal, and that its presence is not the result of selection alone.

Nevertheless, many biologists would surely be uninterested in these possibilities or actualities, regarding X as having the function associated with the selective process, even if it were possibly, even definitely, suboptimal. There are various ways of weakening the requirement that X's fitness be greater than those of alternatives. We might demand that X be fitter than *most* alternatives, that it be fitter than the *most frequently occurring* alternatives, and so forth. It requires only a little imagination to devise scenarios in which an entity is inferior in fitness to most of its rivals and/or to its most frequently occurring rivals, even through it may still be ascribed the function associated with the selection process.

Imagine that there is a species of moth that is protected from predatory birds through a camouflaging wing pattern that renders it hard to perceive when it rests on a common environmental background. We observe the population and discover a number of rival wing colorations, none of which ever occurs in substantial numbers. Less than half of these alternatives are absolutely disastrous, and organisms with them are vulnerable to predation, and quickly eliminated. Investigating the other, we find, to our surprise, that they prove slightly superior to the prevalent form, in affording improved camouflage, without any deleterious side effects. However, as the result of various events that we can identify— disruptions of habitat, increased concentrations of predators in areas in which there is a high frequency of the mutants—these alternatives are eliminated as the result of chance. Nonetheless, although it is somewhat inferior to most of its rivals, the common wing pattern still has the function of protecting the moth from predation.

I think that it is obvious what we should say about this and kindred scenarios. The impulse to recognize X as having a function can stem from recognition that X is a response to an identifiable selection pressure, *whether or not the presence of X is completely explicable in terms of selection*. Thus, instead of trying to weaken the conditions on etiological conceptions of function, I suggest that we can accommodate cases that prove troublesome by drawing on the distinctions of sections II and III. I shall now try to show how this leads to a rich account of functional ascriptions that will cover practice in physiology as well as in those areas in which the etiological conception finds its most natural home.

V

Entities have functions when they are designed to do something, and their function is what they are designed to do. Design can stem from the intentions of a cognitive agent or from the operation of selection (and, perhaps, recognizing how unintuitive the notion of design without a designer would have seemed before 1859, from other sources that we cannot yet specify). The link between function and the source of design may be direct, as in instances of agents explicitly intending that an entity perform a particular task, or when the entity is present because of selection for a particular property (that is, its presence is completely explained in terms of selection for that property). Or the link may be indirect, as when an agent intends that a complex system perform some task and a component entity makes a necessary causal contribution to the performance, or when organisms experience selection pressure that demands some complex response of them and one of their parts, traits, or behaviors makes a needed causal contribution to that response. As noted in the previous section, there are also ambiguities about the time period throughout which the selection process is operative. It would be easy to tell a parallel story about agents and their intentions.

I have noted that the strong etiological conception—that based on a direct link between function and the underlying source of design (in this case, selection)—is very demanding. While some ecologists undoubtedly aim to find functions in the strong sense, much functional discourse within ecology, as well as in other parts of biology is more relaxed. Imagine practicing biologists accompanied by a philosophical Jiminy Cricket, constantly chirping doubts about whether selection is *entirely* responsible for the presence of entities to which functions are ascribed. Many biologists would ignore the irritating cavils, contending that attribution of function is unaffected by the possibilities suggested by philosophical conscience. It is enough, they would insist, that genuine demands on the organism have been identified and the entities to which they attribute functions make causal contributions to the satisfaction of those demands. What is wrong with the relaxed attitude?

Functional attributions in the strong sense have clear explanatory work to do. They indicate the lines along which we should account for the presence of the entities to which functions are ascribed. To say that the function of X is F is to propose that a complete explanation of the presence of X (at the appropriate time) should be sought in terms of selection for F. Once we relax the demands on functional ascriptions, the role of selection is no longer clear; indeed, a biologist may explicitly allow that selection has not been responsible for maintaining X (or, at least, not completely responsible). But there is a different type of explanatory project to which the more lenient attributions contribute. They help us to understand the causal role that entities play in contributing to complex effects.

Here we encounter a central theme of the main philosophical rival to the etiological conception, lucidly articulated in an influential article by Robert Cummins.[18] For Cummins, functional analysis is about the identification of constituent causal contributions in complex processes. This style of activity is prominent in physiological studies, where the apparent aim is to decompose a complex "organic function" and to recognize how it is discharged. I claim that Cummins has captured an important part of the notion of biological function, but that his ideas need to be integrated with those of the etiological approach, not set up in opposition to it.

When we attribute functions to entities that make a causal contribution to complex processes, there is, I suggest, always a source of design in the background. The constituents of a machine have functions because the machine, as a whole, is explicitly intended to do something. Similarly with organisms. Here selection lurks in the background as the ultimate source of design, generating a hierarchy of ever more specific selection pressures, and the structures, traits, and behaviors of organisms have functions in virtue of their making a causal contribution to responses to those pressures.

Without recognizing the background role of the sources of design, an account of the Cummins variety becomes too liberal. Any complex system can be subjected to functional analysis. Thus we can identify the "function" that a particular arrangement of rocks makes in contributing to the widening of a river delta some miles downstream, or the "functions" of mutant DNA sequences in the formation of tumors—but there are no genuine functions here, and no functional analysis. The causal analysis of delta formation does not link up in any way with a source of design; the account of the causes of tumors reveals *dysfunctions*, not functions.

Recognizing the liberality of Cummins-style analyses, proponents of the etiological conception drag evolutionary considerations into the foreground. In doing so they make *all* projects of attributing functions focus on the explanation of the presence of the bearers of those functions. However, important though the theory of evolution by natural selection undoubtedly is to biology, there are other biological enterprises, some even continuous with those that occupied pre-Darwinians, which can be carried out in ignorance of the details of selective regimes. Thus the conscience-ridden biologists who offer more relaxed attributions of function can quite legitimately protest that the niceties of selection processes are not their primary concerns: without knowing what alternatives there were to the particular valves that help the heart to pump blood, they can recognize both that there is a general selection pressure on vertebrates to pump blood and that particular valves make identifiable contributions to the pumping. Selection, they might say, is the background source of design here, but it need not be dragged into the foreground to raise questions that are irrelevant to the project they set for themselves (understanding the mechanism through which successful pumping is achieved).

I believe that the account I have offered thus restores some unity to the concept of function through the recognition that each functional attribution rests on some presupposition about design and a pertinent source of design. But it allows for a number of distinct conceptions of function to be developed, based on sources of design (intention versus selection), time relation between source of design and the present, and directness of connection between source of design and the entity to which functions are ascribed. This pluralism enables us to capture the insights of the two main rival philosophical conceptions of function, and to do justice to the diversity of biological projects.

Does it go too far? In their original form, etiological accounts were vulnerable to counterexample, and the resolution invoked selection *ad hoc*, Am I committed to supposing that the leaky valve that asphyxiates the scientist has the function of so doing? No. For there is no explaining the presence of the valve in terms of selection for ability to asphyxiate scientists, nor is there any selection pressure on a larger system to whose response the action of the valve makes a causal contribution. Even though the account I have offered is more inclusive than traditional etiological conceptions, it does not seem to fall victim to the traditional counterexamples.

VI

I have tried to motivate my account of function and design by alluding to some quickly sketched examples. This strategy helps to elaborate the approach, but invites concerns to the effect that a more thorough investigation of biological practice would disclose less ambiguity than I have claimed. To alleviate such concerns I now want to look at some cases of functional attribution in a little more detail.

I shall start with two examples that are explicitly concerned with evolutionary issues. The first concerns a "functional analysis of the egg sac" in golden silk spiders.[19] The orb-weaving spider *Nephila clavipes* lays its eggs under the leaf canopy, covers them with silk, and weaves a loop of silk around twig and branch which holds the sac in place. The authors of the study (T. Christenson and P. Wenzl) investigate the functions of components of the egg-laying behavior. I shall concentrate on the spinning of the loop.

Christenson and Wenzl write:

The functions of the silk loop around the attachment branch were assessed by examining clutches that fell to the ground. We found 19 of the 59 egg sacs that fell due to naturally occurring twig breakage; 84.2% (16) failed to produce spiderlings, 13 because of ground moisture and subsequent rotting, and 3 because of predation.... The remaining three sacs had fallen a few weeks prior to the normal time of spring emergence; the spiderlings appeared to disperse and inhabit individual orbs.[20] In contrast to those that fell, sacs that remained in the tree were dry and appeared relatively safe from predation. Only 4.5% (15 of 353) showed unambiguous signs of predation, that is, some damage to the silk such as a tear or a bore hole.[21]

I interpret this passage as demonstrating a marked fitness difference between spiders who perform the looping operation that attaches the egg sac to twig and branch and those who fail to do so. Christenson and Wenzl are tacitly comparing the normal behavior of *N. clavipes* with mutants whose ability to weave an attachment loop was somehow impaired. Their emphasis on evolutionary considerations is evident not only in their detailed measurements of survivorships, but also in the framing of their analysis and in their final discussion. The authors begin by noting that "[f]unctional analyses of behaviours are often speculative due to the difficulty of demonstrating that the behavior contributes to the individual's reproductive success, and what the relevant selective agents might be."[22] They conclude by contending that "Female *Nephila* maximize their reproductive efforts, in part, through the construction of an elaborate egg sac."[23] This study is thus naturally interpreted as deploying the strong etiological conception of function, linking function directly with selection and proposing that the entities bearing functions are optimal.

Similarly, a study of the function of roaring in red deer by T. Clutton-Brock and S. Albon explicitly connects the attribution of function to claims about selection.[24] The authors begin by examining a traditional proposal:

A common functional explanation is that displays serve to intimidate the opponent.... This argument has the weakness that selection should favour individuals which are not intimidated unnecessarily and which adjust their behavior only to the probability of winning and the costs and benefits of fighting....[25]

Here it seems that a necessary condition on the truth of an ascription of function is that there should not be possible mutants that would be favored by selection. The same strong conception of function is apparent later in the discussion, when Clutton-Brock and Albon consider the hypothesis that roaring serves as an advertisement enabling stags to assess others' fighting ability. Although their careful observations indicate that stags rarely defeat those by whom they have been out-roared, they recognize that their data leave open other possibilities for the relation between roaring and fighting ability. They suggest that fighting and roaring may both draw on the same groups of muscles, so that roaring serves as an "honest advertisement" to other stags. But they note that this depends on assuming that "selection could not produce a mutant which was able to roar more frequently without increasing its strength or stamina in fights."[26] I interpret the caution expressed in their discussion to be grounded in recognition of the stringent conditions that must be met in showing that a form of behavior maximizes reproductive success, and thus their reliance on the strong version of the etiological conception.

I now turn to two physiological studies in which the connection to evolution is far less evident. Here, there are neither detailed measurements of the fitnesses (or proxies such as survivorships) of rival types of organism (as in the study of golden silk spiders) or connections with mathematical models of a selection process (as in the investigation of the roaring of stags). Instead, the authors undertake a mechanistic analysis of the workings of a biological system. Consider the following discussion of digestion in insects.

Food in the midgut is enclosed in the peritrophic membrane, which is secreted by cells at the anterior end of the midgut in some insects or formed by the midgut epithelium in most. It is secreted continuously or in response to a distended midgut, as in biting flies. It is likely that the peritrophic membrane has several functions, although the evidence is not conclusive. It may protect the midgut epithelium from abrasion by food or from attack by microorganism or it may be involved in ionic interactions within the lumen. It has a curious function in some coleopterous larvae, where, in various ways, it is used to make the cocoon.[27]

The interesting point about this passage is that it could easily be accepted by a biologist ignorant of or hostile to evolutionary theory. So long as one has a sense of the overall life of an insect and of the conditions that must be satisfied for the insect to thrive, one can view the peritrophic membrane as making a causal contribution to the organism's flourishing. Of course, Darwinians will view these conditions as grounded in selection pressures to which insects must respond, but physiology can keep this Darwinian perspective very much in the background. It is enough to recognize that insects must have a digestive system capable of processing food items, that the passage of food through the system must not abrade the cells lining the gut, and so forth. I suggest that this, like so many other physiological discussions, presupposes a background picture of the selection pressures on the organisms under study and analyzes the causal mechanisms that work to meet those pressures, without attending to the fitness of alternatives that would have to be considered to underwrite a claim about the operation of selection.

Finally, I turn to a developmental study of sexual differentiation in *Drosophila*.[28] The problem is to understand simultaneously how an embryo with two X chromosomes becomes a female, how an embryo with one X chromosome becomes a male, and how the organism compensates for the extra chromosomal material found in females. The author (M. Kaulenas) summarizes a complex causal story, as follows:

The primary controlling agent in sex determination and dosage compensation is the ratio between the X chromosomes to sets of autosomes (the X:A ratio). This ratio is "read" by the products of a number of genes; some of which function as numerator elements, while others as denominator elements. Two of the numerator genes have been identified [*sisterless a (sis a)* and *sisterless b (sis b)*] and others probably exist. The denominator elements are less clearly defined. The end result of this "reading" is probably the production of DNA-binding proteins, which, with the cooperation of the *daughterless (da)* gene product (and possibly other components) activate the *Sex lethal (sxl)* gene. This gene is the key element in regulating female differentiation. One early function is autoregulation, which sets the gene in the functional mode. Once functional, it controls the proper expression of the *doublesex (dsx)* gene. The function of *dsx* in female somatic cell differentiation is to suppress male differentiation genes. *Dsx* needs the action of the *intersex (ix)* gene for this function. Female differentiation genes are not repressed, and female development ensues.[29]

Here is a causal story about how female flies come to express the appropriate proteins in their somatic cells. The elements of the story concern the ways in which particular bits of DNA code for proteins that either activate the right genes or block transcription of the wrong ones. In the background is a general picture of how selection acts on sexually reproducing organisms, a picture that recognizes the selectively dis-advantageous effects of failing to suppress one set of genes (those asso-ciated with the distinctive reactions that occur in male somatic cells) and of failing to activate the genes in another set (those whose action is responsible for the distinctive reactions of female somatic cells). The functions of the specific genes identified by Kaulenas are understood in terms of the causal contributions they make in a complex process. There is no attempt to canvass the genetic variation in *Drosophila* populations or to argue that the specific alleles mentioned are somehow fitter than their rivals. The discussion takes for granted a particular type of selection pressure—thus adopting the environment-centered perspective on evolu-tion—and considers only the causal interactions that result in a re-sponse to that selection pressure. The causal analysis is vividly presented in a diagram (reproduced in the figure 18.1), which shows the kinship between the type of mechanistic approach adopted in this study and the analysis of complex systems designed by human beings. Selection fur-nishes a context in which the overall design is considered, and, within that context, the physiologist tries to understand how the system works.

I offer these four examples as paradigmatic of two very common types of biological practice offering ascriptions of function. I hope that it is evi-dent how introducing the strong etiological conception within the last two would distort the character of the achievement, rendering it vulner-able to skeptical worries about the operation of selection that are in fact quite irrelevant. By the same token, it is impossible to appreciate the line of argument offered in the explicitly evolutionary studies without recog-nizing the stringent requirements that the strong etiological conception imposes. There are undoubtedly many instances in which the notion of function intended is far less clear. I believe that keeping our attention focused on paradigms will be valuable in the work of disambiguation.

VII

Philosophical discussions of function have tended to pit different analyses and different intuitions against one another without noting the pluralism inherent in biological practice.[30] On the account I have offered here,

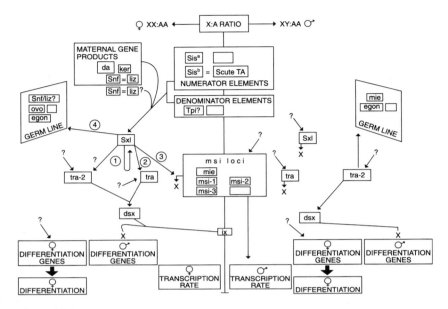

Figure 18.1
Diagram illustrating the interrelationships of the genes involved in the control of sexual differentiation and dosage compensation in *Drosophila* (from M. Kaulenas, *Insect Accessory Reproductive Structures: Function, Structure and Development*, New York: Springer, 1992, 18).

there is indeed a unity in the concept of function, expressed in the connection between function and design, but the sources of design are at least twofold and their relation to the bearers of function may be more or less direct. This means, I believe, that the insights of the main competitors, Wright's etiological approach and Cummins's account of functional analysis, can be accommodated (and, as the discussion in section IV indicates, variants of the etiological approach can also be given their due).

The result is a general account of functions that covers both artifacts and organisms. I believe that it can also be elaborated to cover the apparently mixed case of functional ascriptions to social and cultural entities, in which both explicit intentions and processes of cultural selection may act together as sources of design. But working out the details of such impure cases must await another occasion.

Notes

I am extremely grateful to the Office of Graduate Studies and Research at the University of California–San Diego for research support, and to Bruce Glymour for research assistance. My thinking about functional attributions in biology has been greatly aided by numerous conversations with Peter Godfrey-Smith. Despite important residual differences, I have been much influenced by Godfrey-Smith's careful elaboration and resourceful defense of an etiological view of functions.

1. I shall sometimes identify the bearers of functions simply as "entities," sometimes, for stylistic variety, talk of traits, structures, organs, behaviors as having functions. I hope it will be obvious throughout that may usage is inclusive.

2. This aspect of Darwin's accomplishment is forcefully elaborated by Richard Dawkins in *The Blind Watchmaker* (London, 1987). Although I have reservations about Dawkins's penchant for seeing adaptation almost everywhere in nature, I believe that he is quite correct to stress Darwin's idea of design without a designer.

3. The fact that the intentions of the Creator are in the remote background in much pre-Darwinian physiological work is one of the two factors that allow for continuity between pre-Darwinian physiology and the physiology of today. As I shall argue later, appeals to selection as a source of design are kept in the remote background in contemporary physiological discussions.

4. Richard Lewontin, "Organism and Environment" (manuscript), and Lewontin and Richard Levins, *The Dialectical Biologist* (Cambridge, Mass., 1987).

5. Wright, "Functions," *Philosophical Review* 82 (1973): 139–168. For further elaboration, see Ruth Millikan, *Language, Thought, and Other Biological Categories* (Cambridge, Mass., 1984); Karen Neander "Functions as Selected Effects: The Conceptual Analyst's Defense," *Philosophy of Science* 58, 168–184; and Peter Godfrey-Smith, "Functions," forthcoming.

6. This example stems from Christopher Boorse, "Wright on Functions," *Philosophical Review* 85 (1976): 70–86.

7. This way of evading the trouble is due to Millikan, *op. cit.*

8. These issues are broached by Godfrey-Smith in his forthcoming paper. He and I are in broad agreement about questions of timing and diverge in our approaches to the second cluster of questions.

9. Here, and in the ensuing discussion, I permit myself an obvious shorthand. In speaking of the origination of an entity in an organism I do not, of course, mean to refer to the mutational and developmental history that lies behind the emergence of the entity in an individual organism but in the process that culminates in the initial fixation of that entity in members of the population. I hope that this abbreviatory style will not cause confusions.

10. See N. Tinbergen, "On War and Peace in Animals and Man," *Science* 160 (1968): 1411–1418.

11. Gould and Vrba, "Exaptation—A Missing Concept in the Science of Form." *Paleobiology* 8 (1982): 4–15.

12. Mayr, "The Emergence of Evolutionary Novelties" in *Evolution and the Diversity of Life* (Cambridge, Mass., 1976).

13. The point that biologists often ignore in practice the strictures on adaptationist claims that they recognize in theory is very clearly expressed in Gould and Lewontin "The Spandrels of San Marco and the Panglossian Paradigm: A Critique of the Adaptationist Programme," *Proceedings of the Royal Society* B 205 (1979): 581–598.

14. Bigelow and Pargetter, "Functions," *Journal of Philosophy* 84 (1987); 181–196.

15. See his "Functions" (forthcoming). For my own commitments to a similar view see "Why Not The Best?" in John Dupre, *The Latest on the Best: Essays on Evolution and Optimality* (Cambridge, Mass., 1988), and "Developmental Decomposition and the Future of Human Behavioral Ecology," *Philosophy of Science* 57 (1990): 96–117.

16. See, for example, the discussion of Geoffrey Parker's ingenious and sophisticated work on copulation time in male dungflies in my *Vaulting Ambition: Sociobiology and the Quest for Human Nature* (Cambridge, Mass., 1985), chapter 5.

17. See the examples given in section VI below.

18. Robert Cummins, "Functional Analysis," *Journal of Philosophy* 72 (1975): 741–765.

19. T. Christenson and P. Wenzl, "Egg-Laying of the Golden Silk Spider, *Nephila cavipes* L. (Araneae, Araneidae): Functional Analysis of the Egg Sac," *Animal Behaviour* 28 (1980): 1110–1118.

20. I should note here that spiderlings typically overwinter in the egg sac, so that the period of a few weeks represents a fall only a *short* time before the usual time of emergence. Thus the successful instances are those in which the normal course of development is only slightly perturbed.

21. Christenson and Wenzl, *op. cit.*, 1114.

22. Ibid., 1110.

23. Ibid., 1115.

24. T. Clutton-Brock and S. Albon, "The Roaring of Red Deer and the Evolution of Honest Advertisement," *Behaviour* 69 (1979): 145–168.

25. Ibid., 145.

26. Ibid., 165.

27. J. McFarlane, "Nutrition and Digestive Organs," in M. Blum, *Fundamentals of Insect Physiology* (Chichester, 1985), 59–90. The quoted passage is from p. 64.

28. M. Kaulenas, *Insect Accessory Reproductive Structures: Function, Structure, and Development* (New York, 1992), Section 2.3 "Genetic Control of Sexual Differentiation."

29. Ibid., 17.

30. As I have argued elsewhere, biological practice is pluralistic in its employment of concepts of gene and species and in its identification of units of selection. See my essays, "Genes" (*British Journal for the Philosophy of Science* 33 [1982]: 337–359) and "Species" (*Philosophy of Science* 51 [1984]: 308–333), and Kim Sterelny and Philip Kitcher, "The Return of the Gene" (*Journal of Philosophy* 85 [1988]: 335–358).

V

Design

19

Historical Biology and the Problem of Design

George V. Lauder

1   Introduction

The problem of biological design is one of the oldest topics in biology. It has been clear, at least from the time of Aristotle, that organisms possess a "unity of plan" or underlying commonality of structure, and that the function of biological features bears some relationship to their form. E. S. Russell in his classic book (1916) has interpreted the history of morphology as an attempt to understand the relationship between form and function.

The concepts of commonality of structure and functional design play prominent roles in modern evolutionary morphology. Yet we seem to have made little progress toward a synthetic approach to biological structure, or towards a "science of form" (Gould, 1971). In this paper I will recast the dichotomy between form and function into a somewhat different framework, that of historical vs. equilibrium analysis, suggest how a synthesis might be achieved (and why it is necessary), and outline a set of outstanding problems that must be examined if a general theory about the transformation of biological design is to be developed. Two related general questions can be considered to underly this discussion. (1) Why does the range of extant phenotypes, when mapped onto a theoretical "morphospace," fill so little of it (Raup, 1966; McGhee, 1980)? (2) Is it possible to produce testable explanations for why certain morphologies (such as vertebrates with wheeled appendages) have not evolved?

To approach these questions, two classes of morphological analysis may be distinguished: equilibrium and historical. Equilibrium analysis (Lewontin, 1969; Lauder, 1981) is by far the dominant approach to the analysis of biological design in the fields of functional morphology, ecology, and population genetics. Organisms are considered to be in equilibrium with the environment and present-day environmental correlates of structure are sought in an attempt to explain morphology (e.g. Wiens and Rotenberry, 1980). An equilibrium approach may also have a time axis in that changes in the environment through time are related to the morphology of those organisms inhabiting the environments, and structural change is described as an adaptive modification to changing external conditions (e.g. Valentine, 1973, 1975).

Historical analysis is concerned with the evolutionary transformation of intrinsic organizational features (Lauder, 1981) and not with the relationship between form and the (extrinsic) environment. A crucial element in analyzing structural transformation is the phylogenetic reconstruction of nested sets of homologies which indicate the historical sequence in which new morphological features were acquired in a lineage (Eldredge and Cracraft, 1980). A phylogenetic approach to the analysis of design reveals the historical pattern by which any particular combination of structural features was constructed. (In this paper I define *design* as the organization of biological structure in relation to an hypothesized function. *Adaptation* [following Lewontin, 1978; Cracraft, 1981] is restricted to features that have arisen by means of natural selection.)

If an understanding is ever to be achieved of the pathways by which form has diversified to fill the occupied phenotype space, or of the mechanics underlying these evolutionary patterns, then a synthesis of both extrinsic environmental limits to form and intrinsic structural and functional properties must be achieved within a phylogenetic (historical) framework.

2   Extrinsic Determinants of Design

There is little doubt that the environment poses at least general constraints on the design of organisms (Lewontin, 1978) and the fields of biomechanics and functional morphology have been largely devoted to analyzing the relationship between biological structure and environmental forces experienced by organisms (e.g. Gans, 1974; Wainwright, 1980). For the most part, this work falls under the equilibrium method defined above. Two main contributions have emerged from this approach: (1) a reasonably precise characterization of the structural principles and systems necessary to meet particular functional situations, and (2) the delimitation of boundary conditions to biological design. As an example of a structural property found widely in biological systems, consider helically-wound collagen fibers. In many organisms in which resistance by the skin or body surface to radial expansion is necessary, while retaining body flexibility and/or the ability to elongate, helically-wound fibers occur in the body surface (Wainwright *et al.*, 1975). A crossed helix is formed with fibers at a helical angle of from 30° to 70°. This arrangement permits lateral bending and elongation but resists radial expansion and buckling due to internal pressure changes. The diversity of organisms in which this system has been found (sharks, teleost fishes, squid, earthworms, nematodes, plants) attests to its generality as a constructional principle, and Wainwright *et al.* (1975) propose this concept as one of their six design principles for biological systems. As a second example, Webb (1978) has examined the functional tradeoffs between rapid accelerations and steady swimming in teleost fishes. The theoretical optima for each locomotor mode is definable by a mechanical analysis. Rapid accelerations are best achieved (in part) be increasing depth along a flexible body, while steady swimming efficiency increases with body stiffness, caudal fin aspect ratio, and reduced total caudal fin area. Webb (1978) was able to demonstrate experimentally that fishes of different body shape have very similar fast acceleration performances due to the tradeoffs in different species between lateral body profile and the mass of body musculature. This type of analysis is of interest because of the relatively well-defined mechanical theory underlying the interpretation of design, and the consequent utility of these analyses in making precise comparative predictions about the relative performance of unstudied species.

Boundary conditions on the range of possible morphologies may also be revealed by a biomechanical analysis. McGhee (1980) has shown that much of the "morphospace" available to brachiopods is unfilled because of geometrical constraints on shell growth. McGhee defined the theoretical morphospace available to brachiopod shells by elaborating a geometrical model of shell growth and ascribed the limited distribution of actual shapes within the theoretical space to two factors: the fact that the two valves of brachiopods must articulate and thus cannot overlap, and small size imposed by the ratio between surface area and internal volume.

While these examples illustrate the utility of an extrinsic explanatory framework for understanding the relationship between structure and function and limits to design, there are important restrictions on the explanatory power of this approach. What of the unoccupied phenotype space that is not prohibited by mechanical factors? How do we explain the absence of certain morphologies when there appears to be no mechanical, functional, or structural reason why such an organism might not have existed? Finally, how is an observed lack of correlation between the structures possessed by organisms and environmental factors to be interpreted? If an organism predicted on the basis of a mechanical argument to have a helically-wound fiber system does not, how can this discrepancy be explained? The traditional escape is to interpret the absence of predicted features, the lack of a predicted structure-environment correlation, or an unfilled portion of the morphospace in terms of the original equilibrium premise. An additional unanalyzed component of the system has prevented the close concordance of structure and environment; or in the case of the morphospace, the requisite historical pattern of environmental change has not been present to generate the selective forces necessary to produce the new morphologies (Raup and Stanley, 1971). But these explanations beg the historical question: might there not be intrinsic, phylogenetic constraints on form that canalize or facilitate certain evolutionary transformations? In order to evaluate this possibility, we need a testable program for historical research.

3   Historical Morphology: Intrinsic Determinants of Design

The study of evolutionary morphology has been dominated by the search for explanations of unique structural or functional aspects of design. The difficulty with this approach is that while a reasonable hypothesis may be formulated for why any particular unique structural feature has evolved, it is difficult if not impossible to test the hypothesis. The foundation of a scientific approach to historical events lies in testing by comparison. Hypotheses of structure-function relationship in any group of organisms can be tested by finding similar structures in another group that has acquired them independently, and determining if the predicted functional correlate is present. If not, then the hypothesis is refuted. But this procedure is impossible if the structures or functions selected for analysis are unique.

The most clear-cut cases in evolutionary biology of attempts to explain unique features involve discussions of the causal basis for the origin and "adaptive" radiation of major lineages. The usual procedure is to identify some morphological or physiological feature that seems to play an important role in the biology of a group, and then to maintain that this feature is causally related to both the morphological change and subsequent speciation pattern of the taxon. One example is the analysis of the adaptive radiation of multicellular organisms provided by Stebbins (1973). Stebbins identified the "most significant determining characteristics" of adaptive radiation such as a rigid cell wall and the presence of cellular polarization. These structural properties are then invoked as being causally related to the adaptive radiation of multicellular organisms. A similar approach to other taxa may be found in Liem (1973), Stanley (1968), and Lewis (1972). The difficulty with these discussions is the untestable nature of the premise: that a particular structural feature can be identified as having caused adaptive radiation. How would one choose between two different structural attributes of a taxon, each cited by different authorities as being *the* feature that caused adaptive radiation?

An alternative approach to the historical analysis of design is possible, however, and I suggest three principles for a scientific (testable) approach to the evolutionary transformation of structure. (1) Explanation must be for general properties of a structure or functional complex, and not for unique features. (2) A phylogenetic hypothesis of relationship that depicts the nested hierarchy of structural features in the historical sequence they were acquired (a cladogram) is fundamental to the analysis of historical patterns. (3) Historical hypotheses are tested by the examination of related monophyletic lineages. Hypotheses about classes of evolutionary events thus are tested by repeated attempts at refutation via the phylogenetic history of individual lineages.

As an example of how these concepts might be applied, consider the interesting hypothesis of Vermeij (1973*a*) that

the potential versatility of a given taxon or body plan is determined by
the number and range of independent parameters controlling form.

This hypothesis has stimulated considerable interest and yet has not been explicitly tested, although Vermeij (1973b) has listed several cases which seem to support it. In order to test Vermeij's explanation for morphological diversity, it is necessary to realize that the hypothesis is fundamentally historical in nature. That is, a specific evolutionary pattern (an increase in structural versatility) is predicted to occur as a consequence of a change in structural organization. This is exactly the general type of hypothesis that can be tested with respect to nested sets of structural features in a class of monophyletic clades. As increasingly restricted sets of taxa (cladistically more derived forms) are considered, an increase in structural diversity should occur for those clades which primitively possess an increased number of parameters controlling form. To state Vermeij's hypothesis explicitly: primitive members of morphologically diverse monophyletic lineages possess an increased number of independent parameters controlling form compared to closely related (sister) lineages whose primitive members lack the increased parameters. The hypothesis is refuted if these related lineages show similar patterns of structural diversification to the taxa with a primitively greater number of morphogenetic pathways. When stated in this way, it is clear that a test of this hypothesis is possible by the repeated examination of sister lineages (lineages more closely related to each other than to any other lineage) which differ in the number of parameters controlling form. If the initial relationship is corroborated, then there should be a positive correlation between the number of controlling parameters and the degree of body-plan versatility as expressed in morphological diversity. Ideally, such tests should span a wide range of diverse taxa differing in general bauplan to determine the generality of the phenomenon.

The basic theme underlying this approach to historical morphology is that general aspects of structural organization such as the number of functional linkage systems, the number of epigenetic pathways controlling form, or complexity of organization, canalize historical change in morphology and restrict (or, possibly, increase) the potential structural and functional diversity of descendent taxa. Thus, when a potential morphospace is found to be partially unoccupied, the explanation may not lie with a lack of appropriate selection pressures or with mechanical reasons for why morphologies in the unoccupied area would not work, but rather in the phylogenetic constraints on possibilities for morphological change. This possibility is usually acknowledged (e.g. Rensch, 1959; Riedl, 1978; Seilacher, 1973), but a testable research program in historical morphology has not yet been applied to the distribution of structure in the theoretical phenotype space.

An interesting consequence of a phylogenetic/historical approach to biological design is the possibility of elucidating general laws of structural change in organisms. To the extent that a repeated examination of the historical consequences of intrinsic organizational properties of organisms in many monophyletic lineages reveals substantial regularity, then a law-like deterministic view of structural evolution is corroborated. This result would be particularly important in the light of recent stochastic simulations of morphological evolution (Raup and Gould, 1974; Raup, 1977; Gould *et al.*, 1977; see Schopf, 1979). These investigations have shown that simple stochastic models of morphological evolution in randomly generated clades produces patterns of clade organization, extinction, and morphology that are extremely similar to those observed in biological systems. The consequence of these results for deterministic explanations of morphological evolution is severe: if observed patterns in nature cannot be convincingly distinguished from stochastically generated ones, then the dominant explanatory framework in evolutionary biology today would appear to have limited utility. Despite this challenge, there does not appear to have been a convincing reply. Evolutionary morphologists have continued to explain structural patterns within a deterministic framework.

But congruent patterns of structural transformation in unrelated monophyletic lineages provide the possibility of testing the importance of stochastic vs. deterministic factors. The repeated similarity of evolutionary structural modification in unrelated groups corroborates intrinsic phylogenetic constraints as important determinants of biological design as it refutes both stochastic and environmental (extrinsic) deterministic explanations.

The analogy between this approach to historical patterns of form and the emerging science of historical biogeography is very close. Historical biogeography (Nelson, 1969; Nelson and Platnick, 1981) is concerned with the distribution of organisms in space as related to both historical events of earth geography and the phylogenetic (genealogical) relationships of the earth's biota. Congruence between the pattern of environmental fragmentation in space and time (vicariance) with the pattern of genealogical branching (Rosen, 1978) supports a causal connection between genealogical diversification and environmental change. Thus, to the extent that genealogical splitting is congruent with patterns of environmental fragmentation, then unique dispersalist scenarios for how each lineage came to occupy its present range are rendered unparsimonious. A stochastic explanation of biotic distribution is also rejected with increasing confidence with higher congruence between geology and the phylogenetic histories of unrelated lineages. Invoking dispersal to explain the distribution of each species while failing to see if a general pattern that is congruent with geological and environmental history is present, is comparable to adducing unique selection pressures and historical sequences of environmental change to explain the forms of each evolutionary lineage. Such interpretations are rendered unparsimonious if general historical patterns common to taxa with widely differing bauplans are found.

4   Conclusions and Prospectus

The problem of biological design is fundamentally a historical one. Without an understanding of the past history of organisms, there is little hope of effectively analyzing the relationship between form and function and the interaction between organisms and the environment. The past constrains and determines future directions of structural change and organisms carry with them, through the retention of primitive characters, a record of their past history.

Yet it is precisely the historical element that has been missing from explanations for biological design. To be sure, the existence of historical constraints is widely mentioned, but there have been few attempts to directly grapple with the testing of historical hypotheses, or the influence of intrinsic factors on the design of biological systems. (Nowhere is the lack of a historical framework more evident than in D'Arcy Thompson's, 1942, *On Growth and Form*, and the widespread influence of his approach to design is evidence of the minor role historical factors have played in the explication of organic form.) The failure to consider adequately this class of explanation is particularly apparent in two areas of research: (1) optimal modelling of structure, ecology, or behavior, and (2) the "deductive" approach to morphology (Dullemeijer, 1974; Gutmann, Vogel and Zorn, 1978).

Optimal models explicitly neglect the historical component of biological systems and treat the organism-environment relationship as though the system were in equilibrium, evidenced most clearly by the widespread use of the *ceteris paribus* assumption (Lewontin, 1978). The "constraints" inherent in optimal models reflect the initial assumptions about the equilibrium situation, and not historical factors or intrinsic phylogenetic limits to optimality. In so doing, these models reveal little about the potential transformation of design, and the conclusions tend to be a reflection of the initial factors selected for analysis (Lewontin, 1978).

"Deductive" models for the evolution of design (e.g. Gutmann, 1977; Gutmann *et al.*, 1978) utilize an *a priori* functional model or mechanical analysis to predict a morphological transition between different structural types. The method of this approach (see Gutmann, 1977) is to (1) hypothesize selective factors that have governed morphological change, (2) propose a gradual series of "adaptive" structural and functional modifications that will transform one bauplan into another (annelid into brachiopod, for example), and (3) claim that this morphological series of "adaptive" changes reflects the actual process by which evolutionary transformation took place. The basis of this approach is the assumption that selective forces can be identified in the historical record and that an alternative model, equally consistent with simple structural and mechanical principles, will not provide as convincing a demonstration of structural transformation. Without a phylogenetic hypothesis independent of the hypothesized transformation series of designs, the model cannot be tested. Thus, this approach is circular because the test of the proposed biomechanical model requires the very phylogeny the model was used to construct.

With this background of what I perceive to be several of the methodological problems and logical difficulties with some current approaches to the analysis of biological design, three inter-related issues may be defined that may provide a focus for future research.

1. Are there historical laws of morphologic change? To the extent that clades of differing bauplan exhibit historical congruence in basic pathways of structural and functional transformation attributable to a common general organizational property (such as complexity), then stochastic or individual selectionist explanations become of less generality. Simpson (1964, p. 128) maintained that

the search for historical laws is ... mistaken in principle ... historical events are unique, usually to a high degree, and hence cannot embody laws defined as recurrent repeatable relationships.

This characterization of history is a recipe for the continued search for individual explanations and untestable historical scenarios as have characterized biogeography for much of this century. Historical events are not unique, organisms exhibit a commonality of structure as a result of conservative epigenetic pathways (Alberch, 1980), and thus general properties of biological design are available for analyzing the occurrence of repeated transformations of structure and function.

2. What is the null hypothesis for extrinsic explanations of design? One of the values of stochastic modelling of morphological change (Raup, 1977) is that a null hypothesis is provided from which deviations in structure can be determined. In a similar fashion, general patterns of structural transformation in lineages that are shown to be the consequence of intrinsic organizational properties, become the null hypothesis for extrinsic environmental explanations for structure. These explanations must be assessed against the "expected" properties of design that result from intrinsic phylogenetic determinants of form. The demonstration that intrinsic properties have played an important role in governing the transformation of design does not rule out the possibility that selection and changing extrinsic conditions ultimately drive change. However, the historical record provides only limited access to extrinsic data, while it does provide the opportunity of examining intrinsic factors.

Finally, 3. is it possible to test explanations for why certain designs have *not* evolved? To some extent, yes. Overlapping valves in brachiopods appear to be a prohibited morphology for simple mechanical reasons. Similarly, coiled cephalopods do not look like submarines (despite the close similarity in functional requirements) because of constraints associated with accretionary growth of shells (Raup, 1972). On the other hand, a large portion of the unoccupied morphospace is not prohibited by mechanical restraints, and I can conceive of only three possible explanations for this: chance, environmental determinism, or intrinsic constraints. Intrinsic explanations may reveal that canalization of the direction of morphological change in evolution is severe and that chance plays a relatively limited role. Invoking the lack of proper selection pressures and environmental change to produce morphologies in the unoccupied morphospace produces an untestable explanation, as our ability to analyze these factors even in present-day environments is very restricted.

If intrinsic constraints are revealed to be very course and only to provide broad limits to the potential pathways of structural change, then chance factors may have played an important role in generating the range of extant phenotypes.

The key to discovering the limits to deterministic explanation in the historical record will be the extent to which general historical pathways in the transformation of biological design are revealed by a phylogenetic analysis of structural and functional patterns.

References

Alberch, P. (1980). *Am. Zool.* 20, 653.

Cracraft, J. (1981). *Am. Zool.* 21, 21.

Dullemeijer, P. (1974). *Concepts and Approaches in Animal Morphology.* The Netherlands: Van Gorcum.

Eldredge, N. and Cracraft, J. (1980). *Phylogenetic Patterns and the Evolutionary Process.* New York: Columbia.

Gans, C. (1974). *Biomechanics, an Approach to Vertebrate Biology.* Philadelphia: J. B. Lippincott.

Gould, S. J. (1971). *New Lit. Hist.* 2, 229.

Gould, S. J., Raup, D. M., Sepkoski, J. J., Schopf, T. J. M. and Simberloff, D. (1977). *Paleobiol.* 3, 23.

Gutmann, W. F. (1977). In: *Major Patterns in Vertebrate Evolution* (M. K. Hecht, P. C. Goody and B. M. Hecht, eds), pp. 645–669, New York: Plenum Press.

Gutmann, W. F., Vogel, K. and Zorn, H. (1978). *Science* 199, 890.

Lauder, G. V. (1981). *Paleobiol.* 7, 433.

Lewis, O. J. (1972). In: *The Functional and Evolutionary Biology of Primates* (R. Tuttle, ed.), pp. 207–222, Chicago: Aldine.

Lewontin, R. C. (1969). *J. Hist. Biol.* 2, 35.

Lewontin, R. C. (1978). *Sci. Am.* 239, 156.

Liem, K. F. (1973). *Syst. Zool.* 22, 425.

McGhee, G. R. (1980). *Paleobiol.* 6, 57.

Nelson, G. (1969). *Syst. Zool.* 18, 243.

Nelson, G. and Platnick, N. (1981). *Systematics and Biogeography, Cladistics and Vicariance.* New York: Columbia.

Raup, D. M. (1966). *J. Paleon.* 40, 1178.

Raup, D. M. (1972). In: *Models in Paleobiology* (T. J. M. Schopf. ed.), San Francisco: Freeman, Cooper and Co., pp. 29–44.

Raup, D. M. (1977). In: *Patterns of Evolution, as Illustrated by the Fossil Record* (A. Hallam, ed.), New York: Elsevier, pp. 59–78.

Raup, D. M. and Gould, S. J. (1974). *Syst. Zool.* 23, 305.

Raup, D. M. and Stanley, S. M. (1971). *Principles of Paleontology.* San Francisco: Freeman.

Rensch, B. (1959). *Evolution above the Species Level.* New York: Columbia.

Riedl, R. (1978). *Order in Living Organisms.* New York: Wiley.

Rosen, D. E. (1978). *Syst. Zool.* 27, 159.

Russell, E. S. (1916). *Form and Function.* London: John Murray.

Schopf, T. J. M. (1979). *Paleobiol.* 5, 337.

Seilacher, A. (1973). *Syst. Zool.* 22, 451.

Simpson, G. G. (1964). In: *This View of Life.* New York: Harcourt, Brace and World, pp. 121–148.

Stanley, S. M. (1968). *J. Paleont.* 42, 214.

Stebbins, G. L. (1973). *Syst. Zool.* 22, 478.

Thompson, D. W. (1942). *On Growth and Form.* Cambridge: Cambridge University Press.

Valentine, J. W. (1973). *Evolutionary Paleoecology of the Marine Biosphere.* Englewood Cliffs, New Jersey: Prentice Hall.

Valentine, J. W. (1975). *Am. Zool.* 15, 391.

Vermeij, G. J. (1973a). *Proc. Natn. Acad. Sci. U.S.A.* 70, 1936.

Vermeij, G. J. (1973b). *Syst. Zool.* 22, 466.

Wainwright, S. A. (1980). In: *The Mechanical Properties of Biological Materials* (J. F. V. Vincent and J. D. Curry, eds), London: Society for Experimental Biology, pp. 437–453.

Wainwright, S. A., Biggs, W. D., Currey, J. D. and Gosline, J. M. (1975). *Mechanical Design in Organisms.* London: Edward Arnold.

Webb, P. W. (1978). *J. Exp. Biol.* 74, 211.

Wiens, J. A. and Rotenberry, J. T. (1980). *Ecol. Monogr.* 50, 287.

20

Exaptation—A Missing Term in the Science of Form

Stephen Jay Gould and Elisabeth S. Vrba[1]

I Introduction

We wish to propose a term for a missing item in the taxonomy of evolutionary morphology. Terms in themselves are trivial, but taxonomies revised for a different ordering of thought are not without interest. Taxonomies are not neutral or arbitrary hat-racks for a set of unvarying concepts; they reflect (or even create) different theories about the structure of the world. As Michel Foucault has shown in several elegant books (1965 and 1970, for example), when you know why people classify in a certain way, you understand how they think.

Successive taxonomies are the fossil traces of substantial changes in human culture. In the mid 17th century, madmen were confined in institutions along with the indigent and unemployed, thus ending a long tradition of exile or toleration for the insane. But what is the common ground for a taxonomy that mixes the mad with the unemployed—an arrangement that strikes us as absurd. The "key character" for the "higher taxon," Foucault argues, was idleness, the cardinal sin and danger in an age on the brink of universal commerce and industry (Foucault's interpretation has been challenged by British historian of science Roy Porter, MS). In other systems of thought, , at seems peripheral to us becomes central, and distinctions essential to us do not matter (whether idleness is internally inevitable, as in insanity, or externally imposed, as in unemployment).

II   Two Meanings of Adaptation

In the vernacular, and in sciences other than evolutionary biology, the word adaptation has several meanings all consistent with the etymology of *ad* + *aptus*, or towards a fit (for a particular role). When we adapt a tool for a new role, we change its design consciously so that it will work well in its appointed task. When creationists before Darwin spoke of adaptation—for the term long precedes evolutionary thought—they referred to God's intelligent action in designing organisms *for* definite roles. When physiologists claim that larger lungs of Andean mountain peoples are adapted to local climates, they specify directed change for better function. In short, all these meanings refer to historical processes of change or creation for definite functions. The "adaptation" is designed specifically for the task it performs.

In evolutionary biology, however, we encounter two different meanings—and a possible conflation of concepts—for features called adaptations. The first is consistent with the vernacular usages cited above: a feature is an adaptation only if it was built by natural selection for the function it now performs. The second defines adaptation in a static, or immediate way as any feature that enhances current fitness, regardless of its historical origin. (As a further confusion, adaptation refers both to a process and a state of being. We are only discussing state of being here— that is, features contributing to fitness. We include some comments about this further problem in section VIE.)

Williams, in his classic book on adaptation, recognized this dilemma and restricted the term to its first, or narrower, meaning. We should speak of adaptation, he argues, only when we can "attribute the origin and perfection of this design to a long period of selection for effectiveness in this particular role" (1966, p. 6). In his terminology, "function" refers only to the operation of adaptations. Williams further argues that we must distinguish adaptations and their functions from fortuitous effects. He uses "effect" in its vernacular sense—something caused or produced, a result or consequence. Williams' concept of "effect" may be applied to a character, or to its usage, or to a potential (or process), arising as a consequence of true adaptation. Fortuitous effect always connotes a con-

Table 20.1
A taxonomy of fitness.

Process	Character		Usage
Natural selection shapes the character for a current use—adaptation	adaptation		function
A character, previously shaped by natural selection for a particular function (an adaptation), is coopted for a new use—cooptation	exaptation	aptation	effect
A character whose origin cannot be ascribed to the direct action of natural selection (a nonaptation), is coopted for a current use—cooptation			

sequence following "accidentally," and not arising directly from construction by natural selection. Others have adopted various aspects of this terminology for "effects" *sensu* Williams (Paterson 1981; Vrba 1980; Lambert, MS). However, Williams and others usually invoke the term "effect" to designate the *operation* of a useful character *not* built by selection for its current role—and we shall follow this restriction here (table 20.1). Williams also recognizes that much haggling about adaptation has been "encouraged by imperfections of terminology" (1966, p. 8), a situation that we hope to alleviate slightly.

Bock, on the other hand, champions the second, or broader, meaning in the other most widely-cited analysis of adaptation from the 1960s (Bock and von Wahlert 1965; Bock 1967, 1979, 1980). "An adaptation is, thus, a feature of the organism, which interacts operationally with some factor of its environment so that the individual survives and reproduces" (1979, p. 39).

The dilemma of subsuming different criteria of historical genesis and current utility under a single term may be illustrated with a neglected example from a famous source. In his chapter devoted to "difficulties on theory," Darwin wrote (1859, p. 197):

The sutures in the skulls of young mammals have been advanced as a beautiful adaptation for aiding parturition, and no doubt they facilitate, or may be indispensable for this act; but as sutures occur in the skulls of young birds and reptiles, which have only to escape from a broken egg, we may infer that this structure has arisen from the laws of growth, and has been taken advantage of in the parturition of the higher animals.

Darwin asserts the utility, indeed the necessity, of unfused sutures but explicitly declines to label them an adaptation because they were not built by selection to function as they now do in mammals. Williams follows Darwin and would decline to call this feature an adaptation; he would designate its role in aiding the survival of mammals as a fortuitous effect. But Bock would call the sutures and the timing of their fusion an adaptation, and a vital one at that.

As an example of unrecognized confusion, consider this definition of adaptation from a biological dictionary (Abercrombie et al. 1951, p. 10): "Any characteristic of living organisms which, in the environment they inhabit, improves their chances of survival and ultimately leaving descendants, in comparison with the chances of similar organisms without the characteristic; natural selection therefore tends to establish adaptations in a population." This definition conflates current utility with historical genesis. What is to be done with useful structures not built by natural selection for their current role?

III A Definition of Exaptation

We have identified confusion surrounding one of the central concepts in evolutionary theory. This confusion arises, in part, because the taxonomy of form in relation to fitness lacks a term. Following Williams (see Table 20.1), we may designate as an *adaptation* any feature that promotes fitness and was built by selection for its current role (criterion of *historical genesis*). The operation of an adaptation is its *function*. (Bock uses the term function somewhat differently, but we believe we are following the biological vernacular here.) We may also follow Williams in labelling the operation of a useful character not built by selection for its current role as an *effect*. (We designate as an effect only the usage of such a character, not the character itself, see p. 521.) But what is the unselected, but useful character itself to be called? Indeed it has no recognized name (unless we accept Bock's broad definition of adaptation—the criterion of current utility alone—and reject both Darwin and Williams). Its space on the logical chart is currently blank.

We suggest that such characters, evolved for other usages (or for no function at all), and later "coopted" for their current role, be called *exaptations*. (See VIA on the related concept of "preadaptation.") They are fit for their current role, hence *aptus*, but they were not designed for it, and are therefore not *ad aptus*, or pushed towards fitness. They owe their fitness to features present for other reasons, and are therefore *fit* (*aptus*) *by reason of* (*ex*) their form, or *ex aptus*. Mammalian sutures are an exaptation for parturition. Adaptations have functions; exaptations have effects. The general, static phenomenon of being fit should be called aptation, not adaptation. (The set of aptations existing at any one time consists of two partially overlapping subsets: the subset of adaptations and the subset of exaptations. This also applies to the more inclusive set of aptations existing through time; see Table 20.1.)

IV   The Current Need for a Concept of Exaptation

Why has this conflation of historical genesis with current utility attracted so little attention heretofore? Every biologist surely recognizes that some useful characters did not arise by selection for their current roles; why have we not honored that knowledge with a name? Does our failure to do so simply underscore the unimportance of the subject? Or might this absent term, in Foucault's sense, reflect a conceptual structure that excluded it? And, finally, does the potential need for such a term at this time indicate that the conceptual structure itself may be altering?

Why did Williams not suggest a term, since he clearly recognized the problem and did separate usages into functions and effects (corresponding respectively to adaptations and to the unnamed features that we call exaptations)? Why did Bock fail to specify the problem at all? We suspect that the conceptual framework of modern evolutionary thought, by continually emphasizing the supreme importance and continuity of adaptation and natural selection at all levels, subtly relegated the issue of exaptation to a periphery of unimportance. How could nonadaptive aspects of form gain a proper hearing under Bock's definition (1967,

p. 63): "On theoretical grounds, all existing features of animals are adaptive. If they were not adaptive, then they would be eliminated by selection and would disappear." Williams recognized the phenomenon of exaptation and even granted it some importance (in assessing the capacities of the human mind, for example), but he retained a preeminent role for adaptation and often designated effects as fortuitous or peripheral— "merely an incidental consequence" he states in one passage (p. 8).

We believe that the adaptationist program of modern evolutionary thought (Gould and Lewontin 1979) has been weakening as a result of challenges from all levels, molecules to macroevolution. At the biochemical level, we have theories of neutralism and suggestions that substantial amounts of DNA may be nonadaptive at the level of the phenotype (Orgel and Crick 1980; Doolittle and Sapienza 1980). Students of macroevolution have argued that adaptations in populations translate as effects to yield the patterns of differential species diversification that may result in evolutionary trends (Vrba's effect hypothesis, 1980). If nonadaptation (or what should be called nonaptation) is about to assume an important role in a revised evolutionary theory, then our terminology of form must recognize its cardinal evolutionary significance —cooptability for fitness (see Seilacher 1972, on important effects of a nonaptive pattern in the structure and coloration of molluscs).

Some colleagues have said that they prefer Bock's broad definition because it is more easily operational. We can observe and experiment to determine what good a feature does for an organism now. To reconstruct the historical pathway of its origin is always more difficult and often (when crucial evidence is missing) intractable.

To this we reply that we are not trying to dismantle Bock's concept. We merely argue that it should be called aptation (with adaptation and exaptation as its modes). As aptation, it retains all the favorable properties for testing enumerated above.

Historical genesis is, undoubtedly, a more difficult problem but we cannot therefore ignore it. As evolutionists, we are charged, almost by definition, to regard historical pathways as the essence of our subject. We cannot be indifferent to the fact that similar results can arise by different historical routes. Moreover, the distinction between ad- and exaptation, however difficult, is not unresolvable. If we ever find a small running dinosaur, ancestral to birds and clothed with feathers, we will know that early feathers were exaptations, not adaptations, for flight.

V Examples of Exaptation

A. Feathers and Flight-sequential Exaptation in the Evolution of Birds

Consider a common scenario from the evolution of birds. (We do not assert its correctness, but only wish to examine appropriate terminology for a common set of hypotheses.) Skeletal features, including the sternum, rib basket and shoulder joint, in late Jurassic fossils of *Archaeopteryx* indicate that this earliest known bird was probably capable of only the simplest feats of flight. Yet it was quite thoroughly feathered. This has suggested to many authors that selection for the initial development of feathers in an ancestor was for the function of insulation and not for flight (Ostrom 1974, 1979; Bakker 1975). Such a fundamental innovation would, of course, have many small as well as far-reaching, incidental consequences. For example, along no descendant lineage of this first feathered species did (so far as we know) a furry covering of the body evolve. The fixation early in the life of the embryo, of cellular changes that lead on the one hand to hair, and on the other to feathers, constrained the subsequent course of evolution in body covering (Oster 1980).

Archaeopteryx already had large contour-type feathers, arranged along its arms in a pattern very much as in the wings of modern birds. Ostrom (1979, p. 55) asks: "Is it possible that the initial (pre-*Archaeopteryx*) enlargement of feathers on those narrow hands might have been to increase the hand surface area, thereby making it more effective in catching insects?" He concludes (1979, p. 56): "I do believe that the predatory design of the wing skeleton in *Archaeopteryx* is strong evidence of a prior predatory function of the proto-wing in a cursorial proto-*Archaeopteryx*." Later selection for changes in skeletal features and feathers, and for specific neuromotor patterns, resulted in the evolution of flight.

The Black Heron (or Black Egret, *Egretta ardesiaca*) of Africa, like most modern birds, uses its wings in flight. But it also uses them in an interesting way to prey on small fish: "Its fishing is performed standing in shallow water with wings stretched out and forward, forming an umbrella-like canopy which casts a shadow on the water. In this way its food can be seen" (McLachlan and Liversidge 1978, p. 39, Plate 6). This "mantling " of the wings appears to be a characteristic behavior pattern, with a genetic basis. The wing and feather structures themselves do not seem to be modified in comparison with those of closely related species, the individuals of which do not hunt in this way (A. C. Kemp, pers. comm.).

We see, in this scenario, a sequential set of adaptations, each converted to an exaptation of different effect that sets the basis for a subsequent adaptation. By this interplay, a major evolutionary transformation occurs that probably could not have arisen by purely increasing adaptation. Thus, the basic design of feathers is an adaptation for thermoregulation and, later, an exaptation for catching insects. The development of large contour feathers and their arrangement on the arm arise as adaptations for insect catching and become exaptations for flight. Mantling behavior uses wings that arose as an adaptation for flight. The neuromotor modifications governing mantling behavior, and therefore the mantling posture, are adaptations for fishing. The wing per se is an exaptation in its current effect of shading, just as the feathers covering it also arose in different adaptive contexts but have provided much evolutionary flexibility for other uses during the evolution of birds.

B. Bone as Storage and Support   The development of bone was an event of major significance in the evolution of vertebrates. Without bone, vertebrates could not have later taken up life on land. Halstead (1969) has investigated the question: granting its subsequent importance as body support in the later evolution of vertebrates, why did bone evolve at such an early stage in vertebrate history? Some authors have hypothesized that bone initially arose as an osmoregulatory response to life in freshwater. Others, like Romer (1963), postulate initial adaptation of bony "armor" for a protective function. Pautard (1961, 1962) pointed out that any organism with much muscular activity needs a conveniently accessible store of phosphate. Following Pautard, and noting the seasonal cycle of phosphate availability in the sea, Halstead (1969) suggested the following scenario: Calcium phosphates, laid down in the skin of the earliest vertebrates, evolved initially as an adaptation for storing phosphates needed for metabolic activity. Only considerably later in evolution did bone replace the cartilaginous endoskeleton and adopt the function of support for which it is now most noted.

Thus, bone has two major uses in extant vertebrates: support/protection and storage/homeostasis (as a storehouse for certain mineral ions, including phosphate ions). The ions in vertebrate bone are in equilibrium with those in tissue fluids and blood, and function in certain metabolic activities (Scott and Symons 1977). For instance, in humans, 90% of body phosphorus is present in the inorganic phase of bone (Duthie and Ferguson 1973).

Following Halstead's analysis, the deposition of phosphate in body tissues originally evolved as an adaptation for a storage/metabolic function. The metabolic mechanism for producing bone per se can thus be interpreted as an exaptation for support. The metabolic mechanisms for depositing an increased quantity of phosphates and for mineralization, as well as the arrangement of bony elements in an internal skeleton, are then adaptations for support.

C. The Evolution of Mammalian Lactation

Dickerson and Geis (1969) recount how Alexander Fleming, in 1922, discovered the enzyme lysozyme. He had a cold and, for interest's sake, added a few drops of nasal mucus to a bacterial culture. To his surprise he found, after a few days, that something in the mucus was killing the bacteria: the enzyme lysozyme, since found in most bodily secretions and in large quantities in the whites of eggs. Lysozyme destroys many bacteria by lysing, or dissolving, the mucopolysaccharide structure of the cell wall. The amino acid sequence of α-lactalbumin, a milk protein of previously unknown function, was then found to be so close to that of lysozyme, that some relationship of close homology must be involved. Dickerson and Geis (1969, pp. 77–78) write:

α-Lactalbumin by itself is not an enzyme but was found to be one component of a two-protein lactose synthetase system, present only in mammary glands during lactation.... The other component (the "A" protein) had been discovered in the liver and other organs as an enzyme for the synthesis of N-acetyllactosamine from galactose and NAG. But the combination of the A protein and α-lactalbumin synthesizes the milk sugar lactose from galactose and glucose instead. The non-catalytic α-lactalbumin evidently acts as a control device to switch its partner from one potential synthesis to another.... It appears that when a milk-producing-system was being developed during the evolution of mammals, and when a need for a polysaccharide-synthesizing enzyme arose, a suitable one was found in part by modifying a pre-existing polysaccaride-cutting enzyme.

Thus, lysozyme, in all vertebrates in which it occurs, is probably an adaptation for the function of killing bacteria. Further evolution in mammals (alteration of a duplicated gene according to Dickerson and Geis, 1969) resulted in α-lactalbumin, an adaptation (together with the A protein) for the function of lactose synthesis and lactation. Human lysozyme, in this scenario, is an adaptation for lysing the cell walls of bacteria, and an exaptation with respect to the lactose synthetase system.

D. Sexual "Mimicry" in Hyenas   Females of the spotted hyena, *Crocuta crocuta*, are larger than males and dominant over them. Pliny, and other ancient writers, had already recognized a related and unusual feature of their biology in calling them hermaphrodites (falsely, as Aristotle showed). The external genitalia of females are virtually indistinguishable from the sexual organs of males by sight. The clitoris is enlarged and extended to form a cylindrical structure with a narrow slit at its distal end; it is no smaller than the male's penis and can also be erected. The *labia majora* are folded over and fused along the midline to form a false scrotal sac (though without testicles of course), virtually identical in form and position with the male's scrotum (Harrison Matthews 1939).

The literature on this sexual "mimicry" is full of speculations about adaptive meaning. Most of these arguments have conflated current utility and historical genesis in assuming that the demonstration of modern use (Bockian adaptation) specifies the path of origin (adaptation as used by Williams and Darwin, and as advocated in this paper). We suggest that the absence of an articulated concept of exaptation has unconsciously forced previous authors into this erroneous conceptual bind.

Kruuk (1972), the leading student of spotted hyenas, for example, notes that the enlarged sexual organs of females are used in an important behavior know as the meeting ceremony. Hyenas spend long periods as solitary wanderers searching for carrion, but they also live in well integrated clans that defend territory and engage in communal hunting. A mechanism for reintegrating solitary wanderers into their proper clan must be developed. In the meeting ceremony, two hyenas stand side to side, facing in opposite directions. Each lifts the inside hind leg, exposing an erect penis or clitoris to its partner's teeth. They sniff and lick each other's genitals for 10 to 15 seconds, largely at the base of the penis or clitoris and in front of the scrotum or false scrotum.

Having discovered a current utility for the prominent external genitalia of females, Kruuk (1972, pp. 229–230) infers that they must have evolved for this purpose:

> It is impossible to think of any other purpose for this special female feature than for use in the meeting ceremony.... It may also be, then, that an individual with a familiar but relatively complex and conspicuous structure sniffed at during the meeting has an advantage over others; the structure would often facilitate this reestablishment of social bonds by keeping partners together over a longer meeting period. This could be the selective advantage that has caused the evolution of the females' and cubs' genital structure.

Yet another hypothesis, based upon facts known to every Biology I student, virtually cries out for recognition. The penis and clitoris are homologous organs, as are the scrotum and labia majora. We know that high levels of androgen induce the enlargement of the clitoris and the folding over and fusion of the labia until they resemble penis and scrotal sac respectively. (In fact, in an important sense, they *are* then a penis and scrotal sac, given the homologies.) Human baby girls with unusually enlarged adrenals secrete high levels of androgen, and are born with a peniform clitoris and an empty scrotal sac formed of the fused labia.

Female hyenas are larger than males and dominant over them. Since these features are often hormonally mediated in mammals, should we not conjecture that females attain their status by secreting androgens and that the peniform clitoris and false scrotal sac are automatic, secondary by-products. Since they are formed anyway, a later and secondary utility might ensue; they may be coopted to enhance fitness in the meeting ceremony and then secondarily modified for this new role. We suggest that the peniform clitoris and false scrotal sac arose as nonaptive consequences of high androgen levels (a primary adaptation related to the unusual behavioral role of females). They are, therefore, exaptations for the meeting ceremony, and their effect in enhancing fitness through that ceremony does not specify the historical pathway of their origin.

Yet this obvious hypothesis, with its easily testable cardinal premise, was not explicitly examined until 1979 after, literally, more than 2000 years of speculation in the adaptive mode (both ancient authors and medieval bestiaries tried to infer God's intent in creating such an odd beast). Racey and Skinner (1979) found no differences in levels of androgen in blood plasma of male and female spotted hyenas. Female fetuses contained the same high level of testosterone as adult females. In the other two species of the family Hyaenidae, however, androgen levels in blood plasma are much lower for females than for males. Females of these species are not dominant over males and do not develop peniform clitorises or false scrotal sacs.

We do not assert that our alternative hypothesis of exaptation must be correct. One could run the scenario in reverse (with a bit of forcing in our judgment): females "need" prominent genitalia for the meeting ceremony; they build them by selection for high androgen levels; large size and dominance are a secondary by-product of the androgen. We raise, rather a different issue: why was this evident alternative not considered, especially by Kruuk in his excellent exhaustive book on the species? We suggest that the absence of an explicitly articulated concept of exaptation has constrained the range of our hypotheses in subtle and unexamined ways.

E. **The Uses of Repetitive DNA** For a few years after Watson and Crick elucidated the structure of DNA, many evolutionists hoped that the architecture of genetic material might fit all their presuppositions about evolutionary processes. The linear order of nucleotides might be the beads on a string of classical genetics: one gene, one enzyme; one nucleotide substitution, one minute alteration for natural selection to scrutinize. We are now, not even 20 years later, faced with genes in pieces, complex hierarchies of regulation and, above all, vast amounts of repetitive DNA. Highly repetitive, or satellite, DNA can exist in millions of copies; middle-repetitive DNA, with its tens to hundreds of copies, forms about one quarter of the genome in both *Drosophila* and *Homo*. What is all the repetitive DNA for (if anything)? How did it get there?

A survey of previous literature (Doolittle and Sapienza 1980; Gould 1981) reveals two emerging traditions of argument, both based on the selectionist assumption that repetitive DNA must be good for something if so much of it exists. One tradition (see Britten and Davidson 1971 for its *locus classicus*) holds that repeated copies are conventional adaptations, selected for an immediate role in regulation (by bringing previously isolated parts of the genome into new and favorable combinations, for example, when repeated copies disperse among several chromosomes). We do not doubt that conventional adaptation explains the preservation of much repeated DNA in this manner.

But many molecular evolutionists now strongly suspect that direct adaptation cannot explain the existence of all repetitive DNA: there is simply too much of it. The second tradition therefore holds that repetitive DNA must exist because evolution needs it so badly for a flexible future—as in the favored argument that "unemployed," redundant copies are free to alter because their necessary product is still being generated by the original copy (see Cohen 1976; Lewin 1975; and Kleckner 1977, all of whom also follow the first tradition and argue both sides). While we do not doubt that such future uses are vitally important consequences of repeated DNA, they simply cannot be the cause of its existence, unless we return to certain theistic views that permit the control of present events by future needs.

This second tradition expresses a correct intuition in a patently nonsensical (in its nonpejorative meaning) manner. The missing thought that supplies sense is a well articulated concept of exaptation. Defenders of the second tradition understand how important repetitive DNA is to evolution, but only know the conventional language of adaptation for expressing this conviction. But since utility is a future condition (when the redundant copy assumes a different function or undergoes secondary adaptation for a new role), an impasse in expression develops. To break this impasse, we might suggest that repeated copies are nonapted features, available for cooptation later, but not serving any direct function at the moment. When coopted, they will be exaptations in their new role (with secondary adaptive modifications if altered).

What then is the source of these exaptations? According to the first tradition, they arise as true adaptations and later assume their different function. The second tradition, we have argued, must be abandoned. A third possibility has recently been proposed (or, rather, better codified after previous hints): perhaps repeated copies can originate for no adaptive reason that concerns the traditional Darwinian level of phenotypic advantage (Orgel and Crick 1980; Doolittle and Sapienza 1980). Some DNA elements are transposable; if these can duplicate and move, what is to stop their accumulation as long as they remain invisible to the phenotype (if they become so numerous that they begin to exert an energetic constraint upon the phenotype, then natural selection will eliminate them)? Such "selfish DNA" may be playing its own Darwinian game at a genic level, but it represents a true nonaptation at the level of the phenotype. Thus, repeated DNA may often arise as a nonaptation. Such a statement in no way argues against its vital importance for evolutionary futures. When used to great advantage in that future, these repeated copies are exaptations.

VI   Significance of Exaptation

A.   A Solution to the Problem of Preadaptation   The concept of preadaptation has always been troubling to evolutionists. We acknowledge its necessity as the only Darwinian solution to Mivart's (1871) old taunt that "incipient stages of useful structures" could not function as the perfected forms do (what good is 5% of a wing). The incipient stages, we argue, must have performed in a different way (thermoregulation for feathers, for example). Yet we traditionally apologize for "preadaptation" in our textbooks, and laboriously point out to students that we do not mean to imply foreordination, and that the word is somehow wrong (though the concept is secure). Frazzetta (1975, p. 212), for example, writes: "The association between the word 'preadaptation' and dubious teleology still lingers, and I can often produce a wave of nausea in some evolutionary biologists when I use the word unless I am quick to say what I mean by it."

Indeed, the word is wrong and our longstanding intuitive discomfort is justified (see Lambert, MS). For if we divide that class of features contributing to fitness into adaptations and exaptations, and if adaptations were constructed (and exaptations coopted) for their current use, then features working in one way cannot be pre*a*daptations to a different and subsequent usage: the term makes no sense at all.

The recognition of exaptation solves the dilemma neatly, for what we now incorrectly call "preadaptation" is merely a category of exaptation considered before the fact. If feathers evolved for thermoregulation, they become exaptations for flight once birds take off. If, however, with the hindsight of history, we choose to look at feathers while they still encase the running, dinosaurian ancestors of birds, then they are only potential exaptations for flight, or *preaptations* (that is, *aptus*—or fit—before their actual cooptation). The term "preadaptation" should be dropped in favor of "preaptation." Preaptations are potential, but unrealized, exaptations; they resolve Mivart's major challenge to Darwin.

B.   Primary Exaptations and Secondary Adaptations   Feathers, in their basic design, are exaptations for flight, but once this new effect was added to the function of thermoregulation as an important source of fitness, feathers underwent a suite of secondary adaptations (sometimes called post-adaptations) to enhance their utility in flight. The order and arrangement of tetrapod limb bones is an exaptation for walking on land; many modifications of shape and musculature are secondary adaptations for terrestrial life.

The evolutionary history of any complex feature will probably include a sequential mixture of adaptations, primary exaptations and secondary adaptations. Just as any feature is plesiomorphic at one taxonomic level and apomorphic at another (torsion in the class Gastropoda and in the phylum Mollusca), we are not disturbed that complex features are a mixture of exaptations and adaptations. Any coopted structure (an exaptation) will probably not arise perfected for its new effect. It will therefore develop secondary adaptations for the new role. The primary exaptations and secondary adaptations can, in principle, be distinguished.

C.   The Sources of Exaptation   Features coopted as exaptations have two possible previous statuses. They may have been adaptations for another function, or they may have been non-aptive structures. The first has long been recognized as important, the second underplayed. Yet the enormous pool of nonaptations must be the wellspring and reservoir of most evolutionary flexibility. We need to recognize the central role of "cooptability for fitness" as the primary evolutionary significance of ubiquitous nonaptation in organisms. In this sense, and at its level of the phenotype, this nonaptive pool is an analog of mutation—a source of raw material for further selection.

Both adaptations and nonaptations, while they may have non-random proximate causes, can be regarded as randomly produced with respect to any potential cooptation by further regimes of selection. Simply put: all exaptations *originate* randomly with respect to their effects. Together, these two classes of characters, adaptations and nonaptations, provide an enormous pool of variability, at a level higher than mutations, for cooptation as exaptations. (Lambert, MS, has discussed this with respect to preadaptations only—preaptations in our terminology. He explored the evolutionary implications of the notion that for any function, resulting directly from natural selection at any one time, there may be multiple effects.)

If all exaptations began as adaptations for another function in ancestors, we would not have written this paper. For the concept would be covered by the principle of "preadaptation"—and we would only need to point out that "preaptation" would be a better term, and that etymology requires a different name for preaptations after they are established. Exaptations that began as nonaptations represent the missing concept. They are not covered by the principle of preaptation, for they were not adaptations in ancestors. They truly have no name, and concepts without names cannot be properly incorporated in thought. The great confusions of historical genesis and current utility primarily involve useful features that were not adaptations in ancestor—as in our examples of sexual "mimicry" in hyenas and the uses of middle-repetitive DNA.

D.   The Irony of Our Terminology for Nonaptation   It seems odd to define an important thing by what it is not. Students of early geology are rightly offended that we refer to 5/6 of earth history as Precambrian. Features not now contributing to fitness are usually called nonadaptations. (In our terminology they are nonaptations.) This curious negative definition can only record a feeling that the subject is "lesser" than the thing it is not. We believe that this feeling is wrong, and that the size of the pool of nonaptations is a central phenomenon in evolution. The term "nonadaptive" is but another indication of previous—and in our view false—convictions about the supremacy of adaptation. The burden of nomenclature is already great enough in this paper and we do not propose a new term for features without current fitness. But we do wish to record the irony.

E. Process and State-of-being Evolutionary biologists use the term adaptation to describe both a current state-of-being (as discussed in this paper) and the process leading to it. This duality presents no problem in cases of true adaptation, where a process of selection directly produces the state of fitness. Exaptations, on the other hand, are not fashioned for their current role and reflect no attendant process beyond cooptation (Table 20.1); they were built in the past either as nonaptive by-products or as adaptations for different roles.

Perhaps we should begin our analysis of process with a descriptive approach and simply focus upon the set of features that increase their relative or absolute abundance within populations, species or clades by the only general processes that can yield such "plurifaction," or "more making": differential branching or persistence (see Arnold and Fristrup, MS). This descriptive process of plurifaction has two basic causes. First, features may increase their representation actively by contributing to branching or persistence either as adaptations evolved by selection for their current function, or exaptations evolved by another route and coopted for their useful effect. Secondly, and particularly at the higher level of species within clades, features may increase their own representation for a host of nonaptive reasons, including causal correlation with features contributing to fitness, and fortuitous correlation found at such surprisingly high frequency in random simulations by Raup and Gould (1974). These nonaptive features establish an enormous pool for potential exaptation.

VII Conclusion

The ultimate decision about whether we have written a trivial essay on terminology or made a potentially interesting statement about evolution must hinge upon the importance of exaptation, both in frequency and in role. We believe that the failure of evolutionists to codify such a concept must record an inarticulated belief in its relative insignificance.

We suspect, however, that the subjects of nonaptation and cooptability are of paramount importance in evolution. (When cooptability has been recognized—in the principle of "preadaptation"—we have focussed upon shift in role for features previously adapted for something else, not on the potential for exaptation in nonapted structures.) The flexibility of evolution lies in the range of raw material presented to processes of selection. We all recognize this in discussing the conventional sources of genetic variation—mutation, recombination, and so forth—presented to natural selection from the genetic level below. But we have not adequately appreciated that features of the phenotype themselves (with their usually complex genetic bases) can also act as variants to enhance and restrict future evolutionary change. Thus the important statement of Fisher's fundamental theorem consider only genetic variance in relation to fitness: "The rate of increase in fitness of any organism at any time is equal to its genetic variance in fitness at that time" (Fisher 1958). In an analogous way, we might consider the flexibility of phenotypic characters as a primary enhancer of or damper upon future evolutionary change. Flexibility lies in the pool of features available for cooptation (either as adaptations to something else that has ceased to be important in new selective regimes, as adaptations whose original function continues but which may be coopted for an additional role, or as nonaptations always potentially available). The paths of evolution—both the constraints and the opportunities—must be largely set by the size and nature of this pool of potential exaptations. Exaptive possibilities define the "internal" contribution that organisms make to their own evolutionary future.

A. R. Wallace, a strict adaptationist if ever there was one, nonetheless denied that natural selection had built the human brain. "Savages" (living primitives), he argued, have mental equipment equal to ours, but maintain only a rude and primitive culture—that is, they do not use most of their mental capacities and natural selection can only build for immediate use. Darwin, who was not a strict adaptationist, was both bemused and angered. He recognized the hidden fallacy in Wallace's argument: that the brain, though undoubtedly built by selection for some complex set of functions, can, as a result of its intricate structure, work in an unlimited number of ways quite unrelated to the selective pressure that constructed it. Many of these ways might become important, if not indispensable, for future survival in later social contexts (like afternoon tea for Wallace's contemporaries). But current utility carries no automatic implication about historical origin. Most of what the brain now does to enhance our survival lies in the domain of exaptation—and does not allow us to make hypotheses about the selective paths of human history. How much of the evolutionary literature on human behavior would collapse if we incorporated the principle of exaptation into the core of our evolutionary thinking? This collapse would be constructive because it would vastly broaden our range of hypotheses, and focus attention on current function and development (all testable propositions) instead of leading us to unprovable reveries about primal fratricide on the African savanna or dispatching mammoths at the edge of great ice sheets—a valid subject, but one better treated in novels that can be quite enlightening scientifically (Kurtén 1980).

Consider also the apparently crucial role that repeated DNA has played in the evolution of phenotypic complexity in organisms. If each gene codes for an indispensable enzyme (or performs any necessary function), asks Ohno (1970) in his seminal book, how does evolution transcend mere tinkering along established lines and achieve the flexibility to build new types of organization. Ohno argues that this flexibility must arise as the incidental result of gene duplication, with its production of redundant genetic material: "Had evolution been entirely dependent upon natural selection, from a bacterium only numerous forms of bacteria would have emerged.... Only the cistron which became redundant was able to escape from the relentless pressure of natural selection, and by escaping, it accumulated formerly forbidden mutations to emerge as a new gene locus" (from the preface to Ohno 1970).

We argued in section VE that much of this repetitive DNA may arise for nonaptive reasons at the level of the individual phenotype (as in the "selfish DNA" hypothesis). The repeated copies are then exaptations, coopted for fitness and secondarily adapted for new roles. And they are exaptations in the interesting category of structures that arose as nonaptations, when the "selfish DNA" hypothesis applies.

Thus, the two evolutionary phenomena that may have been most crucial to the development of complexity with consciousness on our planet (if readers will pardon some dripping anthropocentrism for the moment) —the process of creating genetic redundancy in the first place, and the myriad and inescapable consequences of building any computing device as complex as the human brain—may both represent exaptations that began as nonaptations, the concept previously missing in our evolutionary terminology. With examples such as these, the subject cannot be deemed unimportant!

In short, the codification of exaptation not only identifies a common flaw in much evolutionary reasoning—the inference of historical genesis from current utility. It also focusses attention upon the neglected but paramount role of nonaptive features in both constraining and facilitating the path of evolution. The argument is not anti-selectionist, and we view this paper as a contribution to Darwinism, not as a skirmish in a nihilistic vendetta. The main theme is, after all, cooptability for *fitness*. Exaptations are vital components of any organism's success.

Acknowledgments

The following have commented on the manuscript: C. K. Brain, C. A. Green, A. C. Kemp, H. E. H. Paterson. One of us (E.S.V.) owes a debt to Hugh Paterson for an introduction, during extensive discussions, to the terminology of effects (*sense* Williams). We both thank him for referring us to the examples of mantling behavior in the Black Heron and lysozyme/ α-lactalbumin evolution. D. M. Lambert has given us access to an unpublished manuscript, and has discussed with us the ubiquitous presence, and enormous importance, in evolution of what he and others call preadaptation.

Note

1. An equal time production; order of authorship was determined by a transoceanic coin flip.

Literature Cited

Abercrombie, M., C. H. Hickman, and M. L. Johnson. 1951. A Dictionary of Biology. 5th edition, 1966. Hunt Bernard and Co. Ltd., Aylesbury, Great Britain.

Arnold, A. J. and K. Fristrup. 1982. The hierarchical basis for a unified theory of evolution. Paleobiology, in press.

Bakker, R. T. 1975. Dinosaur renaissance. Sci. Am. 232(4):58–78.

Bock, W. 1967. The use of adaptive characters in avian classification. Proc. XIV Int. Ornith. Cong., pp. 66–74.

Bock, W. 1979. A synthetic explanation of macroevolutionary change—a reductionistic approach. Bull. Carnegie Mus. Nat. Hist. No. 13:20–69.

Bock, W. J. 1980. The definition and recognition of biological adaptation. Am. Zool. 20:217–227.

Bock. W. J. and G. von Wanlert. 1965. Adaptation and the form-function complex. Evolution. 10:269–299.

Britten, R. J. and E. H. Davidson. 1971. Repetitive and nonrepetitive DNA sequences and a speculation on the origins of evolutionary novelty. Q. Rev. Biol. 46:111–131.

Cohen, S. N. 1976. Transposable genetic elements and plasmid evolution. Nature. 263:731–738.

Darwin, C. 1859. On the Origin of Species. J. Murray: London.

Dickerson, R. E. and I. Geis. 1969. The Structure and Action of Proteins. Harper and Row; New York.

Doolittle, W. F. and C. Sapienza. 1980. Selfish genes, the phenotype paradigm, and genome evolution. Nature. 284:601–603.

Duthie, R. B. and A. B. Ferguson. 1973. Mercer's Orthopaedic Surgery. 7th edition. Edward Arnold; London.

Fisher, R. A. 1958. Genetical Theory of Natural Selection. (2nd revised edition). Dover; New York.

Foucault, M. 1965. Madness and Civilization. Random House; New York.

Foucault, M. 1970. The Order of Things. Random House; New York.

Frazzetta, T. H. 1975. Complex Adaptations in Evolving Populations. 267 pp. Sinauer Associates; Sunderland, Massachusetts.

Gould, S. J. 1981. What happens to bodies if genes act for themselves? Nat. Hist. November.

Gould, S. J. and R. C. Lewontin. 1979. The spandrels of San Marco and the Panglossian Paradigm: a critique of the adaptationist programme. Pp. 147–164. In: Maynard Smith, J. and R. Holliday, eds. The Evolution of Adaptation by Natural Selection. R. Soc. London.

Halstead, L. B. 1969. The Pattern of Vertebrate Evolution. Oliver and Boyd; Edinburgh.

Harrison Matthews, L. 1939. Reproduction in the spotted hyena *Crocuta crocuta* (Erxleben). Phil. Trans. R. Soc. (B) 230:1–78.

Kleckner, N. 1977. Translocatable elements in procaryotes. Cell. 11:11–23.

Kruuk, H. 1972. The Spotted Hyena, a Study of Predation and Social Behavior. Univ. Chicago Press; Chicago, Illinois.

Kurtén, B. 1980. Dance of the Tiger. Pantheon; New York.

Lewin, B. 1975. Units of transcription and translation. Cell. 4:77–93.

McLachlan, G. R. and R. Liversidge. 1978. Roberts' Birds of South Africa. 4th edition (first publ. in 1940). John Voelcker Bird Book Fund; Cape Town.

Mivart, St. G. 1871. On the Genesis of Species. MacMillan; London.

Ohno, S. 1970. Evolution by Gene Duplication. 160 pp. Springer; New York.

Orgel, L. E. and F. H. C. Crick. 1980. Selfish DNA: the ultimate parasite. Nature. 284:604–607.

Oster, G. 1980. Mechanics, morphogenesis and evolution. Address to Conference on Macroevolution, October 1980, Chicago.

Ostrom, J. H. 1974. *Archaeopteryx* and the origin of flight. Q. Rev. Biol. 49:27–47.

Ostrom, J. H. 1979. Bird flight: how did it begin? Am. Sci. 67:46–56.

Paterson, H. E. H. 1982. Species as a consequence of sex, in press. Am. Sci.

Pautard, F. G. E. 1961. Calcium, phosphorus, and the origin of backbones. New Sci. 12:364–366.

Pautard, F. G. E. 1962. The molecular-biologic background to the evolution of bone. Clin. Orthopaed. 24:230–244.

Porter, R. MS. Problems in the treatment of 'madness' in English science, medicine and literature in the eighteenth century.

Racey, P. A. and J. C. Skinner. 1979. Endocrine aspects of sexual mimicry in spotted hyenas Crocuta crocuta. J. Zool. London. 187:315–326.

Raup, D. M. and S. J. Gould. 1974. Stochastic simulation and evolution of morphology—towards a nomothetic paleontology. Syst. Zool. 23:305–322.

Romer, A. S. 1963. The 'ancient history' of bone. Ann. N.Y. Acad. Sci. 109:168–176.

Scott, J. D. and N. B. B. Symons. 1977. Introduction to Dental Anatomy. Churchill Livingstone; London.

Seilacher, A. 1970. Arbeitskonzept zur Konstruktionsmorphologie. Lethaia. 3:393–396.

Seilacher, A. 1972. Divariate patterns in pelecypod shells. Lethaia. 5:325–343.

Vrba, E. S. 1980. Evolution, species and fossils: how does life evolve? S. Afr. J. Sci. 76:61–84.

Williams, G. C. 1966. Adaptation and Natural Selection. Princeton University Press; Princeton, New Jersey.

21

Adaptation and the Form-Function Relation[1]

Carl Gans

Form and function are related!

Ever since Aristotle we have known that the idea of a form-function correlation implies more than that the form has been shaped by function. The correlation is of interest because it implies some benefit for the individual that displays it; *i.e.*, wings are assumed to be beneficial for a hawk and fins for a herring (Williams, 1966). The concept that form is of utility to the individual is subsumed under the heading of adaptation, examples of which are extremely obvious; however, the idea has led to some confusing misinterpretations.

About ten years ago we were faced with a major attack on the reality of adaptation. The initial statements (*cf.* Lewontin, 1978) listed some unequivocally adaptive aspects, but mainly noted the difficulty of documenting that a particular phenotypic aspect had indeed arisen by selection for its present function. This was followed by arguments for structuralist explanations in the famous "Spandrels of San Marco" Paper in which adaptationists were explicitly branded Panglossian (Gould and Lewontin, 1979, 1982). Repeatedly during the next decade, the popular and semi-popular literature was used to imply that adaptationist explanations of structure were false or incomplete, and that the practitioners of the adaptationist "programme" were fools, "vulgar" Darwinians, panselectionists or worse (Gould, 1980, 1985; Lewontin, 1981*a, b*).

Particularly Gould has written widely and often skillfully in the semi-popular literature (which he and others cite in more technical papers). A variety of later articles and studies have picked up the argument (*e.g.*, Jaksic, 1981; Reid, 1985), sometimes branding students of adaptation as holding most simplistic views. The attack has led to enough confusion to justify the present review of what to some of us should be deemed a restatement of the obvious.

It is important to note that in this literature the explanation of cases seems to oscillate as does the definition of terms; this makes it difficult at any time to determine what the concepts were intended to mean when they were used. Most important, adapted structures are defined as only those that arose historically for their present roles. This allows the argument that adaptation is a rare phenomenon, with many of the cases in the literature representing something else, even accidental matching. Complaints refer to the documentation and logic underlying arguments for adaptation but counter arguments often are accepted on even flimsier evidence. Developmental plasticity and behavioral variants tend to be ignored as is the variability of the system. Furthermore, adaptationists are assumed to believe that everything is "optimized," a term which inappropriately becomes equivalent to "perfected."

The initial comments suggesting caution in the application of adaptationist explanations were actually well received by professional biologists. The long-term acceptance of adaptation as a mechanism had led to lack of explanatory rigor in the popular, semipopular and even professional literature; evolution and selection were seen as the ultimate causes underlying animal form. One encountered statements such as "buffalo are adapted to eat grass" or "buffalo evolved to eat grass," neither of which contains more information than does the simpler "buffalo eat grass." Such lack of rigor commonly involved students for whom theoretical concepts of adaptation are peripheral. Some, who had only incidental acquaintance with the area, did not know the risk of misunderstanding and paid no attention to the nicety of definition. However, even professional statements about adaptation often omitted key arguments in the middle of a sequence as being "obvious" to the reader, a didactic practice that occasionally masked error.

However, the recent sequence of antiadaptationist arguments seemed to represent more and more of an overkill and the entire body of such literature deliberately destructive. The key point is best exemplified by Hull (1986), who notes that the mistakes due to facile use of adaptationist explanations "would be only of parochial interest to evolutionary biologists if it were not for the extension of adaptationist explanations to the human species." Indeed, the most trenchant examples in the Spandrel's paper and assorted articles concern applications to human biology (see also Kitcher, 1985). Many of the recent anti-adaptationist arguments seemingly represent attempts to state positions relating to human evolution. One feels that a justified concern for the misapplication of their science has led some authors to a general negation of adaptive explanations. The ends of this anti-adaptationist project may be commendable, but they do not justify the means.

I hope in the present review to document that adaptation exists and represents the major basis of the match of form and function. In short, "Organisms are designed to do something" (Trivers, 1985). However, I shall also stress that the underlying adaptive mechanisms are much more complex than sometimes assumed, so that the form–function relation should not be taken for granted. The complexity has led to misunderstanding, witness the spurious conflict between developmental and functional constraints. Most important, there is no reason to expect perfection or optimization in any simple match. Whatever their ultimate aim, the Gould/Lewontin arguments have led to more rigorous analysis of the terminology of adaptation.

Examples

Adaptation was a key aspect of biology, long before Darwin (1859) documented that selection was a process that would lead to a form–function match in which the function was advantageous to the organism. One could easily assemble more than 10,000 references addressing the concept in more than passing (*cf.* Grant, 1963; Gans, 1966; Williams, 1966; Stern, 1970; Krimbas, 1984). This seems inappropriate in the present framework! I propose instead to offer some examples and then to provide definitions of the way in which adaptation has recently been used and how these usages may underlie some of the confusion referred to above.

First, let us consider some examples of the potential complexity of the form–function match. We all know that bats use sonar to orient themselves in the dark. The echoes of the sound pulses they emit also serve to tell them of the instantaneous position of their insect prey. Yet many species of moths, for whom these bats are major predators elegantly avoid their predators. Moths generally use sonic signalling, producing and hearing sounds (Blest *et al.*, 1963). As soon as the ears of some moths detect the sonar pulses, the animals dive in a characteristically erratic pattern and often succeed in hiding among vegetation until the predator has passed. However, these moths in turn are parasitized by small mites that deposit their eggs in the moth's ear cavity. However, this usage as a Kinderstube incapacitates the ear and likely reduces the bat-detecting ability, which will have a negative effect on the ability of the moth to escape bat predation and in turn on the survival of the mite's brood.

Dr. Treat, a student of these mites, was interested in the frequency of moths with parasitized ears (1975). To his surprise, he noticed that, when the gravid female mite climbed onto a moth, she did not deposit her eggs in the first ear reached. Rather she "rummaged" about, perhaps searching for a mite pheromone trail. Only if this was absent, both on the first and on the second ear did she oviposit. If she found either ear already parasitized, the female mite either used the already occupied chamber or got off at the next flower; there waiting for the passage of another moth. Clearly, the behavior mediated by the form of the mite's nervous system kept it from generating deaf moths that would not be able to keep the mite's brood from bat predation; it thus promoted a role that substantially increases the survival of its brood.

The integument of many animals camouflages them by duplicating the reflective coloration of the environment they occupy. It has long been assumed that this matching of the background color makes it more difficult for the predator to discover potential prey. Two decades ago, Norris (1967) attempted to test this for some lizards occupying rocky outcrops in the sand deserts of California. To quantify the color matching of animal and substrate, he used reflectant spectrophotometry. The matching was beautiful and the animals always reflected the color of the particular substrate. However, the most convincing support for the camouflage hypothesis was that the color matching was strongest in the visual range of the predators. Outside of this range the color differed from that of the background.

In East Africa, there are local populations of butterflies some of which feed on noxious plants and sequester the poisons these produce (Clarke and Sheppard, 1963; Sheppard, 1975). This endows them generally distasteful, and observations indicate that bird predators avoid these species. The butterflies apparently facilitate the bird's recognition of their noxious aspects; they display bright colors in characteristic patterns. Local ranges of these species are overlapped by those of a much more widely ranging swallowtail butterfly, which lacks noxious properties and hence makes desirable prey. Wherever the range of this species overlaps that of the noxious butterflies, the harmless species mimics the color and pattern of the noxious one. Experiments indicate that many predators cannot distinguish model from mimic and so the coloration presumably achieves substantial protection. The fact that the association is not accidental is indicated, among other things, by the observation that multiple distinct noxious species are involved. Also only part of the population (generally only one sex) of the very common, palatable species plays any one mimicry game; part of the remainder may have a different coloration. Sometimes these butterflies may mimic a second local species and they also change the model across their range. The number of mimics is restricted to a fraction of that of models (the exact ratio reflecting the nastiness thereof); this assures that attacks by predators on animals bearing the characteristic warning pattern will only rarely result in the capture of individuals that are tasty rather than noxious. (Another such case involves the color resemblance among Central American coral snakes and various local colubrids [Greene and McDiarmid, 1981].)

Such cases of a form–function match represent elegant examples; however, the conditions they characterize are hardly unique. For hundreds of years, people have observed and commented on the "beautifully elegant" way in which the structure of organisms allowed them to make their living or to survive in the circumstances in which they were observed in nature. Such observations once formed the basis of natural theology, popular in pre-Darwinian days (Gillespie, 1979). This viewpoint is exemplified in Kipling's poem (Shiva and the grasshopper, 1893) about the contest between the god Shiva and his wife, Parvati, who hid a grasshopper to prove that Shiva would miss the provision of food for at least one creature. Yet when the grasshopper was disclosed, lo, it was chewing on a new-grown leaf provided by Shiva, presumably to match its feeding mechanism!

Our explanation of the basis for the phenomenon has changed. We now talk about predators matching available prey, rather than food being provided to predators. Our operative mechanisms are natural and represent variants of Darwinian natural selection.

Terminology

Terms such as function and adaptation often have been assigned different meanings. Hence, I begin by offering simple definitions of form, function, role, and adaptation. These will be followed by considerations of the way adaptation originates, of biological variability and the limitations of optimization. I will conclude with operational approaches to the test for adaptation.

We don't have to worry much about complicated definitions of "form." By and large, the term implies some aspect of the phenotype of the organism (*cf.* Anonymous, 1980). Phenotype in this sense implies structure, but we must remember that structure may define, for instance, metabolic pathways and neuronal architecture, thus constraining physiology and behavior. Furthermore, discussions of "form–function complexes" or the "form" of a species, should always keep in mind that the form must be specified for more than a single individual; members of a population will differ depending on their sex and even more on their age. Many populations (such as the snakes discussed above) display polymorphism, the genetic basis of which currently is still subject to debate (Boag, 1987; Smith, 1987). Variations and differences among the phenotypes of a species will affect the degree of the form-function match as this must be established independently for the form of each and every "typical phenotype."

The term "function" is more problematical as it implies performance, the multiple attributes of a phenotype or those things which it does. Depending upon the way it is examined, any phenotypic element will have a variety of attributes. Most obvious to a human observer will be its appearance noted by the light it radiates or reflects. The appearance differs depending upon the wavelength, witness floral colors that may or may not incorporate ultraviolet so that flowers have a different "form" to the eye of an insect which perceives UV (Silverglied, 1979). The appearance also changes because of magnification, whether 1,000 or 100,000 times, and also because of sectioning displaying its internal, microscopic texture. Other attributes will be the odor of the phenotype, and its resistance to steady and to dynamic loads, to mention only a few. The reaction of the phenotype to each such set of external influences represents a functional attribute.

It has long been known that only a fraction of such "functions" are of biological interest to the organism (Bock and von Wahlert, 1965). These are the functions that allow the animals to survive and their offspring in turn to reproduce. This *useful* subset of the total function contributes to the congruence of the form-function pair. It should be referred to as the "role" of the structure (initially biological role [Bock and von Wahlert, 1965]), an ideal which is very close to Aristotle's concept of a "final cause" (Moore, this symposium). Hence, form-function is really phenotype-role and is so interpreted hereafter. We will see below that other attributes of the form might be incidental to the way that the phenotype was formed during its development or in history (see developmental and phylogenetic constraints).

We see the form-function, *i.e.*, "phenotype-role," match most clearly whenever the role of a structure is extremely important to the survival of the organism. Examples are seen in the ability of the males of many species of insects to detect at very low densities the pheromones produced by their females, and in the specialization of the rod cells of the mammalian eye to respond electrically to the input of a single photon (Hecht *et al.*, 1942). Hardly anyone would argue that these phenotypes match their roles beautifully!

The next term is "adaptation" which unfortunately means at least two quite distinct things (Fisher, 1985; Regal, 1985*b*). The first definition makes adaptation the "functional" aspect of the structure, it is the aspect that allows the present organism to survive and in turn to produce fit offspring (Ridley, 1986). Hence it is another aspect of current role. A second and quite different usage of the word assumes that adaptation refers to the process of *becoming* adapted to the present role (*cf.*

Cracraft, 1981). (This meaning is hidden in the etymology of the term adaptation, leading to the term aptation; Gould and Vrba, 1982.)

In the first definition, adaptation is obviously a current state, and it is often argued that adaptation in this sense represents a tautology, *i.e.*, the animal is surviving, hence it must be adapted! However, this is inappropriate as adaptation will be seen not to be an absolute concept. Rather it is relative and there are degrees of adaptation and degrees to which observed adaptations are heritable (Clutton-Brock and Harvey, 1979). Consideration of different individuals shows that for any set of circumstances some sets of adaptations will be more effective or efficient than others. See, for instance, the variation in holes bored by predatious gastropods. A look at the recent, and even the fossil record, has suggested that many of the attacks were unsuccessful (Benton, 1986). In short, not all predators could overcome all prey. Still, adaptation in this first sense represents a present condition and few would argue that it does not exist.

In the second definition, adaptation means *becoming* adapted to the present role and this is where many of the recent troubles have arisen. Certainly, many and perhaps most adaptive structures arose in other habitats. Their forms then had different roles. However, the term adaptation defined as a process, refers to fitting organism to environment, *i.e.*, the development of forms that will perform a more useful role.

Hence, this second usage of the term adaptation is intrinsically quite different from the first. It does not ask one to describe a present condition, but treats a historical process. It is unlikely that the phenotypes seen today arose in Recent time and, of course, it is almost certain that the environment has changed in the interval. Consequently, the second usage of adaptation demands information about the way in which the phenotypes and presumably, their roles, were initially generated, how they were modified and refined as well as how they are maintained. Approaches for such study are given in the last section.

This usage of adaptation as a process has often been misunderstood. First, even people who understand that current adaptation does not necessarily imply development of the form by the influence of the current role tend to imply such an origin by use of a kind of shorthand which may be misleading or even wrong. Second, there is general acceptance that natural selection has formed phenotypes but less general understanding of the mechanisms involved and of their intrinsic limitations. As noted before, the linkage of current (adaptation, role) and past adaptation (evolution) occurs in every day speech, in popular writing and even more in introductory text books; most regrettably, it often creeps into some kinds of papers that note evolutionary topics only peripherally. (Such remarks tend to sneak in because it is sometimes appropriate to add some incidental generalizations at the end of an observational or experimental study. By then, one may be loath to think through all possible implications; one's primary investment seems to lie in the generalizations as opposed to the data.)

The definitions should make it clear that understanding of the system, and resolution of the problems that seem to be appearing in the literature, require answers to the question of how present phenotypes arose, and how the adaptation we now see was generated.

The Process of Adaptation

The starting point should be the observation that current animals are more or less adapted (in the first sense) to the particular situations they occupy; from this we may deduce that past animals also were. This means that discussion of any change, leading to a shift of adaptedness, has to start with a precursor population of individuals, members of which were adapted to the circumstances in which they occurred. Implicit in any such statement must be that each such population likely would have been variable both phenotypically and genotypically.

At least two kinds of adaptive change may occur *in situ*. The first would reflect a variant in genotype due to factors such as mutation, recombination, and interbreeding with another species (*i.e.*, introgression), with some resulting phenotypes being better than those previously existing. The probability of improved adaptation due to such a phenomenon is finite but rather small. A second kind of *in situ* change, that may be more common, results from change of the overall environment; for instance, changes of tectonic, climatic and ecological factors will impose changed selection on the population. Extreme examples of such environmental change may be the rise of coastlines, an ice age with associated glaciation, and the advent of new predators, food objects, competitors and parasites. Smaller changes may result from changes in the weather, which is notoriously fickle so that we always hear about the hottest summer or the wettest summer-fall-winter and spring, thus keeping the TV forecasters in small talk. The population may respond by evolutionary change or may go extinct. Change may follow any of several theoretically possible trajectories. The smaller the geographic range of a species, the more likely that its adaptive level will respond to local change.

In situ change is likely to represent an intraspecific event. The population size and range need not change. The genetic bases of such changes are complex, and the adaptive values of different allelic combination are likely to. Also, *in situ* change may, of course, produce difficulties for paleontologists. If they only sample such a changing population every hundred, thousand or million years, they will encounter quantum phenotypic change, leading to discussions as to whether or not such samples should be referred to as distinct species.

A third category of adaptive change involves only some members of an initial population without change in adaptive conditions for the remaining group. This is presumably the most common kind of change. Two variants may be seen. In the first, the species occupies a large range in space or time, so that diverse subpopulations likely may encounter diverse circumstances. As local selection may be unique, genetic exchange across the range may be slow enough to permit local differentiation, which represents local change in adaptation. In the second, the range of the species may change due to its invasion of adjacent zones and resultant change of adaptation resulting from exposure to a different but usable habitat.

As the latter kind of adaptive change seems most important, it may be useful to present some examples. Organisms always test the limits of the portion of the environment (habitat) they occupy, although such dispersal may occur at different rates depending on the age of the individual. For instance, seeds may be broadcast by a tree, and most will fall nearby and likely in fertile soil; however, some of them will spread much farther and reach regions that for most years lack sufficient moisture for germination. Can any of these seeds survive there? Will the colonizing population become self-replicating? How many of the hooked seeds that catch the fur coats of some mammals actually result in viable plants?

In another example, newly emergent toadlets of many species march outward from the site of metamorphosis; the distance they travel and their rate of survival is affected by weather conditions of the year. Most migrating toadlets likely will die as they never encounter a situation that their phenotype can tolerate. However, some of the best travellers may encounter a previously unoccupied patch of suitable environment (perhaps a nutrient-rich temporary pond). The process of testing may, for instance, allow expansion of the range of aquatic species during wet years and its constriction during dry ones. Similarly, species may change their latitudinal ranges, for instance coincident with glaciation (Coope, 1977).

Change of habitat may be temporal as well. Various insects and frogs breed annually with fertilization restricted to a relatively brief period. A minor change in the time during which members of a subpopulation are receptive to mating, whether triggered temporally or cued by different temperatures, would result in so-called sympatric speciation. It is easy to forget that such temporal separation will be as absolute as a geographic isolation or a phenotypic difference more obvious to us.

Such testing of the limits of the habitat may modify the range, but it may also lead to the discovery of a situation incorporating substantial new resources of a kind usable by the invaders (Bock, 1959). The invasion of the new habitat then will be followed by an initial build-up in the numbers of the invading population. Simultaneously, the new environment will, of course, expose the invaders to its characteristic selective regime, which is likely to differ from the original one. Consequently, stability of the new set of resources for a number of generations, might lead to differentiation of the invaders from their ancestral population. We may see change of adaptation which will often be sufficient later to lead to speciation away from the parental population. The greater the difference between the original conditions and those in the invaded zone, the greater will be the tendency to change the genotype frequency of the invaders.

This pattern of invasion and colonization of a new environment may be modelled by study of real invasions (Simberloff, 1969; Pickett and

White, 1985; Drake and Williamson, 1986; Regal, 1986; Lewin, 1987). There are invasions of empty regions (*i.e.*, volcanic islands and those emptied by catastrophies) and invasions of zones that already have some kind of biota (*i.e.*, islands and mountain tops). However, the frequency of success is very low; indeed, success results from repeated attempts by potential invaders and often the initial success will be due to behavioral change.

Success and the trajectory established depend partly on the phenotype of the invading individuals and the extent to which this allows their survival during their initial discovery of the new set of opportunities. It also reflects the characteristic of the intermediate zone. For example, the probability of a shift from an environment with low salinity to one with medium salinity will be reduced if the two environments are separated by a zone of high salinity. Thus, species adapted to estuarine situations can only invade other estuaries if they can tolerate or bypass the oceanic region between the river mouths.

We may, in retrospect, refer to members of populations that were capable of colonizing such a different area, as having been "protoadapted" for it. Such protoadaptation (once preadaptation—which had earlier referred to orthogenesis, *cf.* Gans, 1979; and more recently exaptation, Gould and Vrba [1982]; Bock's [1976] paraadaptation refers to an almost identical phenomenon) only implies a successful historical accident. Stout fins may allow fishes with lungs to shift to a terrestrial life, but fins and lungs had different origins. Very rarely one notes that a successful invasion does more than to expand a simple habitat; then, selection in the newly discovered zone discloses "key innovations" ("evolutionary novelties," Mayr [1960]; "broad adaptations," Schaeffer [1965]) that facilitate the colonization of other related habitats (and perhaps further evolution therein). Naturally, protoadaptations and "key innovations" only take their meaning *ex post facto*; furthermore, it is likely that their initial appearance represents a simpler state than that noted in later descendants. Many so-called macroevolutionary changes fit into this category. The next point then must be determination of why the phenotypes of most members of any population allow them to invade new environments.

Variability and Constraints

Obviously, the phenotypes (and genotypes) that permit the successful colonization of the new environment are a subset of those carried by the environment's discoverers. What factors give individual members of the original population a chance to colonize, *i.e.*, to survive under changed circumstances? Two overlapping phenomena need to be discussed, namely intrapopulational variability and excessive construction.

Discussions of a form-function, *i.e.*, "phenotype-role," match often tend to ignore the existence of the former, but intrapopulational "variability" is one of the key characteristics of organisms (Bennett, 1988; Feder, 1988). We have already noted the differences between the sexes, the changes that occur during ontogeny and other polymorphisms (as in the butterfly example). Consider the differences between a newly born, blind puppy and an older and playful puppy, and between these and a mature adult dog. The roles of their changing phenotypes also change. It is important to note that the match of structure and role must change during ontogeny; one would expect the match to be loosest at the time of transition of phenotype, *i.e.*, at times of hatching and metamorphosis. Witness that although most of the phenotypic shift from caterpillar to butterfly takes place within the pupal case, there is the brief interval during which the emergent butterfly has to spread and dry its wings.

Also, we see individual variation for all such categories. First of all, any phenotype is obviously influenced partly by genetic and partly by environmental factors. Even a population with a constant genotype (such as a clonal one) will display some phenotypic variability, because the environment is always variable. The top and bottom of a nest will differ in temperature and humidity, as will the chemistry and structure of its walls. Growth and structure of the posthatching individual are likely to be affected by the weather. Then, there are also accidental variants, scars encountered in falls, phenotypic changes due to encounters with parasites or predators. All represent one category of environmental effect, although their expression may be genetically determined and they may represent statistically predictable phenomena. To the extent that the phenotypes differ, they may permit the individuals to maintain themselves in slightly different habitats.

"Excessive construction" is just a fancy name for something that is always with us (Gans, 1979). It refers to the fact that most individual phenotypes will at any one time be able to tolerate more severe conditions than the minimum seemingly required of members of the parent population, *i.e.*, most of the time the organism has "reserve" capacity and is able to do more than what is required at the moment. Such excessive construction has several causes. First of all, genotypic instructions always are read off against a poorly predictable environment. If there is selection for a target value, genetic and developmental variability both will eventually produce phenotypes greater or lesser than this; the degree of excess is based on the variation likely in the environment. Next, organisms often use their characteristics at a relatively low level; for instance, individuals may have the metabolic scope for fast and sustained locomotion, but mostly travel at a slower and less persistent pace (Hertz *et al.*, 1988). An extreme of these causes for excessive construction is the influence of rare (but important) events, such as the demands of the mating season or the advantage of resistance to infrequent fires in the bark of trees (Gans, 1979). Excessive construction suggests that the vast majority of individuals of any species have "reserve capacity" for some aspects and currently will be able to survive conditions more stringent than those which the parental population encounters.

Probably, the occurrence of excessive construction has the byproduct of allowing organisms to widen their adaptive situation. It also facilitates more fundamental environmental shifts and the invasion of new habitats discovered during the regular process of testing the current limits of their range. Excessively constructed variants may then be able to pass the barriers defining the normal range of any population and occasionally to locate, invade and occupy new habitats. They facilitate protoadaptations for different ways of making a living.

Any individual successful invaders will incorporate only a fraction of the population genotype leading to the important concepts of population bottleneck, founder effects and genetic drift (Wright, 1948, 1955). As the successful invaders immediately encounter selection for the unusual and perhaps drastically different situation, genotypic change may follow rapidly. This very simple set of ideas characterizes a system in which selection can fairly rapidly lead to rather profound changes in phenotype; for the vast majority of cases, it removes the need to look for evolution by special saltational or other mechanisms.

How Good Is the Fit?

The concept of form and function and the often obvious role of the phenotype has tended to suggest idealistic perfection. Such ideas of phenotypic perfection may well be a holdover from natural theology when it was argued that the form–function match must be perfect because it reflected the "wisdom ... manifested in the works of creation" (Ray, 1691). Also witness Lyell's (1832) pre-Darwinian reference to the "fitness, harmony and grandeur" of nature, and statements such as "nature ... gives us ... objects of interest, or images of beauty" (Bell and Wyman, 1902, written 50 years earlier), which just substitutes the concept of "Nature" for a deity. However, we will see clear indication that perfection is neither common nor necessary to the organism. Early discussions leave it unclear whether the perfection applies to one particular aspect (characteristic) or the totality of the phenotype.

Demand for perfection has led to prolonged objections of the concept that some aspects of organisms might be adaptively neutral (Kimura, 1983). (The acceptance of the concepts of neutrality and drift should not imply that every claim for these is indeed valid. Nor should their occasional existence have much bearing on the reality of natural selection.) There are many reasons why perfection in any of its guises is both unlikely and a non-operational concept.

Certainly, any naturalist encounters individual organisms that are obviously inadequate, due to damage by accident, disease or parasitism or are in the process of being eaten by a predator. Then there are the transitions in metamorphosis, the above-mentioned pupa to butterfly and the tadpole to froglet. Also, phenotypes will be damaged by events that occur routinely during life and yet the organism functions well. Witness the scars on the head of dominant male ungulates or on the trunk of male walrus. The apologia that the "typical" (undamaged or invariant) individuals might have been perfectly adapted is inappropriate. For instance, the feet of a population of adult arboreal lizards collected in South America proved to have from three to four (rather than five) claws, the remainder having been damaged and lost (Rand, 1965); yet such animals seemed to move without handicap. Hence, the five-toed condition was not a state perfectly matched to its role; rather the foot of such a lizard and the head skin of a male ungulate can perform their respective roles even though damaged by common biological events.

Several other factors would limit approximation of the form-role match to a perfect state. Genes are pleiotropic (each genetic unit contributing to several characters); also many genes will be involved in forming any phenotype. Then there are factors, such as genetic linkage, so that selection for any genetically affected aspect will involve others. The adaptive improvement of the first aspect likely will be delayed (or may be limited) while other expressions of the genes are matched to the overall genotype. Another factor is adaptive compromise, defined as the phenomenon that each structure is likely to participate in more than one role; these roles may conflict, keeping each from being optimized.

An example of adaptive compromise may be seen in the ventilatory system of adult frogs and some salamanders. Lung ventilation in postmetamorphic animals depends on a broad, buccal frame, the soft tissues of which can be lifted rapidly to drive air into the lung. In most of these species, the buccal volume is then related to the requirement for a single lung-filling cycle which establishes a very definite minimum buccal size. However, the feeding patterns of these animals may produce quite different demands on the buccal cavity. Many frogs then rely on various anterior kinds of projectile tongues that are anchored to the mandibular symphysis; also, no frog appears to have developed the inertial feeding patterns so common and successful in Recent lizards and crocodilians (Gans, 1974). One family of salamanders which has lost their lungs and restricted gas exchange to their exterior, was able to resolve the compromise in a different way; the floor of the liberated buccal cavity serves into a base for a projectile tongue. Another kind of constraint is suggested by the compromise seen in elongate reptiles between the cost of diametric increase on locomotion and the demands on coelomic space for reproduction and for feeding (*cf.* Gans *et al.*, 1978).

Perhaps the best example for an adaptive compromise is sexual selection. The various secondary sexual characteristics may have costs in terms of mechanical optimization of feeding, locomotion and other daily activities. Peacocks and male bowerbirds are not optimized for aerodynamic ability. The bright red head of the sexually mature males of some agamid lizards hardly represent good camouflage. Improved mating success seems more important in these species than is flight effectiveness or predator avoidance.

Recently, there has been an argument for a so-called symmorphosis, which suggests that when a physiological process involves multiple steps, the factor of safety will be equal for each (Taylor and Weibel, 1981); the conceptual difficulties with this idea suggest that further tests would be desirable (Garland and Huey, 1987). In some such cases we deal with a situation analogous to what is in engineering practice referred to as a factor of "safety"; the causes are different although the term seems to be instantly recognizable by students.

One byproduct of the assumption of perfect matching may be seen in the discussion of various kinds of biological constrains, phylogenetic, developmental and constructional. For instance, only a limited number of the geometrically or mechanically possible phenotypes are observed (Alberch, 1980; Alexander, 1985); however, this should be expected rather than providing grounds for astonishment. Thus, the ancestral organisms had a particular genetic-developmental mechanism the nature of which determined the variation seen in their immediate offspring. Of course, such mechanisms limit the phenotypes we observe! Similarly, the genetic instructions of the ancestral organisms limit the materials synthesized in animal bodies, as well as the sequence in which these form; hence, these limitations certainly constrain the forms we observe. No animal has yet synthesized materials equivalent to titanium crystals or fiberglass. Enamel, dentine and fibrous cartilage are the closest to such materials. For any general body pattern (*e.g.*, fish, eel, whale, snake) we see improvisations on common themes. (The discovery of such similarity led to the concept of archetype as one of a number of underlying plans which all members of a group of organisms, whether fishes or crabs, are trying to achieve. However, the commonality probably reflects mainly the limits up to which the genetic heritage and biological materials of an organism allow relatively facile change.)

It is useful to note that survival in a habitat demands that the organism meet certain necessary minimum conditions, but particular roles need not establish requirements beyond sufficiency. Certain amphibians have become miniaturized and prove to be very successful in particular habitats; their feet may require only a small contact area and multiple species retain only four toes. The toe loss apparently occurs differently in salamanders and frogs, the former losing digit five and the latter digit one (Shubin and Alberch, 1986). Apparently, the reduced surface area is both necessary and sufficient; however, the pattern of toe reduction does not matter from a functional viewpoint and each group presumably utilizes its own developmental sequence as being that easiest to generate.

The concept of sufficiency (or adequacy, Bartholomew, 1986) has been nicely illustrated in a recent book by Bradshaw (1986). He notes that the great success of desert lizards reflects less their special adaptations to desertic conditions, than the general capacity of all lizards for dealing with problems of heat, water and a variable internal environment. He sees this as an argument against specific adaptation; however, lizard characteristics seem to represent key innovations for terrestriality and desert lizards displaying them need compete only against other lizards. Hence, specific adaptation appears on this level.

One only has to look at horse and elephant dung to see that these animals are not perfect. They, and also tadpoles, "high grade" their food, extracting only a small fraction of the potentially available nutrients from the plant material they are sampling (many sparrows and dung beetles make an excellent living out of what is left over). In contrast, cows and monitor lizards as well as many reptilian herbivores extract a much higher fraction of the original resource. The high-graders do obtain an additional advantage by spreading the seeds of their prey plants; however, the key is that even the "inefficient" utilization of a very large resource provides sufficient resources for maintaining very substantial populations over extensive areas.

The concept of perfection is even less useful in application to the nature of ecological colonizations. This is true despite the popular assumption that speciation implies improvement, in common parlance interpreted as greater efficiency. The critical concept for a successful invasion is that the new situation need only provide a pool of resources sufficient so that the available space-energy-nutrients allow support of a population at the efficiency level at which the organism operates.

Success in a new habitat may require only that the invaders be able to survive a temporarily restricted period of adverse circumstances. In many parts of the world, fresh water habitats include highly productive swampy zones. However, such environments commonly suffer partial desiccation, often associated with reduction in the amount of dissolved oxygen. This makes them briefly, but absolutely uninhabitable for those fishes that must obtain all of their oxygen from the water via their gills. Whatever the duration of the anoxic conditions, the event will wipe out all purely gilled fishes. However, any fish that also manages temporarily to utilize gaseous oxygen, no matter how inefficient, suboptimal and imperfect its oxygen extraction process, can maintain itself in the area and gain access to its substantial resources. The initial steps may well be behavioral (Gans, 1970; Bartholomew, 1988). Among Recent teleosts, we see experiments in utilizing bubbles of air and extracting their oxygen in the mouth, branchial pouches, and gut diverticula (Gans, 1971). Such devices will be useful immediately, long before the circulation has been modified or the gas extraction process has attained the efficiency routinely seen in fish gills. This concept, that initial invaders may well be inefficient, may provide the ecological dimension for survival of the small populations in which founder effects and genetic drift as well as selection lead to new adaptive modes (Wright, 1955).

Naturally, the invasion and protracted occupancy of a new situation changes the selective regime and one would expect to see selection for the multiple specializations that might improve the effectiveness of the phenotype. However, such subsequent improvement within the population will be constrained by compromises with other biological roles, and the constraints of existing genetic, developmental and structural aspects. All of these could be influenced by the initial conditions of the organism, the environmental circumstances which affect it and the length of time that this system remains stable or oscillates. Whether a population adapts along one trajectory or another is in effect random—nature does not prescribe or predetermine the role that the organism fills.

How to Study Adaptation

Study of the form-function, *i.e.*, "phenotype–role," relation then requires determination of the past and present biological roles of the phenotype.

This obviously poses problems because we cannot examine the organisms at the time at which the biological changes presumably occurred (Rudwick, 1964; Regal, 1985a). Hence, we tend to operate by studying the current adaptations of organisms and extrapolating from these to the historical changes. Sometimes we discover fossils that help in this process. However, the analysis is difficult because the intrinsic imperfection of structure does not allow detailed extrapolation of roles from current or past structure. The phenotype only tells us what the animals could and could not have done, but does not indicate the fraction of this array of aspects they actually perform(ed). With this, we are left to a consideration of the major functional aspects and face some degree of uncertainty about their minor tuning.

Even within the Recent, the determination of role poses some substantial problems (Gans, 1974, 1988b; Bock, 1980). Initially, one needs a time budget, preferably over the life cycle of several members of the species to know how often parts of the phenotype may be used and how important they then are to the organism. In this, one is reminded of the fact that constraints affect not only the phenotype but the time dimension as well. Observation of a statistically representative sample of the stages and sexes of an organism throughout the organism's life cycle presents a major task. Hence, analyses tend to be subdivided and several comparative techniques are used to compensate for this difficulty. One of these is the study of more or less related forms occupying distinct habitats, another the study of species in which roles of interest are highly specialized and involve an extreme solution (Gans, 1974); the selective effect of role on phenotype than facilitates recognition of the particular adaptation.

Should a particular phenotype-role (*ex.* form-function) match occur repeatedly in association with a particular behavioral or environmental aspect, one may test the assumption that there be a causal explanation so that one is dealing with a role. The greater the number of species being compared, the less likely that the result will be biased by selection of an aberrant species. As the analysis is designed to partition among multiple genetic and environmental factors one must compare multiple species differing in the factors involved. The situation is analogous to the mathematician's need for at least as many simultaneous equations as there are unknowns (Gans, 1985).

The need to compare forms among species of organisms requires the utilization of morphological terminology (Gans, 1985). The phenotypes of any recent organism allow us to study their form and materials, the developmental causes by which these are generated (indeed the heritability of each such) and the roles these perform. These aspects may be examined in ever greater detail by modern techniques utilized by students of anatomy, development and functional morphology. However, the determination of the causes of similarity (and dissimilarity) requires very careful phylogenetic comparison. Aspects of the phenotype of various organisms may be similar because of shared ancestry (homology), accident (homoplasy), and because of shared function (analogy). Analogy may involve both homologous and homoplasic structures, so that one refers to homologous and homoplasic analogy. Some kinds of the former reflect parallelism, *i.e.*, equivalent genotypic change in closely related lines. Homoplasic analogy often results from convergence (the similar shaping of unrelated organisms by common environment) and its argued that it provides the best of role. As noted above, analogy as a cause of similarity is a difficult explanation as the similarity may have arisen in the past (historical, rather than current analogy), so that the operant influences can no longer be observed.

This scheme seems simply logical; however, the decisions required depend on the rigor of the associated classification. Placement of animals on cladograms or their equivalents tend to precede the analysis of adaptations (but see Reid, 1985). Phylogenetic arrangements provide opportunity to test whether similar structures represent parallelisms or convergences. More important, they provide a basis for determining and testing the direction (polarity) of evolutionary change within a particular lineage (*cf.* Lauder, 1981; Huey, 1988). This is critical as we repeatedly see equivalent changes in multiple parallel lines; cladograms provide tests for the causative features of historical analogies (*cf.* Shine, 1987).

A separate set of questions asks about the units being selected for, a condition therefore being that the functional characters involved in roles should be heritable; we now see successful efforts at approaching this in particular cases (Arnold, 1986). The possibility of resolving the pattern of historical analogy is of course most facile in those few cases in which its role has remained operative. An example is the nice correlation suggesting that the vertical migration behavior of zooplankton is a response to the existence of certain predatory fishes (Gliwicz, 1986; Huntington and Metcalfe, 1986).

It must also be noted that there is a suite of phenotypic plasticity which reflects environmental effects, such as the temperature at a particular stage of development (*cf.* Stearns, 1982). Physiologists refer to these as acclimations or adaptations (Prosser, 1986), this being a completely different usage of the latter term. The response often is advantageous to the organism, but it does not need to be this (Bradshaw, 1986). Advantageous or not the capacity for acclimation must be genetically determined and acclimated characters must have been subject to selection. Physiologists also refer to a useful modification of the idea of perfection, namely optimization, which tends to represent the best level of improvement possible under the evolutionary circumstances (Gans, 1983; Lindstedt and Jones, 1988). Hence it takes into account developmental and other constraints, acclimation, excessive constructions and adaptive compromises.

Biologists have long known that recognition of role demands study of the time budget and other aspects for the organism as a whole, indeed for many members of its population. However, each activity tends to involve multiple aspects of the phenotype and each aspect of the phenotype may be involved in multiple activities. This leads to arguments for a "holistic" approach. Holism is a vague term, variously defined, suggesting that all aspects of the organism need to be looked at simultaneously. In the real world, this is, of course, quite impossible unless one chooses to shift ever more closely to the status of a superficial dilettante. As documented by the talks that follow, most biologists practice some level of reduction (*cf.* Alexander, 1987, for a discussion of such concepts). They study one process and in enough detail to allow them to utilize modern techniques and provide a satisfactory level of reliability. However, it is hoped that such students will always keep in mind the possibility of alternate roles and that they will attempt to determine whether the conditions being studied are limiting.

Recently we have seen attempts to subdivide the tasks of evolutionary biologists into structural and functional aspects (*cf.* Seilacher, 1970; Dwyer, 1984; Raup, 1972). A variant of this is the subdivision of morphological features into structural and Darwinian factors (Wake and Larson, 1987). To a very large extent, form is that with which organisms are endowed and which can only be modified in the long term (involving several generations); initial modification will likely be limited (unless one considers the special case of acclimation or physiological adaptation). However, the organism presumably does not see itself this way and one has to be careful that a convenient subdivision, that may facilitate some analyses, does not lead one into the trap of assuming that structure and function are selected for independently.

Any analysis must allow for the possibility of misinterpretations. A potential one derives from the fact that our interpretation of animals is in the eye of the (human) beholder. Witness, the protracted discussion about the possible role of the beautiful iridescence seen whenever the skin of subterranean uropeltid snakes is viewed in the sunshine. Reanalysis (Gans and Baic, 1977), shows that the colors are a byproduct of the surface architecture of the beta-keratin. The pattern inhibits wetting of the snake's surface and reduces its frictional coefficient allowing such animals to move through tunnels with minimal drag.

Also important is the requirement that the tools of functional analysis be appropriate and carefully used. For example, force applications about joints have traditionally been analyzed by considering muscles to be force-generating systems so that the moment arms of their insertion sites seemed to be important descriptors. However, this comparison is not appropriate if the physiological properties of shortening fibers are taken into account. Each sarcomere is then seen to generate a unit of moment because the excursion and velocity also reflect the distance of the insertion from the fulcrum; they reduce the force as the moment increases (Gans, 1988a). Hence, comparisons of moment arm without consideration of excursion test inappropriate questions. Our increasing understanding of biomechanical factors has to be incorporated in analysis.

Most important for the identification of roles is the reminder that the possibility of falsifying a particular biological role, will never be equivalent to proof of an alternative one. Nature is complex and we must expect more than a single pair of potential roles. The alternative may have been properly selected, but it remains critical that other possibilities continue to be tested. For this phase, experimentalists concentrating on the techniques of physiology and functional anatomy must pay attention to, note the results of and perhaps become expert in, studies of behavior and ecology, *i.e.*, of the natural history of their animals. Field observations may be the closest we come to time budget analyses and to an understanding of the potentially possible roles that may be of interest to the animal.

Conclusion

We can see that the form-function (*i.e.*, "phenotype-role") analysis derives historically from an attempt to understand the diversity of organisms. This led to the recognition that the observed form of organisms facilitates functions useful to them, *i.e.*, organisms are generally adapted at the present time. The matching of the most conspicuous aspects of diversity to the current needs of the respective organisms was already known to Aristotle. However, it took more than two millenia to reach a powerful and operational explanation. The Lamarckian view reflects a kind of striving to match environmental demands, whereas the Darwinian approach argued for natural selection and provided a more mechanistic generative pattern. Obviously natural selection had to be refined as the nature of inheritance became clarified. As we have seen, in spite of the complications due to phylogenetic, developmental and structural constraints, and genetic variants and other potential sources of noise, recognition of adaptation remains the key to an understanding of organismic diversity. It explains more than any other concept and its abandonment would sterilize biology.

Acknowledgments

I am grateful to several colleagues with whom I was able to discuss these matters over the years. At risk of slighting others I mention, R. B. Huey, P. Pridmore and R. Wassersug. Instars of this manuscript were read by Richard D. Alexander, K. Andow G., Gary Belowsky, David R. Carrier, Daniel C. Fisher, Raymond B. Huey, Elazar Kochva, John Olson, Philip J. Regal, and Linda Trueb, whose sometimes conflicting comments were most helpful. Preparation of the manuscript was supported by NSF grant G-BSR-850940.

Notes

1. From the Symposium on *Science as a Way of Knowing—Form and Function* presented at the Annual Meeting of the American Society of Zoologists, 27–30 December 1987, at New Orleans, Louisiana.

References

Alberch, P. 1980. Ontogenesis and morphological diversification. Amer. Zool. 20:653–667.

Alexander, R. D. 1987. *The biology of moral systems.* Aldine de Gruyter, New York.

Alexander, R. NcN. 1985. The ideal and the feasible: Physical constraints on evolution. Biol. J. Linn. Soc. 26:345–348.

Anonymous. 1980. Symposium: Anaysis of form. Some problems underlying most studies of form. Amer. Zool. 20:619–722.

Arnold, S. J. 1986. Laboratory and field approaches to the study of adaptation. *In* M. E. Feder and G. V. Lauder (eds.), *Predator-prey relationships: Perspectives and approaches from the study of the lower vertebrates,* pp. 157–179. University of Chicago Press, Chicago.

Bartholomew, G. A. 1986. The role of natural history in contemporary biology. BioScience 36:324–329.

Bartholomew, G. A. 1988. Interspecific comparison as a tool for ecological physiologists. *In* M. E. Feder, A. F. Bennett, W. Burggren, and R. B. Huey (eds.), *New directions in ecological physiology,* pp. 11–37. Cambridge Univ. Press. Cambridge, U.K.

Bell, C. and J. Wyman. 1902. *Animal mechanics.* Riverside Press, Cambridge.

Bennett, A. F. 1988. Interindividual variability: An underutilized resource. *In* M. E. Feder, A. F. Bennett, W. Burggren, and R. B. Huey (eds.), *New directions in ecological physiology,* pp. 147–169. Cambridge Univ. Press, Cambridge, U.K.

Benton, M. J. 1986. Predation by drilling gastropods. Nature 321:110–111.

Blest, A. D., T. S. Collett, and J. D. Pye. 1963. The generation of ultrasonic signals by a New World arctiid moth. Proc. R. Soc. London (B) 158(971): 196–207.

Boag, P. 1987. Adaptive variation in bill size of African seed-crackers. Nature 329:669–670.

Bock, W. J. 1959. Preadaptation and multiple evolutionary pathways. Evolution 13:194–211.

Bock, W. J. 1976. Adaptation and the comparative method. *In* M. K. Hecht, P. C. Goody, and B. M. Hecht (eds.), *Major patterns in vertebrate evolution,* pp. 57–82. Plenum Press, New York.

Bock, W. J. 1980. The definition and recognition of biological adaptation. Amer. Zool. 20:217–227.

Bock, W. J. and G. von Wahlert. 1965. Adaptation and the form-function complex. Evolution 19(3): 269–299.

Bradshaw, S. D. 1986. *Ecophysiology of desert reptiles*. Academic Press Australia, North Ryde NSW 2113, Australia.

Carroll, R. L. 1986. Physical and biological constraints on the pattern of vertebrate evolution. Geoscience, Canada 13(2):85–90.

Clarke, C. A. and P. M. Sheppard. 1963. Interactions between major genes and polygenes in the determination of the mimetic patterns of *Papilio dardanus*. Evolution 17:404–413.

Clutton-Brock, T. H. and P. H. Harvey. 1979. Comparison and adaptation. Proc. R. Soc. London (B) 205:547–565.

Coope, G. R. 1977. Fossil coleopteran assemblages as sensitive indicators of climatic changes during the Devensian (Last) cold stage. Phil. Trans. R. Soc. London (B) 280:313–340.

Cracraft, J. 1981. The use of functional and adaptive criteria in phylogenetic systematics. Amer. Zool. 21:21–36.

Darwin, C. 1859. *The origin of species by means of natural selection*. J. Murray, London.

Drake, J. A. and M. Williamson. 1986. Invasions of natural communities. Nature 319:718–719.

Dwyer, P. D. 1984. Functionalism and structuralism: Two programs for evolutionary biologists. Notes and comments. Am. Nat. 124:745–750.

Feder, M. E. 1988. The analysis of physiological diversity: The prospects for pattern documentation and general questions in ecological physiology. *In* M. E. Feder, A. F. Bennett, W. Burggren, and R. B. Huey (eds.), *New directions in ecological physiology*, pp. 38–75. Cambridge Univ. Press, Cambridge, U.K.

Fisher, D. C. 1985. Evolutionary morphology: Beyond the analogous, the anecdotal, and the ad hoc. Paleobiology 11(1):120–138.

Gans, C. 1966. Some limitations of and approaches to problems in functional anatomy. *In De anatomia functionali*. Folia Biotheoretica (Leiden) 6:41–50.

Gans, C. 1970. Respiration in early tetrapods—the frog is a red herring. Evolution 24(3):740–751.

Gans, C. 1971. Strategy and sequence in the evolution of the external gas exchangers of ectothermal vertebrates. Forma et Functio 3:66–104.

Gans, C. 1974. *Biomechanics. Approach to vertebrate biology*. Repr. University of Michigan Press, Ann Arbor.

Gans, C. 1979. Momentarily excessive construction as the basis for protoadaptation. Evolution 331(1): 227–233.

Gans, C. 1983. On the fallacy of perfection. In R. R. Fay and G. Gourevitch (eds.), *Perspectives on modern auditory research: Papers in honor of E. G. Wever*, chapter 8, pp. 101–114. Amphora Press, Groton, Connecticut.

Gans, C. 1985. Differences and similarities: Comparative methods in mastication. Amer. Zool. 25(2):291–301.

Gans, C. 1988*a*. Muscle insertions do not incur mechanical advantage. Acta Zool. Cracovienska, Festschrift for M. Mhnarski, 31[pt II] (25) H:193–202.

Gans, C. 1988*b*. Concluding remarks: Morphology, today and tomorrow. Proceeding 2nd Int. Conference Comparative Morphology, Vienna.

Gans, C. and D. Baic. 1977. Regional specialization of reptilian scale surfaces: Relation of texture and biologic role. Science 195(4284):1263, 1348–1350.

Gans, G., H. C. Dessauer, and D. Baic. 1978. Axial differences in the musculature of the uropeltid snakes: The freight-train approach to burrowing. Science 199(4325):189–192.

Garland, T., Jr. and R. B. Huey. 1987. Testing symmorphosis: Does structure match functional requirements? Evolution 4(16):1404–1409.

Gillespie, N. C. 1979. *Charles Darwin and the problem of creation*. University of Chicago Press, Chicago.

Gliwicz, M. Z. 1986. Predation and the evolution of vertical migration in zooplankton. Nature 320:746–748.

Gould. S. J. 1980. *The panda's thumb. More reflections in natural history*. W. W. Norton & Co., New York.

Gould, S. J. 1985. *The flamingo's smile. Reflections in natural history*. W. W. Norton & Co., New York.

Gould, S. J. and R. C. Lewontin. 1979. The spandrels of San Marco and the Panglossian paradigm: A critique of the adaptationist programme. Proc. R. Soc. London (B) 205:581–598.

Gould, S. J. and R. C. Lewontin. 1982. L'adaptation biologique. La Recherche (139):1494–1502.

Gould, S. J. and E. S. Vrba. 1982. Exaptation—a missing term in the science of form. Paleobiology 8(1):4–15.

Grant, V. 1963. *The origin of adaptations*. Columbia University Press, New York.

Greene, H. W. and R. W. McDiarmid. 1981. Coral snake mimicry: Does it occur? Science 213(4513): 1207–1212.

Hecht, S., S. Shlaer, and M. H. Pirenne. 1942. Energy, quanta and vision. J. Gen. Physiol. 25:819–840.

Hertz, P. E., R. B. Huey, and T. Garland, Jr. 1988. Time budgets, thermoregulation, and maximal locomotor performance: Are reptiles olympians or boy scouts? American Zoologist 28:927–938.

Huey, R. B. 1988. Phylogeny, history, and the comparative method. *In* M. E. Feder, A. F. Bennett, W. Burggren, and R. B. Huey (eds.), *New directions in ecological physiology*, pp. 76–98. Cambridge Univ. Press, Cambridge, U.K.

Hull, D. L. 1986. Darwinism and dialectics. Nature 320:23–24.

Huntington, F. A. and N. B. Metcalfe. 1986. The evolution of anti-predatory behavior in zooplankton. Nature 320:682.

Jaksic, F. M. 1981. Recognition of morphological adaptations in animals: The hypothetico-deductive method. BioScience 31:667–670.

Kimura, M. 1983. *The neutral theory of molecular evolution.* Cambridge Univ. Press, Cambridge, U.K.

Kipling, R. 1893. *The jungle book.* Harper and Brothers, New York.

Kitcher, P. 1985. *Vaulting ambition. Sociobiology and the quest for human nature.* MIT Press, Cambridge, Massachusetts.

Krimbas, C. B. 1984. On adaptation, Neo-Darwinian tautology, and population fitness. *In* M. K. Hecht, B. Wallace, and G. T. Prance (eds.), *Evolutionary Biology*, Vol. 17, ch. 1, pp. 1–57.

Lauder, G. V. 1981. Form and function: Structural analysis in evolutionary morphology. Paleobiology 7(4):430–442.

Lewin, R. 1987. Ecological invasions offer opportunities. Science 238(4828):752–753.

Lewontin, R. C. 1987. Adaptation. Sci. Amer. 239(3):212–230.

Lewontin, R. C. 1981*a*. Evolution–creation debate. A time for truth. (Editorial) BioScience 31(8): 559.

Lewontin, R. C. 1981*b*. The inferiority complex. New York Rev. Books 28(16):12–16.

Lindstedt, S. L. and J. H. Jones. 1988. Symmorphosis: The concept of optimal design. *In* M. E. Feder, A. F. Bennett, W. Burggren, and R. B. Huey (eds.), *New directions in ecological physiology*, pp. 290–310. Cambridge Univ. Press, Cambridge, U.K.

Lyell, C. 1832. *Principles of geology.* 3 vols. John Murray, London.

Mayr, E. 1960. The emergence of evolutionary novelties. *In* S. Tax (ed.), *Evolution after Darwin*, Vol. 2, pp. 349–380. University of Chicago Press, Chicago.

Norris, K. S. 1967. Color adaptation in desert reptiles and its thermal relationships. *In* William W. Milstead (ed.), *Lizard ecology: A symposium*, pp. 162–229. Univ. Missouri Press, Columbia.

Pickett, S. T. A. and P. S. While. 1985. *The ecology of natural disturbance and patch dynamics.* Academic Press, Inc., Orlando, Florida.

Prosser, C. L. 1986. *Adaptational biology. Molecules to organisms.* John Wiley & Sons, New York.

Rand, A. S. 1965. On the frequency and extent of naturally occurring foot injuries in *Tropidurus torquatus* (Sauria, Iguanidae). Pap. Avul. Depto. Zool., Sao Paulo 17(17):225–228.

Raup, D. M. 1972. Approaches to morphological analysis. *In* Thomas J. M. Schopf (ed.), *Medels in paleobiology*, pp. 28–45. Freeman, Cooper and Co., San Francisco.

Ray, J. 1691. *The wisdom of God manifested in the works of creation*. S. Smith, London.

Regal, P. J. 1985a. Common sense and reconstructions of the biology of fossils: *Archaeopteryx* and feathers. *In*: M. K. Hecht, J. H. Ostrom, G. Viuohl, and P. Wellnhofer (eds.), *The beginnings of birds*. Proc. Intern. *Archaeopteryx* Conference, Eichstätt.

Regal, P. J. 1985b. The ecology of evolution: Implications of the individualistic paradigm. *In* H. O. Halvorson, D. Pramer, and M. Rogul (eds.), *Engineered organisms in the environment: Scientific issues*, pp. 11–19. Amer. Soc. Microbiol., Washington, D.C.

Regal, P. J. 1986. Models of genetically engineered organisms and their ecological impact. *In* H. A. Mooney and J. A. Drake (eds.), *Ecology of biological invasions of North America and Hawaii. Ecological studies*, Vol. 58, pp. 111–129. Springer-Verlag. New York.

Reid, R. G. B. 1985. *Evolutionary theory: The unfinished synthesis*. Cornell Univ. Press, Ithaca, New York.

Ridley, M. 1986. *Evolution and classification. The reformation of cladism*. Longman, London and New York.

Rudwick, M. J. S. 1964. The inference of function from structure in fossils. Brit. J. Phil. Sci. 15:27–40.

Schaeffer, B. 1965. The role of experimentation in the origin of higher levels of organization. Syst. Zool. 14:318–336.

Seilacher, A. 1970. Arbeitskonzept zur Konstructions–Morphologie. Lethaia 3:393–396.

Sheppard, P. M. 1975. *Natural selection and heredity*, 4th ed. Hutchison University Library, London.

Shine, R. 1987. Parental care in reptiles. *In* C. Gans, and R. B. Huey (eds.), *Biology of the Reptilia*, Vol. 16, pp. 275–329. A. R. Liss, Inc., New York.

Shubin, N. H. and P. Alberch. 1986. A morphogenetic approach to the origin and basic organization of the tetrapod limb. Evol. Biol. 20:319–387.

Silverglied, R. 1979. Communication in the ultraviolet. Ann. Rev. Ecol. Syst. 10:373–398.

Simberloff, D. 1969. Experimental zoogeography of islands: A model for insular colonization. Ecology 50(2):296–314.

Smith, T. B. 1987. Bill size polymorphism and intraspecific niche utilization in an African finch. Nature 329:717–719.

Stearns, S. C. 1982. The role of development in the evolution of life histories. *In* J. T. Bonner (ed.), *Evolution and development*, Dahlem Workshop Report No. 22, pp. 237–258. Springer-Verlag, Berlin.

Stern, J. T., Jr. 1970. The meaning of "adaptation" and its relation to the phenomenon of natural selection. Evol. Biol. 17(1):39–66.

Taylor, C. R. and E. R. Weibel. 1981. Design of the mamalian repiratory system. I. Problem and strategy. Resp. Physiol. 44:1–10.

Treat, A. E. 1975. The genus *Dicrocheles*: Gasmasines of the noctuid ear. *In Mites of moths and butterflies*, pp. 141–167. Cornell Univ. Press. Ithaca and London.

Trivers, R. 1985. *Social evolution*. Benjamin/Cummings Publ. Co., Menlo Park, California.

Vitousek, P. M., L. R Walker, L. D. Whiteaker, D. Mueller-Dombois, and P. A. Matson. 1987. Biological invasions by *Myrica faya* alters ecosystem development in Hawaii. Science 238(4828):802–804.

Wake, D. B. and A. Larson, 1987. Multidimensional analysis of an evolving lineage. Science 238(4823): 42–48.

Williams, G. C. 1966. *Adaptation and natural selection: A critique of some current evolutionary thought*. Princeton University Press, Princeton, New Jersey.

Wright, S. 1948. On the roles of directed and random changes in gene frequency in the genetics of populations. Evolution 2:279–294.

Wright, S. 1955. Classification of the factors of evolution. Cold Spring Harbor Symp. Quant. Biol. 20:16–24.

22

Biological Function, Adaptation, and Natural Design

Colin Allen and Marc Bekoff

1 Introduction

The last half century, especially the last quarter century, has seen the accumulation of a large and diverse literature about the use of teleological notions, such as function, design, and adaptation, in biology. Contributors to this literature include both theoretically oriented biologists and biologically oriented philosophers. In a recent survey of this literature (Allen and Bekoff 1995), we developed a classification scheme identifying nine distinct views about teleology in biology, among contributors as diverse as Ernst Mayr, Robert Hinde, George Williams, Richard Dawkins, Ernest Nagel, Larry Wright, Rob Cummins, and Ruth Millikan. These authors disagree over questions such as whether teleological notions are essential or merely heuristic for understanding biological phenomena and whether biological explanation has a fundamentally different form from explanations in other sciences. For example, despite criticizing the way many biologists use the concept of adaptation, Williams (1966, 11) writes, "I have stressed the importance of use of such concepts as biological means and ends because I want it clearly understood that I think that such a conceptual framework is the essence of the science of biology." In contrast, Nagel writes (1961/1984, 346) that "the use of such explanations in biology is not a sufficient reason for maintaining that this discipline requires a radically distinctive logic of inquiry."

Originally, teleology was controversial because it was associated with pre-Darwinian, creationist views about organisms. Mayr (1974/1988, 40) identifies four reasons to be suspicious of teleological notions, namely

(i) vitalism, (ii) incompatibility with mechanistic explanation, (iii) backwards causation, and (iv) mentalism. To Mayr's list we added a fifth category of methodological concerns about the empirical testability of teleological claims (Allen and Bekoff 1995). These categories may be related; for instance, backwards causation might be incompatible with mechanistic explanation and would entail methodological difficulties. Although concerns centering on vitalism and creationism no longer pose a serious worry for most biologists and philosophers, continuing controversy about teleological notions is attested by the high volume of theoretical papers in the literature.

Even a cursory scan of the theoretical literature reveals that biologists have found it difficult and even undesirable to eliminate teleological notions from their discussions of biological phenomena. Despite this, it is relatively difficult to find *explicit* claims about function or design in articles that primarily report empirical data. English, however, provides a variety of ways for making *implicit* functional claims without using the word "function" (Van Parijs 1982) and it is not difficult to find implicit claims about function. For example, in a paper titled "What is the function of encounter patterns in ant colonies?" Gordon et al. (1993) make no explicit statements about function, but they do say (p. 1099) "An ant that suddenly encounters alien ants may be in danger ... the increase in [antennal] contact rate, though short-lived, may be sufficient to generate a defensive response to the intruders." This contains an implicit suggestion about the function of antennal contacts, but the authors exercise notable caution in avoiding any explicit claims. Another example is provided by Holley (1993), who writes about the bipedal stance that brown hares assume when confronted by red foxes. He says (1993, 21), "The functions of this behavior are considered and competing hypotheses of Predator Surveillance and Pursuit Deterrence are examined by testing predictions against results obtained. The results suggest that by standing erect brown hares signal to approaching foxes that they have been detected." In other words, a function of standing erect by brown hares is to deter pursuit by foxes. Implicit claims about function are also relatively common in textbooks of animal behavior (e.g., Drickamer and Vessey 1992; Alcock 1993). In discussing the classic work of Tinbergen and his colleagues, Drickamer and Vessey (1992, 23) note the plausible hypothesis that "Parents remove white eggshells to protect their young;" clearly they could just as well have written that for these birds a function of eggshell removal behavior is protect their offspring.

In both the theoretical and empirical literatures, explicit claims about design are even harder to find than claims about function. However, for numerous mammals it has been hinted that play signals may be designed to initiate and to maintain social play, and that developmental scheduling and structural sequencing of play (e.g., duration, the interval between play-bouts, the different motor patterns that are used, how they are organized in sequence, and where bites or other actions are directed) are designed to fulfill various functions (e.g., Rasa 1973; Bekoff 1977, 1982, 1988, 1989a,b, 1993; Leyhausen 1979; Bekoff and Byers 1981; Fagen 1981; Martin and Caro 1985; Hass and Jenni 1993; Pellis 1993; Watson and Croft 1993).

Reluctance to discuss design may be due to a couple of factors. First, the notion of design may be considered metaphorical and offputting because it suggests a strong directional component in the evolution (or development) of a behavioral phenotype (Dawkins 1986). Second, many authors seem to accept a principle that straightforwardly assimilates the notion of design to that of function such as:

(ND = F)   T is naturally designed for X if and only if X is a biological function of T.

Some authors make their allegiance to something like ND = F explicit. For example, Kitcher (1993, 379) writes: "the function of an entity *S* is *what S is designed to do*" and Millikan (1984, 17) states that "Having a proper function is a matter of having been 'designed to' or of being 'supposed to' (impersonal) perform a certain function." Other authors write as if they implicitly accept ND = F (e.g. Williams 1966, 9). Further below, we shall argue that this identification ought to be rejected.

2   Naturalizing Biological Function

Like many authors, it is our view that successful naturalization of teleological notions in biology requires that one give an account of these notions that does not involve the goals or purposes of a psychological agent. With respect to the term "function," we agree with Millikan (1989; see also Godfrey-Smith 1994) who refers to an ambiguity in its biological uses, and argues that there is room in biology for at least two different notions—etiological functions (Wright 1973, 1976) and Cummins-functions (Cummins 1975). Kitcher (1993, 395) also notes that "Philosophical discussions of function have tended to pit different analyses and different intuitions against one another without noting the pluralism inherent in biological practice."

Although Millikan and Godfrey-Smith endorse the pluralistic view, they believe, and we agree, that an etiological approach to function, based on Wright (1973), gives the best account of the majority of uses of the notion of function within evolutionary biology. The number of philosophers of biology converging on etiological accounts of function justify our labelling it "the standard line" (Allen and Bekoff 1995); in addition to Millikan and Godfrey-Smith, proponents of the standard line include Ayala (1977), Griffiths (1993), Mithchell (1993), and Neander (1991a,b). The standard line has these three components (from Allen and Bekoff 1995):

1. Functional claims in biology are intended to explain the existence or maintenance of a trait in a given population;
2. Biological functions are causally relevant to the existence or maintenance of traits via the mechanism of natural selection;
3. Functional claims in biology are fully grounded in natural selection and are not derivative of psychological uses of notions such as design, intention, and purpose.

Variants of the standard line differ mostly over how to make component (2) precise. There is also some discussion of (1) with respect to the importance of distinguishing, on the one hand, initial spread of a new phenotypic trait in a population from, on the other hand, the maintenance of traits in populations (Gould and Vrba 1982; we discuss their distinction between adaptation and exaptation below).

3   Function and Adaptation

Tinbergen (1963) is widely regarded as having set the agenda for ethology—the study of the evolution of behavior—by identifying the four areas of evolution, causation, adaptation (or function), and development, as the major areas of concern for ethological study. Ethologists and behavioral ecologists typically follow his identification of function and adaptation (see, for example, Marler and Hamilton 1966; Brown 1975; Eibl-Eibesfeldt 1975; Manning and Dawkins 1992; Drickamer and Vessey 1992; Alcock 1993).

The assimilation of adaptation and function is a characteristic not only of behavioral scientists. For example, Sober (1993, 84) offers this definition of adaptation:

Characteristic C is an adaptation for doing task T in a population if and only if members of the population now have C because, ancestrally, there was selection for having C and C conferred a fitness advantage because it performed task T.

If the phrase "is an adaptation for doing task T" is replaced in this definition with the phrase "has task T as a function," then we have an instance of the standard account of function, and in this sense the two notions are assimilated. Sober's definition is cited approvingly by many biologists (but see Reeve and Sherman 1993 for dissent). We believe, however, that it represents a considerable divergence from normal biological usage. Specifically, nothing in Sober's definition captures the idea of modification in form or structure that is an important component of many notions of adaptation. This is not automatically a criticism. It might be argued that biological usage should be reformed. However, we believe that it is *prima facie* preferable to elucidate scientific usage rather than attempt to reform it. We believe that the notion of an adaptation is better understood in the context of an analysis of natural design than by assimilating it to biological function.

4 Naturalizing Natural Design

If most uses of the notion of biological function are successfully naturalized by the standard line, then the principle ND = F provides a cheap way to naturalize the notion of natural design. We share the goal of naturalizing the notion of design, but we do not believe that it is necessary or desirable to conflate function and design in order to do so. We advocate rejecting ND = F while still basing the analysis of natural design on natural selection. This distinguishes our view from that of Ollason (1987, 549) who claims that "Optimal foraging theory has nothing to do with the theory of evolution: it has to do with the science of design," which entails that the science of design has nothing to do with the theory of evolution.

We claim that uses of the notions of design and function in biological contexts should be understood in ways that do not make biological teleology derivative of psychological teleology. Nonetheless, we shall argue that analysis of *psychological* teleology can, with care, motivate distinctions that are useful for understanding the use of teleological notions in the study of evolution. The methodological principle here is that biological practice provides the stronger constraints on analyses of teleological notions in biology, while considerations from psychological usage provide (at best) weak constraints.

5   Design in Psychological Contexts

Our strategy, then, is to apply what we learn about ordinary psychological usage to help develop a useful account of natural design. We start by noticing that psychological uses of "design" are ambiguous; the term "design" has at least two different but related senses when applied to human activity. The first sense, that we label "goal-driven design," coincides with detailed planning before or during a sequence of behaviors geared to a specific goal. In this sense, architects design buildings and football coaches design plays. Goal-driven design involves the attempt to shape an object or behavior in the light of explicit functional desiderata. Products of goal-driven design are properly called artifacts. The process of design frequently involves successively modified versions of a product that are tested by trial and error. The second sense of "design" is close in meaning to "intentional," and is accordingly labeled "intent design." Someone who is rude intentionally can be said to have been rude by design. Actions may be intentional even when little thought has been given to the action's point or consequences; intentional rudeness very often occurs even when one has not considered what objectives are served by it, or what the aftermath will be. Goal-driven design entails intent design, but the converse is not generally true. Goal-driven designers generally try to anticipate factors which threaten the success of the project, whereas this need not be true of intentional actions, which may occur with relatively little forethought. When talking about design in psychological contexts, we should be understood as using "design" in the goal-driven sense, unless specifically noted.

With respect to goal-driven design, (a) design determines function, but (b) not everything that has a function is designed (for that function). Consider (b) first. Many people use natural objects (driftwood, seashells, heads of game animals, etc.) for decorating rooms and buildings. These objects are clearly not designed for that purpose (although they are presumably placed in strategic locations by design, in the sense of intent-design). A rock on a desk may function as a paperweight, but unless the rock has had a flat base chiselled into it, or other similar modification, it is not appropriate to say that this object was designed for the purpose of holding down papers. A taxidermist may design a ferocious pose for a stuffed grizzly bear; in this case the pose is designed to titillate, but it is not true to say that the bear was designed for this purpose. Thus, function does not entail design for that function.

Now consider (a). Knowing what something is designed for, one knows its (intended) function, even if the item does not and cannot perform that function. Prior to the 1903 Wright Flyer, many contraptions were designed for heavier-than-air powered flight, yet none of them flew. Modern aviation did not have to get off the ground (pun by design!) for it to be the case that the function of those remarkable contraptions was to fly. It *was* their function to fly because that is what they were designed (albeit poorly) to do. Biological functions are importantly disanalogous. Millikan is fond of pointing out that individual hearts (e.g., malformed or diseased ones) may fail to pump blood, but that these hearts would not have this function unless some of their predecessors had actually succeeded in pumping blood. For a thing to possess a biological function, at least some (earlier) members of the class must have successfully performed the function. In cases of psychological design, the corresponding claim is not true. Such cases are called "deviant" or "marginal" by some authors (e.g., Wright 1976; Achinstein 1977; Van Parijs 1982). Nevertheless, they *can* arise in psychological teleology, but not in biological teleology.

6   The Analysis of Natural Design

These considerations about goal-driven design and function suggest two questions about the relationship of natural design to biological function: (1) Are there cases where cases where it is appropriate to say that a trait is naturally designed for X even though it does not have X as a biological function?; and (2) Are there examples where it is appropriate to say that a trait has a certain biological function but is not a product of natural design for that function? Answers to these questions will, of course, depend on what account one gives of biological function and natural design. Accepting ND = F leads to the answer "No" in both cases. However, we believe that this result ignores an important distinction in evolutionary history. Our proposal is to analyze natural design according to this schema (Allen and Bekoff 1995):

Trait T is naturally designed to do X if and only if
i. X is a biological function of T, and
ii. T is the result of a process of change of (anatomical or behavioral) structure due to natural selection that has resulted in T being more optimal (or better adapted) for X than ancestral versions of T.

Clause (i) should be understood according to the standard line, and it has the direct consequence that the answer to question (1) is that there are no cases where a trait is naturally designed for X even though it does not have X as a biological function. Clause (ii) is likely to be controversial because it involves the notions of optimality and adaptedness. The chief criticism of a adaptationism in evolutionary biology is that it is Panglossian (Gould and Lewontin 1978)—i.e., it entails the belief that this is the best of all possible worlds. The "Panglossian Paradigm" is defended by Dennett (1983/1987), who claims that it involves an idealizing assumption about natural selection that is probably false, but that is necessary for making predictions using the theory of natural selection. Byers and Bekoff (1990) worry that arguments for optimality frequently involve logical errors and are bothered that in many cases empirical studies make an adaptationist assumption on the basis of inadequate empirical evidence.

Because it stresses *comparative* judgments, our clause (ii) avoids the charge that it is Panglossian. The required comparison is between the traits of extant organisms and the corresponding traits of their ancestors. For instance, the wings of birds evolved from the forelimbs of their land-bound saurian ancestors. It is hypothesized that the lineage involves successive modifications of forelimbs into more aerodynamically efficient wings found in extant species of birds (ratites and penguins excepted). According to our analysis, to say that the wings of (most) birds are de-signed for flying is to say that (i) enabling flight is a biological function of birds' wings and (ii) extant morphological forms of such wings are the result of natural selection for variants that were better adapted for flying than earlier forms.

Comparative judgments to not require the Panglossian assumption. The claim that A is more optimal or better adapted than B with respect to some function does *not* entail that A is optimal or even good with respect to that function. A Rolls-Royce may consume more fuel per kilometer than a Cadillac, but it does not follow that the Cadillac has good, let alone optimal, fuel efficiency. (The theoretical optimum would be complete conversion of all the energy stored in the fuel into distance travelled.) This simple point about the logic of comparative statements is often overlooked. For example, during the confirmation hearings for U.S. Supreme Court Justice Clarence Thomas, one senator missed this point completely when he repeatedly claimed that Professor Anita Hill was being inconsistent by objecting to the nomination, given that she had told a newspaper that she thought Thomas would make a better Justice now than he would have done 10 years previously.

Clause (ii) is committed *only* to a *comparative* claim about traits. Thus, no statement about overall adaptedness or optimality is implied. Ruse (1993) draws a similar distinction between the notions of "com-parative progress" and "absolute progress" (see also the contributions to Nitecki 1988). It would take us too far afield to discuss the notion of evolutionary progress here, but we are sympathetic to Hull (1988, 45) who claims that "biological evolution has not just [one] direction, but lots of them." In other words, there are many bases for comparison, no one of which can be singled out as an absolute standard. Claims about the natural design of wings are assessed by comparing ancestral forms with descendant forms with respect to effectiveness for the function of flying. Such comparisons can be very specific indeed. For example, the glide ratio of eagle wings can be compared to the glide ratio of the wings of eagle ancestors, perhaps the archaeopteryx. If an eagle's wings result in a higher glide ratio than its ancestors' wings and if having a higher glide ratio provides a comparative fitness advantage, then it can be said that the eagle's wings are designed for soaring. Other wings, e.g., those of hummingbirds, may be designed for other aspects of flying, e.g., hovering.

Because natural selection acts on variation within a population it may reasonably be compared to trial and error testing by human designers. Dennett's "Design Stance" approach to complex systems (Dennett 1971, 1983, 1987) also suggests this comparison. We reject, however, that the view that the legitimacy of the notion of natural design rests on direct comparisons to conscious design by psychological agents. Both biological function and natural design can be fully naturalized (without reference to psychological design).

Hypotheses about natural design are, however, more difficult to establish than hypotheses about biological function. Function, on our view, is neutral with respect to the phylogenetic pathway by which a trait acquires a function. Consider again the behavior of a hare confronted by a fox. Holley (1993) argues that this behavior's function is to indicate to the fox that it has been detected. According to the standard line on functions, Holley's hypothesis is justified if it is reasonable to believe that bipedal standing by ancestral hares had this effect on ancestral foxes, and this effect was (partially) responsible for the transmission of this trait from ancestral hares to descendants. A corresponding design claim about bipedal standing would, on our analysis, require showing that this trait is a direct modification of some ancestral trait that was less efficient with respect to its effects on foxes.

These considerations suggest an answer to question (2) above. A trait may have a biological function but not be naturally designed for that function (although it may be a product of natural design for some other function.) But because showing design is more difficult than showing function, examples are likely to be difficult to find. Gould's (1980) discussion of the panda's thumb provides an intuitively plausible example of a nonbehavioral trait with a function—stripping bark from bamboo—for which it is apparently not designed given that the thumb apparently shows no special modifications for bark stripping, although the maintenance of this trait can presumably be (at least partially) explained by its contribution to bark stripping. However, the comparative evidence needed to support this claim is not readily available.

Our distinction between *design* and *function* is usefully contracted with the widely discussed distinction between *adaptation* and *exaptation* introduced by Gould and Vrba (1982). An adaptation, on their view, is a trait that has been shaped by natural selection for some use; their notion of "shaping" appears similar to our clause (ii) above (and, incidentally, captures the element of change missing from Sober's definition). They reserve the term *function* for cases when such shaping has occurred (thus, like Sober, they conflate function and adaptation, although not in the same way as Sober). They introduce the neologism "exaptation" for cases where either a selected trait of an organism is coopted for a new use (e.g., the use of the tongue to modify speech sounds) or where a characteristic that is produced and maintained by mechanisms other than natural selection is coopted for a current use (e.g., the violinist's use of the chin to hold the instrument). Exapted traits, according to Gould and Vrba merely have effects, they do not have functions. Millikan (1993, 45) notes that "[Gould and Vrba] are aware, of course, that [their] restriction on the term 'function' is stipulative and not just a reflection of general biological usage." Griffiths (1992) applies a version of the standard line on biological function to argue that Gould and Vrba mischaracterize the distinction between function and (mere) effect. He proposes to refer to "unshaped" traits with the right kind of selective history as "exadaptations" rather than adaptations and he proposes to use the term function in both cases. He does not, however, distinguish function from design. Thus our usage differs from both Gould and Vrba's (1982) and Griffiths' (1992) by distinguishing function from design, while it agrees with Griffiths' usage, but not with Gould and Vrba's, on the range of cases to which the notion of function correctly applies. Our view has the added advantage of not proliferating neologisms.

7   Design versus Function: An Ethological Example

The distinction we have drawn between function and design is useful for understanding issues many issues in evolutionary biology. We will illustrate with an example from ethology; specifically we consider the evolution of communicative behavior. Many authors consider transfer of information to be a defining feature of communication (Allen and Hauser 1993). However it seems implausible to treat all information transfer equally as communication. Using the framework developed here, we distinguish three levels of information transfer.

First, the behavior of many organisms provides information even though it is not a biological function of that behavior to do so. For example, migrating birds may provide the information to humans that winter is coming. Plausibly, this effect of migration is not part of the selectional history that accounts for birds migrating. Thus this is a mere effect, not a function (and hence not something that migration is designed to do). Considerations such as these have led many biologists to include selective advantage in their definitions of communication (see, for example, Eibl-Eibesfeldt 1975; Drickamer and Vessey 1992; Alcock 1993); at the second level, then, are those behaviors that have transfer of information as a function. At the third level are those behaviors that have been modified over evolutionary history to better perform this function.

The classical account of the evolution of communicative signals is given by Tinbergen (1952) who considers them to be "ritualized" versions of "intention movements." An intention movement is understood to be a movement from which it is possible to infer what an organism is likely to do next. Ritualization is a process whereby specific features of such movements become progressively exaggerated. Ritualization of a movement is considered likely when the benefits to an organism of conveying information about its future behavior exceeds the costs of doing so. For instance, conveying information about an intention to fight may be beneficial if it prevents the fight from occurring. In this way, for example, behavior patterns that convey information about the likelihood of fight and flight responses, such as the raising of hair due to a hormonal change, may become transformed via ritualization into differentiated expressive behaviors, e.g., hackle raising in dogs, that are clearer and less ambiguous than their ancestral forms (Eibl-Eibesfeldt 1975). Ritualization is the sort of change that would satisfy clause (ii) of our definition of natural design. If measures of communicative efficiency can be derived from notions such as clarity and lack of ambiguity, and if these measures can be applied to behaviors in the sequence from unritualized behavior to ritualized behavior, then it will be possible to assess the claim that ritualized behaviors are signals that are designed for communication. The chief methodological difficulty lies in comparing extant forms of a ritualized behavior to ancestral forms. However, if such comparisons can be made, it is reasonable to consider the possibility that some behaviors have a communicative function without being designed for communication, while other signals are specialized behaviors designed for communication.

8 Methodology and Future Work

On our account, to show that a trait T is naturally designed for some effect X, in addition to showing that X is a function of T one must also show evidence of structural changes in the phylogeny of T so that T is better suited for X than ancestral versions of T. This additional requirement can be very hard to meet, especially when the trait in question is a behavioral trait.

Relatively few anatomical traits fossilize, but fossils can be used to build models for comparison to descendant forms. For example, a model of archaeopteryx wings could be compared aerodynamically to the wings of modern birds. Paleontological evidence of soft tissue changes are extremely difficult to find. Nonetheless, it is sometimes possible to make inferences about soft tissue from fossils—for example, muscle arrangements can be deduced by noticing apparent attachment points on fossilized bones. Comparisons of contemporary species from different taxa can also sometimes be used to draw inferences about ancestral forms. Thus, for example, one might infer natural design in human lungs by comparing membrane oxygen transfer rates to those in the air sacs of certain extant fishes. Such inferences are necessarily very tenuous because they depend on many assumptions about the similarity of those fishes to the common ancestor of terrestrial vertebrates.

For behavioral traits, inferences about design are almost exclusively limited to the comparison of individuals from different extant species and populations. Inferences to the phylogeny of behavior based on such comparisons are difficult but not impoissible. For example, Golani (1992) proposes a scheme for characterizing movement patterns and gradients of movement differences across vertebrate species from different taxonomic groups. He hopes that examination of variations in shared movement patterns of vertebrates will allow specification of what he calls "the ground plan of vertebrate behavior" (ibid., 264). If such a project is successful, it might be possible to compare the relative effectiveness of variations in behavior for specific tasks. It might then be possible to draw further inferences about the evolution of complex behaviors, on which claims about behavioral design could be based. Many methodological difficulties remain unsolved, making this a prime area for future interdisciplinary, comparative research by biologists and philosophers.

Acknowledgments

We thank our colleagues at Texas A&M and the University of Colorado, Boulder, for discussing these issues with us. Conversations with Ruth Millikan and Rob Cummins have also helped us enormously. Finally, we thank Lawrence Shapiro, Elliott Sober, John Fentress, Kim Sterelny, Susan Townsend, Robert Eaton, and Nick Thompson for helpful comments on earlier and less optimal versions of this work. Parts of this essay were adapted from Allen and Bekoff 1995. CA gratefully acknowledges the support of NSF fellowship SBR-9320214 during preparation of this manuscript; MB was supported by a sabbatical leave from the University of Colorado, Boulder.

References

Achinstein, P. (1977), "Function Statements," *Philosophy of Science* 44: 360–376.

Alcock, J. (1993), *Animal Behavior: An Evolutionary Approach*, 5th edition. New York: Sinauer.

Allen, C. and Bekoff, M. (1995), "Function, Natural Design, and Animal Behavior: Philosophical and Ethological Considerations," in N. S. Thompson (ed.), *Perspectives in Ethology* 11, New York: Plenum Press, pp. 1–46.

Allen, C. and Hauser, M. D. (1993), "Communication and Cognition: Is Information the Connection?" *PSA* vol 2: 81–91.

Ayala, F. J. (1977), "Teleological Explanations," in T. Dobzhansky (ed.), *Evolution*. San Francisco: W. H. Freeman and Co., pp. 497–504.

Bekoff, M. (1977), "Social Communication in Canids: Evidence for the Evolution of a Stereotyped Mammalian Display," *Science* 197: 1097–1099.

———. (1982), "Functional Aspects of Play as Revealed by Structural Components and Social Interaction Patterns," *Behavioral and Brain Sciences* 5: 156–157.

———. (1988), "Motor Training and Physical Fitness: Possible Short and Long Term Influences on the Development of Individual Differences in Behavior," *Developmental Psychobiology* 21: 601–612.

———. (1989a), "Social Play and Physical Training: When 'Not Enough' May be Plenty,' *Ethology* 80: 330–333.

———. (1989b), "Behavioral Development of Terrestrial Carnivores," in J. L. Gittleman (ed.), *Carnivore Behavior, Ecology, and Evolution*. Ithaca, NY: Cornell University Press, pp. 89–124.

Bekoff, M. and Byers, J. A. (1981), "A Critical Reanalysis of the Ontogeny and Phylogeny of Mammalian Social and Locomotor Play: An Ethological Hornet's Nest," in K. Immelmann, G. W. Barlow, L. Petrinovich and M. Main (eds.), *Behavioral Development: The Bielefeld Interdisciplinary Project*. Cambridge: Cambridge University Press, pp. 296–337.

Brown, J. L. (1975), *The Evolution of Behavior*. New York: Norton.

Byers, J. A. and Bekoff, M. (1990), "Inference in Social Evolution Theory: A Case Study," in M. Bekoff and D. Jamieson (eds.), *Interpretation and Explanation in the Study of Animal Behavior: Vol. 2, Explanation, Evolution, and Adaptation*. Boulder, CO: Westview Press, pp. 84–97.

Cummins, R. (1975/1984), "Functional Analysis," *Journal of Philosophy* 72: 741–765. Reprinted with minor alterations in E. Sober (ed.), (1984), *Conceptual Issues in Evolutionary Biology*. Cambridge, MA: MIT Press, pp. 386–407.

Dawkins, R. (1986), *The Blind Watchmaker*. New York: Norton.

Dennett, D. C. (1971), "Intentional Systems," *Journal of Philosophy* 68: 87–106.

———. (1983/1987), "Intentional Systems in Cognitive Ethology: The Panglossian Paradigm Defended," *Behavioral and Brain Sciences* 6: 343–390. Reprinted in Dennett, D. C. (1987), *The Intentional Stance*, pp. 237–268.

———. (1987), *The Intentional Stance*. Cambridge, MA: MIT Press.

Drickamer, L. and Vessey, S. H. (1992), *Animal Behavior: Mechanisms, Ecology, and Evolution*, Dubuque, Iowa: Wm. C. Brown.

Eibl-Eibesfeldt, I. (1975), *Ethology: The Biology of Behavior*. Second Edition. New York: Holt, Rinehart, and Winston.

Fagen, R. (1981), *Animal Play Behavior*. Oxford: Oxford University Press.

Godfrey-Smith, P. (1994), "A Modern History Theory of Functions," *Noûs* 28: 344–362.

Golani, I. (1992), "A Mobility Gradient in the Organization of Movement: The Perception of Movement Through Symbolic Language," *Behavioral and Brain Sciences* 15: 249–308.

Gordon, D. M., Paul, R. E., and Thorpe, K. (1993), "What is the Function of Encounter Patterns in Ant Colonies?" *Animal Behaviour* 45: 1083–1100.

Gould. S. J. (1980), *The Panda's Thumb: More Reflections in Natural History*. New York: Norton.

Gould, S. J. and Lewontin, R. (1978), "The Spandrels of San Marco and the Panglossian Paradigm: A Critique of the Adaptationist Programme," *Proceedings of the Royal Society London*. Reprinted in E. Sober (ed.), (1984), *Conceptual Issues in Evolutionary Biology*. Cambridge, MA: MIT Press, pp. 252–270.

Gould, S. J. and Vrba, E. S. (1982), "Exaptation—A Missing Term in the Science of Form," *Paleobiology* 8: 4–15.

Griffiths, P. (1992), "Adaptive Explanation and the Concept of a Vestige," in P. Griffiths (ed.), *Trees of Life*. Netherlands: Kluwer, pp. 111–131.

Griffiths, P. E. (1993), "Functional Analysis and Proper Functions," *The British Journal for the Philosophy of Science* 44: 409–422.

Hass, C. C. and Jenni, D. A. (1993), "Social Play Among Juvenile Bighorn Sheep: Structur, Development, and Relationship to Adult Behavior," *Ethology* 93: 105–116.

Holley, A. J. F. (1993), "Do Brown Hares Signal to Foxes?" *Ethology* 94: 21–30.

Hull, D. L. (1988), "Progress in Ideas of Progress," in M. H. Nitecki (ed.), *Evolutionary Progress*. Chicago: University of Chicago Press, pp. 27–48.

Kitcher, P. (1993), "Function and Design," *Midwest Studies in Philosophy* *XVIII*, Minneapolis: University of Minnesota Press, pp. 379–397.

Leyhausen, P. (1979), *Cat Behavior*. New York: Garland.

Manning, A. and Dawkins, M. S. (1993), *An Introduction to Animal Behaviour*. Fourth Edition. New York: Cambridge University Press.

Marler, P. and Hamilton, W. D. III (1966), *Mechanisms of Animal Behavior*. New York: Wiley.

Martin, P. and Caro, P. M. (1985), "On the Functions of Play and its Role in Behavioral Development," *Advances in the Study of Behavior* 15: 59–103.

Mayr, E. (1974/1988), "The Multiple Meanings of Teleological," reprinted with a new postscript in E. Mayr (1988) *Towards a New Philosophy of Biology*. Cambridge, MA: Harvard University Press, pp. 38–66.

Millikan, R. G. (1984), *Language Thought and Other Biological Categories* Cambridge, MA: MIT Press.

———. (1989), "An Ambiguity in the Notion of Function," *Biology and Philosophy* 4: 172–176.

———. (1993), *White Queen Psychology and Other Essays for Alice*. Cambridge, MA: MIT Press.

Mitchell, S. (1993), "Dispositions or Etiologies? A Comment on Bigelow and Pargetter," *Journal of Philosophy* 90: 249–259.

Nagel, E. (1961/1984), "The Structure of Teleological Explanations," in E. Nagel (ed.), *The Structure of Science*. Indianapolis: Hackett, pp. 401–427. Reprinted in E. Sober (ed.), (1984), *Conceptual Issues in Evolutionary Biology*. Cambridge, MA: MIT Press, pp. 319–346.

Neander, K. (1991a), "The teleological notion of 'function'," *Australian Journal of Philosophy* 69 (4): 454–468.

———. (1991b), "Functions as Selected Effects: The Conceptual Analyst's Defence," *Philosophy of Science* 58: 168–184.

Nitecki, M. H., (ed.), (1988), *Evolutionary Progress*, Chicago: University of Chicago Press.

Ollason, J. G. (1987), "Foraging Theory and Design," in A. C. Kamil, J. R. Krebs, and H. R. Pulliam (eds.), *Foraging Behavior*. New York: Plenum Press, pp. 549–561.

Pellis, S. M. (1993), "Sex and the Evolution of Play Fighting: A Review and Model Based on the Behavior of Muroid Rodents," *Play Theory and Research* 1: 55–75.

Rasa, O. A. E. (1973), "Prey Capture, Feeding Techniques, and their Ontogeny in the African Dwarf Mongoose *Helogale Undulate Rufula*," *Zeitschrift für Tierpsychologie* 332: 449–488.

Reeve, H. K. and Sherman, P. W. (1993), "Adaptation and the Goals of Evolutionary Research," *Quarterly Review of Biology* 68(1): 1–32.

Ruse, M. (1993), "Evolution and Progress," *Trends in Ecology and Evolution* 8(2): 55–59.

Sober, E. (1993), *Philosophy of Biology*. Boulder, Colorado: Westview Press.

Tinbergen, N. (1952), "'Derived' Activities, Their Causation, Biological Significance and Emancipation During Evolution," *Quarterly Review of Biology* 27: 1–32.

———. (1963), "On Aims and Methods of Ethology," *Zeitschrift für Tierpsychologie* 20: 410–429.

Van Parijs, P. (1982), *Evolutionary Explanation in the Social Sciences: An Emerging Paradigm*. Totowa, NJ: Rowman and Littlefield.

Watson, D. M. and Croft, D. B. (1993), "Playfighting in Captive Red-necked Wallabies, *Macropus rofogriseus banksianus*," *Behaviour* 126: 219–245.

Williams, G. C. (1966), *Adaptation and Natural Selection*. Princeton, NJ: Princeton University Press.

Wright, L. (1973/1984), "Functions," *Philosophical Review* 82: 139–168. Reprinted in E. Sober (ed.), (1984), *Conceptual Issues in Evolutionary Biology*. Cambridge, MA: MIT Press, pp. 347–368.

———. (1976), *Teleological Explanations*, Berkeley: University of California Press.

Index

Contributors

Colin Allen, Associate Professor
Department of Philosophy
Texas A&M University

Fred Adams, Professor
Department of Philosophy
University of Maryland

Ron Amundson, Professor
Department of Philosophy
University of Hawaii

Francisco J. Ayala, Professor
Department of Ecology and
Evolution
University of California, Irvine

Mark A. Bedau, Associate Professor
Department of Philosophy
Reed College

Marc Bekoff, Professor
Department of Environmental,
Population, and Organismic Biology
University of Colorado

Walter J. Bock, Professor
Department of Biological Sciences
Columbia University

John Bigelow, Professor
School of Philosophy
La Trobe University, Australia

Robert N. Brandon, Professor
Department of Philosophy
Duke University

Robert Cummins, Professor
Department of Philosophy
University of Arizona

Berent Enç, Professor
Department of Philosophy
University of Wisconsin

Carl Gans, Professor
Department of Biology
University of Michigan

Peter Godfrey-Smith, Associate
Professor
Philosophy Department
Stanford University

Stephen J. Gould, Professor
Museum of Comparative Zoology
Harvard University

Paul Griffiths, Director
Unit for the History and Philosophy of
Science
University of Sydney, Australia

Robert A. Hinde, Professor and Fellow
St. John's College
Cambridge University, England

Philip Kitcher, Professor
Department of Philosophy
University of California, San Diego

George Lauder, Professor
Department of Ecology and
Evolutionary Biology
University of California, Irvine

Ruth Garrett Millikan, Professor
Philosophy Department
University of Connecticut, Storrs

Sandra D. Mitchell, Associate
Professor
Department of Philosophy
University of California, San Diego

Ernest Nagel (deceased)
Professor of Philosophy
Columbia University

Karen Neander, Assistant Professor
Department of Philosophy
Johns Hopkins University

Robert Pargetter, Professor
School of Philosophy
La Trobe University, Australia

Martin Rudwick, Professor
Department of History
University of California, San Diego
and
Department of History and Philosophy
of Science
University of Cambridge, England

Elisabeth Vrba, Professor
Department of Geology and
Geophysics
Yale University

Gerd von Wahlert, Professor
Forschungstelle für Ichthyiologie,
Staatliches Museum für Naturkunde
Stuttgart, Germany

Larry Wright, Professor
Department of Philosophy
University of California, Riverside